Metazoa – Morphology and Evolution of Animals

Metazoa – Morphology and Evolution of Animals

Achim Paululat · Günter Purschke

Metazoa – Morphology and Evolution of Animals

A Practical Guide to the Dissection and Comparative Study of Animals

 Springer

Achim Paululat
Faculty of Biology & Chemistry, Department
of Zoology and Developmental Biology
Osnabrück University
Osnabrück, Lower Saxony, Germany

Günter Purschke
Faculty of Biology & Chemistry, Department
of Zoology and Developmental Biology
Osnabrück University
Osnabrück, Lower Saxony, Germany

ISBN 978-3-662-69903-4   ISBN 978-3-662-69904-1   (eBook)
https://doi.org/10.1007/978-3-662-69904-1

Preface

Since modern biology covers many exciting topics, there is often little space left for a classical zoological dissection course. The number of semester hours spent preparing and dissecting animals has decreased at almost all universities, with so many objects traditionally studied in an introductory zoological course no longer being covered today. Consequently, only a few objects are dealt with, but those chosen are generally thought to represent the most important anatomical concepts, the so-called bauplans. Thus, a more in-depth examination of anatomy and histology, as well as the interplay of anatomical forms, evolution, and the systematic relationships among animal groups, is reserved for advanced courses in zoology. Our book considers the recent teaching requirements and focuses on only a few objects, reflecting metazoan diversity. In each chapter, we provide the essential basic knowledge of animal morphology for all biology students, including our future teachers. We always start with a precise macroscopic external examination of the animal before we proceed with dissection. If available, the living animal is also presented. For an in-depth assessment, specific manual skills are necessary, for example, the correct use of optical instruments such as stereo and compound microscopes. Larger objects must be carefully dissected so as to understand the anatomical organization, and the smaller ones must be examined microscopically or histologically using whole-mount preparations or histological sections. In this book, we suggest a combination of examination methods depending on the size of the animals to be studied.

In addition, our textbook also considers the changed learning behavior of today's students. Unlike many other manuals of dissection instruction, this book focuses primarily on photographs. Schematic drawings are used exceptionally, solely for a better understanding of the preparation steps and observations. Nevertheless, we would like to emphasize the value of drawings made by the students during such a practical course. A drawing made from one's own preparation has a high didactic value. Students who draw precisely accumulate much more knowledge in their long-term memory than those who only take photos or look at objects. Photographs, however, often reflect the natural conditions much better than drawings. Many of the images shown here have been used for teaching purposes in our courses. Our zoology course at the University of Osnabrück is attended by approximately 100 students yearly. The feedback from our students shows us that photos and drawings are a perfect combination for a better understanding of the basic anatomy of animals, ultimately leading to the development of a new concept for the book. Each chapter presents the taxon and the animal to be prepared in a compact text. The most important organs and their function are explained. The focus is on comparative morphology and anatomy to illustrate the evolution of the metazoan organization. This book can supplement other textbooks or be a stand-alone practical book. It will help students gain a sound overview of nature's morphological and evolutionary concepts. We know that each object or species described in our book deserves much more attention than we have offered here. This is because we want to address the aspects that we consider most important for each animal group without losing the compactness of the book.

The animals needed for an introductory zoological dissection course should be as easy to obtain as possible. We order our material mostly from breeding farms that supply, among others, hobby entomologists with cockroaches, pond lovers with pond mussels, zoos with rats, or restaurants with trout and snails, thus excluding any interference with sensitive ecosystems. If the material is collected or fished, for example sea squirts and starfish, it should be done so in a way that does not harm the respective ecosystem. Ethical aspects, of course, play an important role. We have always disclosed our sources in our introductory zoology course and discussed them with our students. The overwhelming majority of our students have rated our concept positively and do not want to do without the dissection of animals, as far as the animals can be made available with due regard for their welfare and without interfering with ecosystems. Modern media, such as video tutorials for dissecting specimens, are, if of high quality, excellent supplements to a zoology course, but they are not an equivalent substitute.

Our book will support as many students as possible in better understanding metazoan morphology. We look forward to feedback from students and teachers, whose suggestions will be taken into account in later editions. We are also open to suggestions for selecting additional animals to be included in the future.

Osnabrück, Germany   Achim Paululat
March 2025   Günter Purschke

Acknowledgments

Numerous colleagues from the Department of Zoology/Developmental Biology at the University of Osnabrück have contributed to the success of our zoology courses, for example, by developing and testing different dissection protocols and fixation techniques. Our thanks go to, among others, Dr. Annette Bergter, Dr. Maik Drechsler, PD Dr. Heiko Harten, Dr. Christine Lehmacher, Dr. Ekaterini Olympia Psathaki, and Prof. Dr. Torsten H. Struck.

We want to express our special thanks to biology student Caici Neerincx for her contribution to this book. Caici showed outstanding commitment in preparing several animals for photographing.

Individual chapters were proofread by Prof. Dr. Hermann Aberle (Düsseldorf), Prof. Dr. Ernst-Martin Füchtbauer (Aarhus), Prof. Dr. Monika Hassel (Marburg), Dr. Peter Heimann (Bielefeld), Dr. Katherina Psathaki (Osnabrück), and Prof. Dr. Torsten Struck (Oslo). We are deeply grateful for this. Living specimens of *Trichoplax adhaerens* were provided by Prof. Dr. Harald Gruber-Vodicka (Kiel).

Some photos were taken during our marine biology practical courses in Roscoff (Station Biologique de Roscoff, Brittany, France) or on Heligoland (Helgoland, Biologische Anstalt Helgoland, Alfred-Wegener-Institut, Schleswig-Holstein, Germany). We were always reliably supplied with animals by our hosts. Our thanks, therefore, go to Ute Kieb, Kathrin Böhmer, Renate Beissner, Uwe Nettelmann, and Dr. Eva Brodte from the BAH/AWI on Heligoland, as well as to Philippe Gaëlle and the staff of the Station Biologique de Roscoff. Furthermore, we thank Mechtild Krabusch and Eva Cordes from our laboratory. They observed the molting of the crayfish and recovered the exuviae. Tomke Ramke made sections of *Astacus astacus*. The beautiful photos of sea urchin fertilization and early development were kindly provided by Prof. Dr. Dietrich Ribbert (University of Münster). Prof. Dr. Ernst-Martin Füchtbauer (University of Aarhus) contributed photos of *Ciona* embryos.

Contents

About the Authors

Achim Paululat, born in 1961, in Soest, Germany, studied biology, chemistry, and educational sciences at the University of Münster. After his first state examination, he decided to pursue a university career, earning his doctorate in 1990, under the supervision of Dietrich Ribbert at the University of Münster. During this time, Paululat was a lecturer at the Pedagogical University Münster and a freelance guide at the Natural History Museum in that same city. He then moved to the Philipps University of Marburg to work in the lab of Prof. Dr. Renate Renkawitz-Pohl as a postdoc on the differentiation of muscle tissue in *Drosophila*. A research internship in Alan Michelson's lab at the HHMI, Harvard Medical School in Boston, was financed by the German Science Foundation (DFG). There, he analyzed the asymmetric cell division of muscle founder cells. Together with Dr. Aria Baniahmad, now a professor of Genetics at the University of Jena, he joined the Collaborative Research Center SFB 397 with an independent research project as a young group leader. After completing his habilitation in the subjects of General Zoology and Developmental Biology at the University of Marburg, Paululat was appointed, in 2004, to the chair of Zoology and Developmental Biology at the University of Osnabrück. As a member of several collaborative research centers (SFB 397, SFB 431, SFB 944, and SFB 1557), as a project leader in international consortia (EU-Myores), and after two terms as faculty dean, he is currently addicted to the research of genetic and cell physiological processes in muscle cells and nephrocytes in the model organism *Drosophila melanogaster*. In addition to his research interests, field biology is particularly close to his heart. He offers avifaunistic excursions to, for example, Heligoland and co-organizes wet lab and field courses at the Biological Station Roscoff in Brittany (France). Since 2020, Paululat has served as an elected member of the DFG Review Board 203 (Zoology) and as co-editor of the German edition of the textbook *Campbell Biology*. He would particularly like to thank his wife Mechtild, who enabled him to work on this book through her great forbearance and tolerance.

Günter Purschke, born in Hamburg in 1954, studied biology and chemistry at the Georg-August-University in Göttingen, with an eye toward becoming a high school teacher. After his first state examination, he received his doctorate in 1984, under Wilfried Westheide. Since his Ph.D. studies, he has been consistently involved in student teaching, first as a student and scientific assistant, after his doctorate as a research associate (assistant) at the University of Osnabrück at the chair for Systematic Zoology, also under Wilfried Westheide. Following his habilitation in zoology in 1997, he was appointed adjunct professor in 2002. After holding the chair in this department in 2003/2004, he continued to work at the current chair for Zoology and Developmental Biology under Achim Paululat. His research focus is the morphology, phylogeny, and systematics of annelids. His scientific interests focus on the structure and evolution of the nervous system and sensory organs, with several studies dedicated to eyes and photoreceptor cells. In addition to numerous original publications, he is co-editor of several scientific books: *Morphology, Molecules, Evolution and Phylogeny in Polychaeta and Related Taxa* in 2005 (with Thomas Bartolomaeus, Bonn), *Structure and Evolution of Invertebrate Nervous Systems* in 2016 (with Andreas Schmidt-Rhaesa, Hamburg, and Steffen Harzsch,

Greifswald), and the 2019, 2020, 2021, and 2022 editions (with Markus Böggemann, Vechta, and Wilfried Westheide, Osnabrück) of the *Handbook of Zoology, Annelida*, each volume of which includes his review contributions. He has also been represented as an author in Westheide/Rieger's *Systematic Zoology* (Spezielle Zoologie) since the first edition in 1996. Since the eighth edition of the German version of the textbook *Campbell Biology*, when it was taken over by biologists from Osnabrück in 2009, he has continuously revised the chapters on phylogeny, diversity, and evolution of Metazoa, comprising both invertebrates and vertebrates. Last but not least, Achim Paululat and he have extensively revised and updated *Wörterbuch der Zoologie* (*Dictionary of Zoology*) in its current form.

1.1   Summary

The vast diversity of animal life on Earth comprises more than 1.5 million described species. This diversity is confusing and incomprehensible for beginners. Since the animals (Metazoa) can be traced back to one last common ancestor, studying their evolutionary history or phylogeny is one key to breaking down this diversity. Although, with the invention of phylogenomic analyses, significant progress has been made in unraveling metazoan phylogeny, open and controversial questions still exist. Nevertheless, the evolutionary history of Metazoa is reflected by comparative morphology, among other things. But what are the reasons for the evolutionary and biological success of Metazoa? This goes back to a few key events in evolution, such as the origin of multicellularity, which was accompanied by the specialization of cells to certain specific functions. After separating into generative and vegetative cells, the latter have been further diversified from less than ten cell types in Placozoa to about 200 in vertebrates. Another key invention was the evolution of true epithelia, sealing the inside from the external environment, a prerequisite for the evolution and function of sensory, nerve, and muscle cells, or nephridia. Several of these innovations have made it into the names of higher taxa, such as Bilateria, Nephrozoa, and Chordata. This practical book focuses on the morphology and the evolutionary context of the major animal lineages. We present the ground patterns and specific morphological adaptations of various animal groups in photos and make them understandable for students in introductory zoological dissection courses.

So, how did we choose the taxa and species presented in this book? Generally, Metazoa are divided into 30–36 so-called phyla, depending on the respective authors. Each represents a different ground pattern (basic design or bauplan). Still, it should be kept in mind that no objective criteria exist for separating a particular group of animals as a separate phylum. Therefore, the term phylum should be used with care and is mainly avoided here. All metazoans go back to a single last common ancestor (LCA) or the stem species of all

Metazoa, see Fig. 1.1. However, most authors of introductory zoology textbooks simplify the phylogenetic tree by omitting certain groups.

For example, one of the best-selling general biology textbooks for beginners, *Campbell's Biology* (Urry et al. 2019), considers only 15 taxa in its figured tree of life, but 23 taxa are mentioned in the main text. The various major groups discussed all represent more or less different ground patterns or apparent differences in their organization. However, many of these major groups will never be observed or examined, even by a biologist. One example is the group of the marine, benthic, deuterostome Hemichordata (acron worms and pterobranchs), which live hidden at or in sea bottoms worldwide. Are they, therefore, less important? Such groups are particularly interesting for zoologists who deal with phylogeny and evolution, especially for understanding evolutionary changes over time. Nevertheless, the education of our biology students must contain at least an idea about the immense diversity of animals in terms of species numbers and their different biology and morphology. In addition, the number of species, including the diversity within these groups, is considerably different, ranging from one or a few in Placozoa to more than one million in Hexapoda. At first sight, the diversity of animals seems to be overwhelming and incomprehensible. This diversity results from variations of a few readily identifiable concepts. These are the ground patterns, the evolutionary history (phylogeny), and the functional morphology and physiology. It follows that a selection must be made because of the number of higher groups and the usually large number of participants in our practical courses. Another criterion for choosing a given species to be presented in this textbook is easy access to sufficient numbers of individuals. However, the selection of animals for a zoology course should always be made with thoughtfulness so that the most significant possible number of different morphological concepts (ground patterns) is included.

The course objects and their probable phylogenetic positions are highlighted in the phylogenetic tree (Fig. 1.1). We propose to study at least one representative from the major lineages,

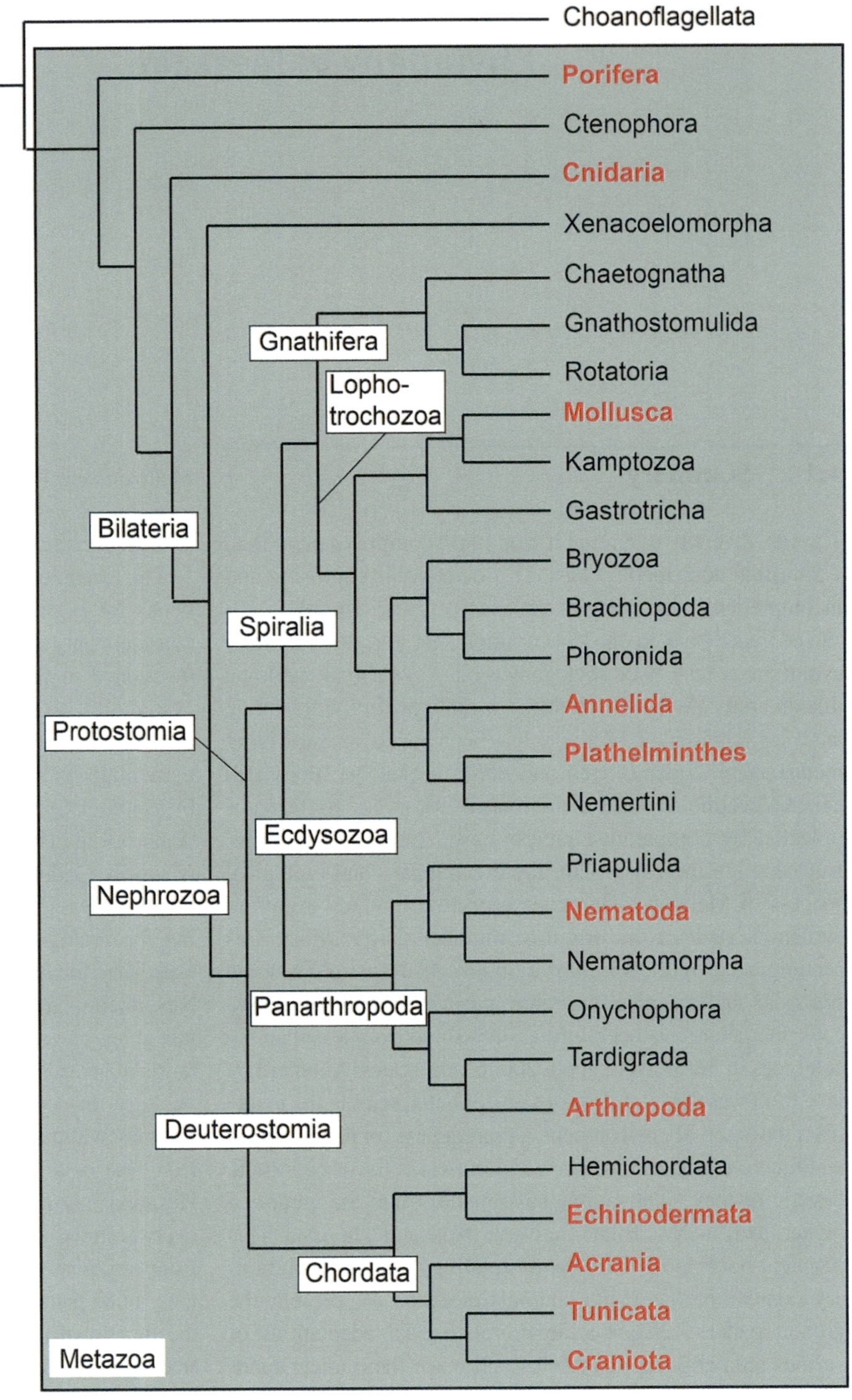

Fig. 1.1 Phylogenetic relationships of the main major taxa of Metazoa; some groups, like Placozoa, Loricifera and Kinorhyncha, are not considered. The topology shown summarizes various phylogenomic studies (see bibliography) and represents only one of several possible hypotheses. The groups usually referred to as animal phyla are the named terminal taxa at the end of the branches. A selection of the more inclusive higher taxa is positioned on the respective branches of the tree. The taxa discussed in this book are highlighted in red and bolded

each representing a different bauplan within the metazoans. We have focused on the more diverse or species-rich groups, considering that our students may come across these candidates later in their studies or during their professional lives. Thus, the basal taxa of Metazoa are represented here by sponges and cnidarians. Within the triploblastic Metazoa, known as Bilateria, representatives of both sister groups, Protostomia and Deuterostomia, have been selected. In each of the two groups, the main lineages are represented: Ecdysozoa with nematodes and arthropods, and Spiralia with flatworms, mollusks, and annelids. Among arthropods, mollusks, and annelids, several objects are presented, one of which may be sufficient for such a general zoology course. Deuterostomia are represented by Echinodermata and all three groups of chordates—Acrania, Urochordata, and Craniota. The latter choice is because we, as humans, belong to the chordates as well. Also, for traditional reasons, vertebrates are often given more space than their species number of about 55,000 would justify, especially compared to well over 1,500,000 arthropod species, meaning that less than 1% of all described species are vertebrates.

The morphology of the animals we treat here always emerges from a combination of phylogenetically ancient structures, some of which arose very early in evolution, and comparatively more recently evolved features. No organism, species, or taxon is entirely primitive or plesiomorphous, but always a mosaic of phylogenetically old and new features, the so-called plesiomorphies and apomorphies. Each major taxon has passed through its irreversible evolutionary history and thus represents a combination of plesiomorphies and apomorphies, referred to as the ground pattern of a given taxon. When we examine a common earthworm or a rainbow trout, we see the result of their individual evolutionary history. From this fact, it inevitably follows that, within these taxa, the selected species never shows the entire set of all group-specific characters or the so-called ground pattern. This is particularly true for the parasitic species, as this lifestyle requires special adaptations to their hosts. For example, most free-living flatworms and the free-living nematodes are microscopically small. Quite different are the parasitic forms, which reach sizes of several centimeters in both taxa, such as the liver fluke and the roundworm treated here, or even meters, like certain tapeworms. Their comparatively large size is thus the main reason why they are selected as objects for a beginner's practical zoology course.

Although it is very likely and beyond discussion that all Metazoa can be traced back to only one last common ancestor, it has yet to become possible to present a phylogenetic system of Metazoa that reflects their phylogenetic relationships with sufficiently high probability at all branches. It should be remembered that a cladogram or a tree of life is always a scientific hypothesis, which, as always in science, can be tested and possibly falsified. It is essential to consider that the relationships of some major taxa, depicted in the phylogenetic tree in Fig. 1.1, are still unclear or controversially discussed in science. This is still the case even after the introduction of phylogenomic analyses, although many contentious or existing problems have been resolved. One reason is that we are dealing with evolutionary events that go back 500 million years that occurred during the so-called Cambrian explosion or even earlier. Among the discussed problems is, for example, which of the following taxa, Porifera, Placozoa (Fig. 1.2), or, most recently brought into discussion, Ctenophora, represents the first branch in the phylogenetic system of extant Metazoa. There is support from both molecular and/or morphological studies for all three hypotheses, so, currently, none can be falsified. However, the respective position of these taxa has severe consequences for the characters to be assumed to have been present in the last common ancestor of Metazoa. For example, at what point did epithelia evolve, or were muscle and nerve cells already present in this ancestor? The tree in Fig. 1.1 represents the "classical" hypothesis that Porifera, the sponges, are the sister group of all other Metazoa. In addition to "new" animal phyla, these modern phylogenomic analyses no longer

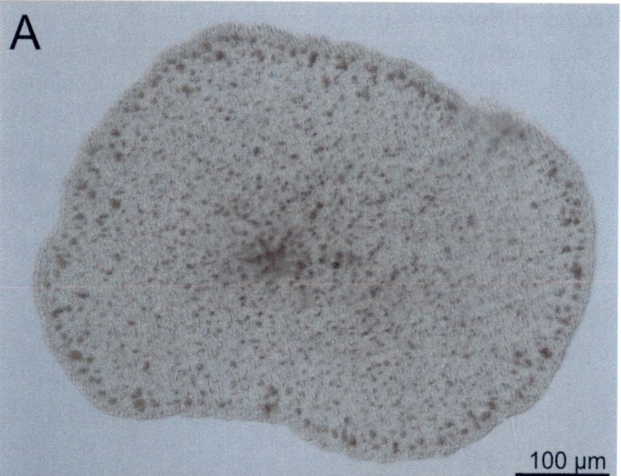
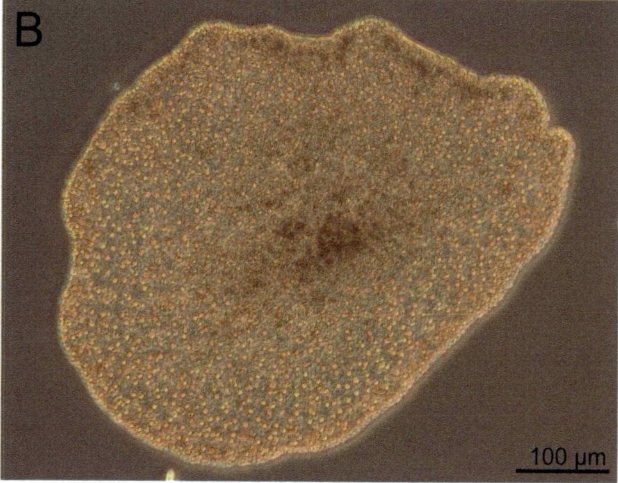

Fig. 1.2 *Trichoplax adhaerens*, Placozoa, live images. (**a**) Bright field, (**b**) Phase contrast. The German zoologist Franz Eilhard Schulze discovered *Trichoplax adhaerens* in 1883, in a sea-water aquarium of the Zoological Institute in Graz. But only works from the 1970s by Karl Grell led to the idea that *Trichoplax adhaerens* is perhaps the simplest of all multicellular animals. About six morphologically distinguishable cell types provide all necessary functions for life. These cells form three layers: an upper and a lower epithelium, and a connective tissue of fiber cells in between. The epithelia of *Trichoplax* differ from the epithelia of other metazoa by the absence of a basal lamina formed by the extracellular matrix (ECM), but not the apical connecting structures, the zonulae adhaerentes, and septate junctions. Like numerous unicellular organisms, *Trichoplax* moves by ciliary beating of their monociliary cells of the lower epithelium by some authors. A movie showing *Trichoplax adherens* in motion can be found at figshare: sn.pub/xr9qhi

mention specific "old" well-known animal phyla in the modern phylogenetic trees, as they have proven to be subgroups of other major taxa or to not be monophyletic (tracing back to a common ancestor unique to them). These groups include, among numerous others, the peanut worms (Sipuncula), as well as beard and tube worms (Pogonophora), but also the articulated animals (Articulata) and roundworms (Nemathelminthes). The latter's members, an assemblage of worm-like animals of uncertain affinities, fell into several clades, such as Ecdysozoa, Gnathifera, and Lophotrochozoa (Fig. 1.1).

Technical Notes

Living animals: We thank all institutions for the opportunity to present these animals here. Other images of living animals, for example, *Sycon raphanus* or *Ciona intestinalis*, were taken during our excursions to Roscoff (Brittany, France) and to the North Sea Island of Sylt or Heligoland (both Schleswig-Holstein, Germany). Supplementary material, including many unique movie clips, can be accessed at fig-share: sn.pub/xr9qhi

Aquariums

Berlin. Aquarium Berlin, Germany (Trout and other fish).

Borkum. Nordseeaquarium Borkum, Germany (Adult moon jellyfish and their polyps).

Delfzijl. MuzeeAquarium Delfzijl, The Netherlands (Spotted ray).

Düsseldorf. Aquazoo Löbbecke Museum, Germany (Pond mussels).

Ibbenbüren. NaturaGart, Germany (Birchirs, Bowfin, Lungfish, Bitterlings and other species).

Lyon. Aquarium de Lyon, France (Alligator gar and other fish).

We thank all aquariums for the opportunity to show the animals here.

Specimens on Display in Museums

Heligoland. Lobsters shed their cuticle many times in the course of their lives. Lost extremities are regenerated during these molts. Exuviae of different old lobsters are on display in the Museum Helgoland, Heligoland, Germany.

Silver Spring. The amputated leg of a patient severely infested with *Wucheria* (elephantiasis) is shown in the National Museum of Health and Medicine, Silver Spring, Maryland, USA. The leg is stored in ethanol.

Toulouse. The coelacanth (*Latimeria chalumnae*), shown in the Craniota chapter is permanently displayed at the Musée de Toulouse in France. It was caught by hand with a horsehair rope off Ajouan Island in 1983, fixed in a formalin bath, and dried.

We want to thank all museums for allowing us to show the exhibits here and providing us with information.

Cat Skeleton We thank the Department of Behavioral Biology at the University of Osnabrück for providing the cat skeleton.

Histological Specimens Some of the microscopic sections shown here are property of the Zoological Teaching Collection of the Philipps University Marburg. The photographs were made possible with the kind support of Prof. Dr. Monika Hassel (Marburg). We would also like to thank Prof. Dr. Hermann Aberle and Dr. Kai Caspar for the opportunity to photograph some specimens from the Zoological Collection of the Heinrich Heine University Düsseldorf. We also thank Patricia and Johannes Lieder, Ludwigsburg, for their generous support. Lieder Laboratory for Microscopy, Johannes Lieder GmbH & Co. KG, Solitudeallee 59, 71636 Ludwigsburg, Germany. Internet: www.lieder.de. All other microscopic sections shown here are property of the Histology Collection of the Department of Zoology and Developmental Biology, University of Osnabrück, and were produced for research and teaching.

Photography The photos of histological sections were mainly taken with a Leica DMLB microscope equipped with Leica HC and HCX lenses and Jenoptik cameras (Progres CF14, Gryphax). The images of sea urchin fertilization and development provided by Prof. Dr. Ribbert (University of Münster, Germany) were taken with a Zeiss Photomicroscope on B/W film. The photos of cilia in the pharynx of *Ciona* were recorded with a Nikon Coolpix 4500 and a Nikon D5200 attached to a Zeiss IM35 at the Biologische Anstalt Helgoland (BAH). For several photos, "focus stacking" was used to achieve an improved depth of field. The images were captured either manually, with an automatic "bracketing" extension ring (Helicon FB Tube), or with an automatic focus stacking rail (WeMacro). Images were then stacked using the "Helicon Focus" software. Focus stacking was also used for several pictures of histological specimens. To achieve high-quality overview pictures, several individual images were usually combined into panoramas (stitching) to achieve the necessary resolution for zooming into the pictures of the ebook edition of this book. Cameras: Nikon Z30, Z50, D300s, D7500, D500, and D800. Lenses: Nikon Z DX 16-50/3,5-6,3 VR, Nikon Z DX 24 mm/1,7, Nikon Z 40 mm/2, Nikon Z DX 12-28 mm/3,5-5,6, Nikon AF-S Micro Nikkor DX 40 mm/2,8 G, Nikon AF-S Micro Nikkor DX 85 mm/3,5 G ED VR, Nikon AF-S DX Nikkor 16–80 mm/2,8–4 ED VR, Schneider Kreuznach Apo-Componon 2,8/40 mm, Laowa 10 mm Cookie, Laowa 65 mm/2,8 CA-Dreamer Ultra Macro Apo 2x, Laowa 25 mm/2,8 Ultra Macro 2,5–5X, Laowa 25 mm/14 2× Macro Probe. All nonmoving objects were aligned and fixed for photography in a dissecting dish. The use of repro stands with calibrated columns (Kaiser and Pentacon), remote releases, polarizing filters, and LED illumination improved the image quality significantly. Software used included Affinity Serife Photo, Affinity Serife Designer, Adobe Photoshop CS and Adobe Illustrator CS, Helicon Focus, and MacroDLSR.

Images and Illustrations

All images and illustrations were captured or drawn by the authors if not stated otherwise. Copyright, 2024, Achim Paululat and Günter Purschke, all rights reserved.

References

Urry, L. A., Cain, M. L., Wasserman, S. A., Minorsky, P. V., & Reece, J. B. (2019). *Campbell biologie* (11. Aufl.). Pearson Studium.

Porifera (Sponges)

► **Summary** Sponges are aquatic and, as adults, sessile and nearly immobile multicellular animals. Their body is called amorphous because no precise body axes can be recognized. Sponges lack distinct organs for circulatory, respiratory, digestive, excretory, and reproductive functions. Instead, the water flow system and the individual cells provide all of these essential physiological tasks. Sponges filter food particles out of the water flowing through them. The water flow is generated by specific flagellated cells called choanocytes. Almost all of the 10,000 known sponge species are marine, and only a few, about 150 species, live in freshwater. However, most people would be surprised to learn that the freshwater sponge *Spongilla lacustris* lives in almost all of our local rivers, ponds, or creeks.

When students collect sponges on a marine biology field trip, many are initially surprised that these often very colorful organisms are indeed multicellular animals and not plants. Sponges (Porifera) belong to the first group of truly multicellular animals (Metazoa) that colonized our earth. Their fossil history can be traced back at least to the early Cambrian 540–480 million years ago; however, they may be much older, since trace chemicals believed to come from sponges indicate an origin 850 million years ago. For this reason, they represent an important milestone in the evolution of all animals living today. Whether they are the recent representatives of the first branching lineage in the phylogenetic tree of all Metazoa (see Fig. 1.1) is a current subject of controversy. In this book, we still follow the *"Porifera first"*-hypothesis. This inevitably results in the assumption that many typical properties assumed for all other metazoans, and often for animals in general, are primarily lacking in Porifera. Accordingly, the typical "animal" characteristics only emerged after the sponges had branched off from the other metazoans. Competing hypotheses on the evolution of the Metazoa claim that the typical features of animals were initially present in sponges but were then lost again during evolution. According to this, sponges' extremely simple body structure only developed secondarily.

Morphology of Sponges

Sponges do not have a fixed body shape defined by the so-called symmetry axes. They show neither a radial symmetry like a sea anemone or jellyfish nor a bilateral symmetry like an insect or a human. Right and left, dorsal and ventral, anterior and posterior are terms not meaningfully applicable to sponges. Instead, the sessile sponges often grow irregularly or adapt their growth to the substrate and environment. Their body form is, therefore, described as amorphous. The simple body organization is reflected in the complete absence of any organs. Sponges have no intestine, kidney, heart, or vascular system. Instead, all body functions are performed at the level of individual cells. In older textbooks, sponges were described as animals lacking true epithelia. However, recent investigations showed that their outer cell layer seals the inside from the exterior, although morphologically distinct cell–cell junctions appear to be absent. Before discussing this important aspect, let's look at how sponges feed. Sponges continuously filter food particles from the water flowing through them. Such a microphagous diet is not unusual for many marine animals, as the seawater contains a multitude of microscopic dead organic particles, prokaryotic or eukaryotic microorganisms (POM, *particulate organic matter*), as well as dissolved organic substances (DOM, *dissolved organic matter*). However, the concentration of these particles in the water is relatively low, so a large amount of water must be filtered to ensure the animals do not starve and stop growing. Some sponges live in symbiosis with microorganisms, which most likely help in the uptake of nutrients. The sponges have uniquely solved this biological problem, and their filtration system is among the most efficient in the animal kingdom. On average, every five seconds, a volume of water that corresponds to a sponge organism's entire body volume passes through it. This is only possible because the body of a sponge bears numerous cavities and channels connected through pores to the external medium. The scientific name of sponges, Porifera, is derived from the Greek and Latin words *ho poros*—"the passage, the pore" and *ferre*—

© The Author(s), under exclusive license to Springer-Verlag GmbH, DE, part of Springer Nature 2025
A. Paululat, G. Purschke, *Metazoa – Morphology and Evolution of Animals*, https://doi.org/10.1007/978-3-662-69904-1_2

"to bear." The internal cavities are referred to in their entirety as the spongocoel. Water flows into the spongocoel through the pores and accumulates in a central cavity, usually called the atrium. From there, the water leaves the spongocoel mainly through a single outflow opening called the osculum (plural oscula). The flagellar movement of choanocytes generates the continuous water flow and flagellated collar cells lining the inner walls of the sponge body (Fig. 2.1). The water flow rate within the sponge body varies greatly. When passing the choanocytes, the water flow is slow, which is essential to provide enough time to exchange nutrients, gasses, and other components. Inside the sponge, the water velocity is regulated by the cross-section diameter of the pipes. So, fluids move more slowly in pipes with large diameters and accelerate in narrower pipes. This is, in the sponge body, delineated by choanocytes, which form the so-called choanoderm. When the water leaves the sponge body via the atrium and oscula, it passes through narrower tubes and thus has a higher speed. A movie showing fluorescein injection into sponges to visualize water flow can be found at figshare: sn.pub/xr9qhi

The complexity of the sponge body can vary greatly depending on the species. Three general body forms have been described, mainly distinguished by the arrangement of choanocytes and the internal channel system. The choanocytes line a single central atrium in sponges of the asconoid design. In contrast, in sponges of the syconoid design (Figs. 2.2, 2.3, and 2.4), all choanocytes are located in radial tubes or folds. The radial folds form as invaginations of the body wall to increase the surface area of the chambers, and thus increase the possible filtration performance. In the leuconoid design, the incoming water first enters several chambers lined with choanocytes before being assembled in a central main chamber and leaving the body from there.

A single cell layer delineates the sponge body from the outside. In sponges, pinacocytes form the outer body layer, called the pinacoderm. The pinacoderm borders the sponge body on the outside and lines the central cavity and the atrium in syconoid and leuconoid sponges. In the choanocyte chambers or—in asconoid sponges—in the atrium, choanocytes replace pinacocytes. The space between the two pinacoderm layers is filled with an extracellular matrix, which

Fig. 2.2 *Sycon raphanus*, radish sponge, live image, Roscoff, France

holds all other cells. This matrix is called mesohyl in sponges, purely for historical reasons. This body structure limits the stability of the body. Because of the thinness of the actual sponge body, asconoid sponges reach a maximum size of only about 1 cm by 1 mm in diameter (Fig. 2.5). Larger sponges of this architecture would become mechanically unstable. In addition, the number of choanocytes would be too small to generate a sufficient filtration current for nutrient supply. In sponges of the syconoid type, the atrium is smaller and, due to their relocation into tube-like chambers, the number of choanocytes is larger. Therefore, syconoid sponges can reach larger body dimensions (Fig. 2.2). In the leuconoid type, the physical ratio becomes even more favorable; thousands of flagellated cell chambers interconnected by a network of thin channels allow for high filtration efficiencies and large body dimensions, which can be in the meter range. Such sponges can contain up to 10,000 small

Fig. 2.1 Choanocyte, with the flagellum and collar built by microvilli. Modified from wall charts from the Biological Collection of the University of Osnabrück

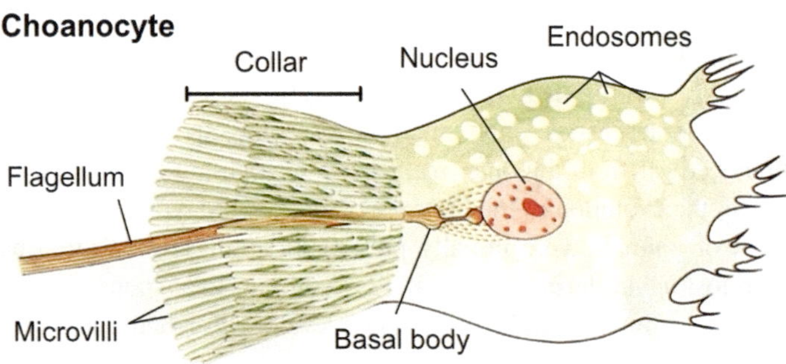

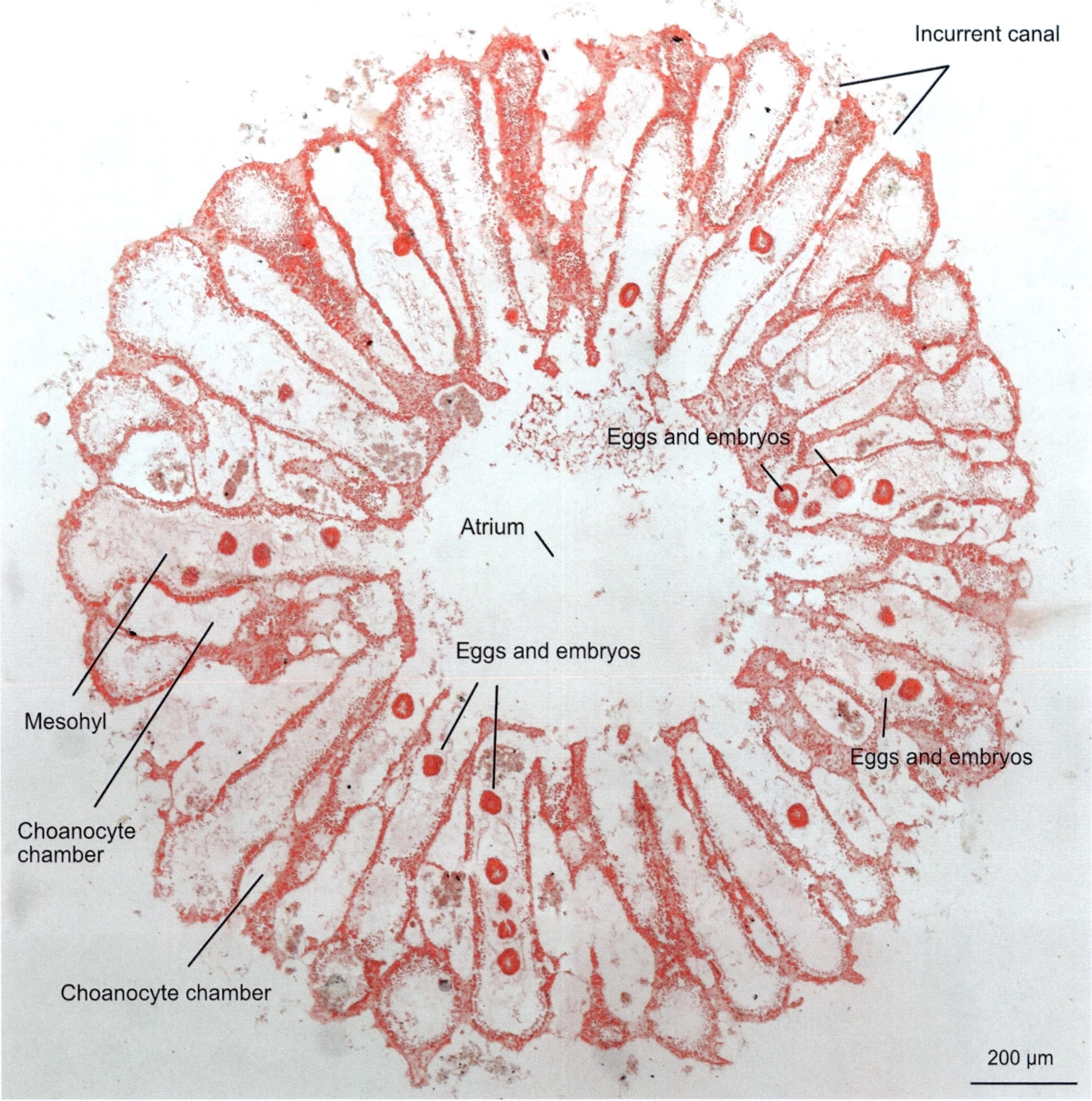

Fig. 2.3 *Sycon raphanus*. Histology. Complete cross-section through the middle region of a sponge. The central atrium has tube-like chambers lined with choanocytes, and incurrent canals with pinacocyte channels; embryos are present in the mesohyl. Specimen from the Zoological Teaching Collection, Philipps University Marburg

chambers per mm^3 lined with choanocytes. Each chamber is about 20–40 µm in diameter.

As already introduced above, choanocytes are essential to generate water movement in the sponge and its food intake. Choanocytes are fascinating cells from an evolutionary point of view, appearing repeatedly in animals in numerous modifications and being crucial for many physiological functions. The basal part of the choanocyte is embedded in the mesohyl, and the apical side faces the atrium or the chambers, through which the water flows. Each choanocyte harbors a single, long flagellum that is surrounded by a ring (collar) of 30–40 long microvilli (Fig. 2.1). Microvilli, unlike flagella or cilia, are not supported by tubulin, but rather by actin filaments, and therefore lack a basal body with axonemal microtubules and are immobile. Thus, choanocytes of sponges structurally resemble the choanoflagellates, unicellular organisms (protists), which are considered to represent the closest relatives of Metazoa (Fig. 1.1). This hypothesis is not only supported by morphological features but also by numerous molecular data. There is also a sensory cell type, the collar receptors,

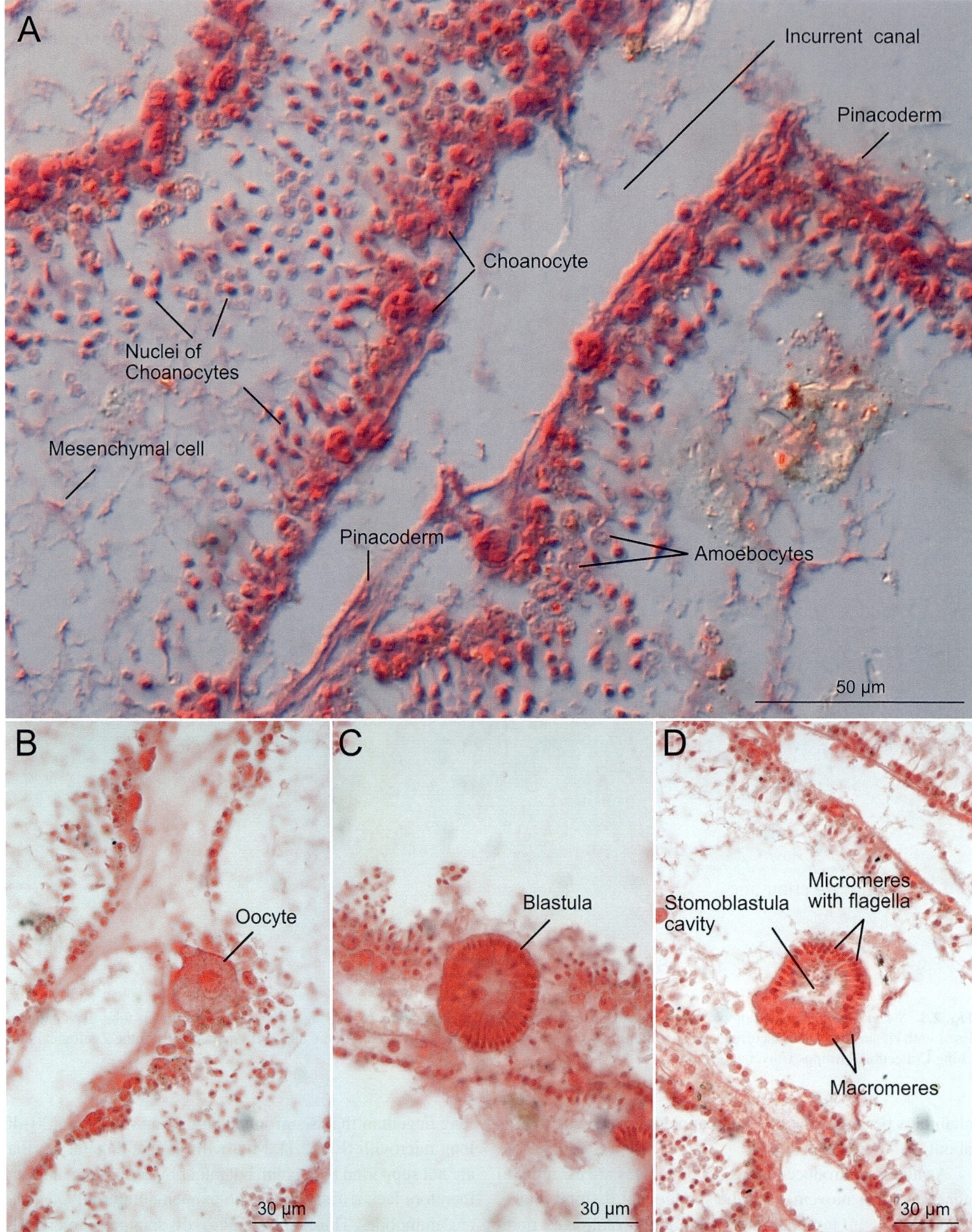

Fig. 2.4 *Sycon raphanus*. (**a**) Enlargement of flagellate chamber and inflow channel, choanocytes, amoebocytes, and pinacocytes. (**b–d**) Developmental stages. (**b**) Oocyte. (**c**) Early blastula. (**d**) Stomoblastula. Specimens from the Zoological Teaching Collection, Philipps University Marburg

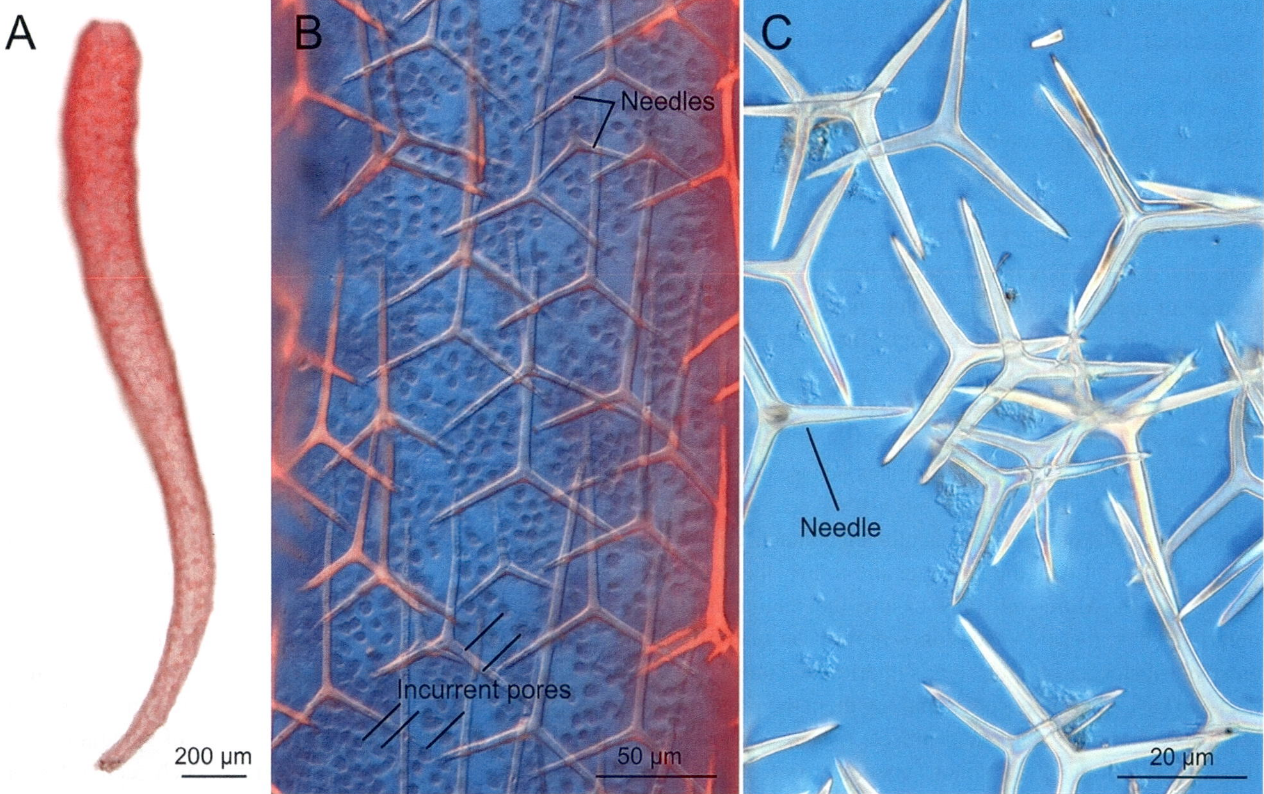

Fig. 2.5 *Leucosolenia blanca*. (**a**) Total view of a specimen approximately 2 mm in size. (**b**) Magnified section. Water flows into the body through ostia, calcareous needles (spicula) embedded in the mesohyl, Nomarski interference contrast. (**c**) Isolated spicula at high magnification. Specimens from the Zoological Teaching Collection, Philipps University Marburg

which are very common in aquatic invertebrates, that is similar to choanocytes. However, both their length and the number of their microvilli (eight or ten) are significantly lower. In sponges, the choanocytes secrete mucus, a particular type of extracellular matrix that remains between the microvilli and forms a viscous film. Since choanocytes form the walls of the central chambers, the permanent movement of the flagella generates a continuous water flow. The water is first directed through the 20–50 µm large pores (ostia), which are formed by porocytes situated between the pinacocytes in the outer wall of the sponge body, into the inflow channels (Fig. 2.4). Thus, larger particles do not even enter the sponge body. In these channels, the diameter continuously decreases, and particles with a diameter of less than 5 µm are phagocytized by the endopinacocytes or archaeocytes; cell types, which we will discuss below. Thus, only the smaller particles are driven through the mucus film of the collar flagellate cells and then taken up by endocytosis of the choanocytes. Sponges, therefore, still feed like the unicellular ancestors of Metazoa; there is no extracellular digestion. The ingested particles are digested intracellularly, and nutrients are taken up by the archaeocytes and passed on to the other body cells. Inorganic particles that enter the sponge are also taken up by archaeocytes and released via the outflow channels.

Recommended Material

- **Collecting sponges:** Sponges can be collected on marine biological or limnological excursions for later use. The animals are fixed for 24 h in Bouin's solution (15 parts saturated, aqueous picric acid, five parts 37% formaldehyde, one part pure acetic acid) and then dehydrated over an ascending ethanol series. Ready-to-use Bouin's solution is sold by several lab suppliers. Afterward, the sponges are washed several times for a few days in 95% ethanol and then stored in 95% ethanol. Spicules can be freed from organic material by bathing sponge tissue in sodium hypochlorite (commercial bleach, e.g., Clorox®). After the tissue has macerated, they are washed with water. Small samples can now be inspected using a microscope (Fig. 2.5).
- **Histological sections:** Science suppliers often sell microscopic sections of *Sycon raphanus*, *Grantia* sp. or related species (*Sycon* sp. = *Scypha* sp. check the World Register of Marine Species at https:// www.marinespecies.org for valid taxon names) (Fig. 2.2).

Do Sponges Possess True Epithelia?

A so-called true epithelium always consists of cells that fundamentally have an apical-basal polarity. The apical side faces the external environment or the lumen of an organ. For instance, the apical membrane is the only one that has the potential to develop cilia or microvilli. In contrast, the basal side is connected to the adjacent tissue via a specialized extracellular matrix, the basal lamina. For this task, morphologically distinct junctions, called hemidesmosomes, are formed in the epithelial cells, which, inside these cells, are linked to the cytoskeleton by specific proteins. Specific band-shaped cell–cell connections, which, in invertebrates, are formed by the so-called belt desmosomes (= zonulae adherentes) followed by septate junctions (the functional pendant to tight junctions in vertebrates), cause the epithelial cells not only to join together but also to seal the interior from the exterior medium. Such cell–cell contacts ensure the mechanical stability of the cell layer and control the exchange of substances. Moreover, they ensure the maintenance of polarity. Thus, these contact zones prevent an uncontrolled entry of substances into the body's interior, and thus form the decisive basis for homeostasis, making them the prerequisite for the evolution of nerve and muscle tissue. Sponges lack these morphologically distinct junctions present in most invertebrate epithelia. However, the cellular properties mentioned are not absent in sponges; instead, sponges can seal their internal milieu from the exterior, i.e., the water surrounding the sponge and passing through its so-called spongocoel. Sponges are capable of maintaining a considerable electrochemical transepithelial potential between the exterior and interior. Pinacocytes also allow for homeostasis, a prerequisite for intracellular communication and excitation. So, the sponge's internal medium differs from the water that bathes the outer membranes of pinacocytes and choanocytes. Although lacking morphologically distinct belt-shaped adherens and septate junctions, as well as hemidesmosomes, they possess the respective proteins that make up these junctions. Moreover, most epithelium-specific genes present in other metazoans have been found in sponges as well. Since the sponge body comprises only two cell layers bordering the mesohyl, sponges do not need organs for osmoregulation or excretion. They release waste material and metabolic byproducts directly into the surrounding water via the pinacocytes. However, sponges are also protected from the outside by these cells, which have an apical-basal polarity as well. So, pinacocytes show all physiological properties of a true epithelium, despite being morphologically distinct from a usual epithelial organization. Thus, the pinacoderm of the sponges may be called an epithelium, although certain authors prefer to call it epithelioid due to the lack of morphologically distinct junctions. These cells also line all inflow and outflow channels of syconoid and leuconoid sponges. Pinacocytes have a flattened plate-like shape and are usually not ciliated.

The respective linings of the inner surfaces and the choanocyte chambers are called the choanoderm, consisting of only one cell type, the already-mentioned choanocytes (Fig. 2.4).

The Extracellular Matrix in Sponges: Home of Several Specialized Cell Types

Between the outer pinacoderm and the inwardly directed choanoderm, we find a highly specialized extracellular matrix, the mesohyl, which has, so to speak, the function of connective tissue (Figs. 2.3 and 2.4). Sponges harbor about 20 different cell types that are embedded in the mesohyl. Many cell types are derived from archaeocytes that move permanently through the matrix like amoebae. They take over the food particles endocytosed by choanocytes and digest and distribute them in the sponge body. Archaeocytes produce digestive enzymes such as lipases, amylases, and proteases and also take on excretory functions. Archaeocytes produce digestive enzymes, such as lipases, amylases, and proteases, while also fulfilling excretory functions. Their unique property lies in their ability to form totipotent stem cells from which pinacocytes, choanocytes, oocytes, and all other sponge cell types originate. Moreover, sclerocytes, which are responsible for the secretion of spicula (skeletal needles) (Fig. 2.5) and spongocytes, which secrete the protein spongin, differentiate from archaeocytes as well. However, sperm originate in sponges from choanocytes. Spongin is related to collagen IV, found in the ECM of all other metazoans. Here, it forms fibers in the mesohyl, which, together with the spicula and skeletal needles, usually constitute the supporting framework of the sponge body. Sponges also produce a "true" collagen IV, secreted as collagen fibers by collenocytes and lophocytes into the mesohyl. The highly variable body shape of sponges is supported by the sum of the embedded skeletal elements of calcium carbonate or silicon dioxide (spicules) and protein fibers (spongin) (Fig. 2.5). Some health and beauty stores sell so-called natural bath sponges. Indeed, they are made from *Spongina officinalis*, the bath sponge. Bath sponges are among the few sponge species that lack spicules. During production, the spongin skeleton is freed of cellular debris by rolling, washing, and exposure to moist air. Natural sponges are almost always quite irregular in shape, and today they come mainly from breeding farms.

Reproduction and Development

As sessile animals, sponges must find new ways to spread and colonize new habitats. The answer to their predicament lies in how sponges and many other sessile animals reproduce in the sea. They disperse via free-moving larvae that drift passively with the currents in the ocean. At that moment, the larvae are an essential component of the plankton. They can only adjust their vertical position in the water column using their outer flagellated cells. When a suitable habitat is reached, the larvae switch to a sessile lifestyle and undergo metamorphosis. Thus, the larvae contribute significantly to the colonization of new habitats. Like most metazoans, sponges primarily reproduce sexually. The gametes are formed meiotically. The germ cells, spermatozoa, and oocytes, are embedded within the mesohyl, the extracellular matrix of sponges, and share the space with many somatic cells, for example, archaeocytes and choanocytes. The immense evolutionary advantage of sexual reproduction lies in creating new genetic variants by fusing two gametes, both of which have received unique genetic material during meiosis. Through the random distribution of maternally and paternally acquired chromosomes, through crossing over, and through frequently occurring spontaneous mutations, genetically unique egg and sperm cells are generated. Of course, this basic principle applies to all sexually reproducing organisms.

Oogenesis and spermatogenesis follow the general pattern found in all Metazoa, with the formation of polar bodies and only one functional oocyte in oogenesis and the specific sperm structure with head, midpiece, and tail. Oocytes, which differentiate from choanocytes or archaeocytes, and sperm, which originate from choanocytes, are built directly in the mesohyl, as sponges do not form gonads (Fig. 2.4). The sperm—and, in some oviparous species, also the oocytes—are released into the open water with the water current and leave the sponge body through the oscula. In oviparous species, therefore, fertilization of the eggs takes place outside the sponge body and, after cleavage, leads to the development of a planktonic larva. The larval forms of sponges are very diverse; there is no typical sponge larva. Some, like the coeloblastula or amphiblastula, resemble the single-layered blastula stage occurring in the ontogeny of many other Metazoa. In contrast, the parenchymula larva, as the name suggests, is filled with cells and has no counterpart in other groups.

In viviparous species, the oocytes remain in the mother animal. Choanocytes endocytose sperm from another animal of the same species. These choanocytes then lose their flagellum and become amoeboid-moving transport cells. They actively bring the sperm to the oocytes, where fertilization occurs. Sperm from foreign species are also endocytosed, but then broken down and digested. The early development of the zygote thus takes place in the parent animal in viviparous species. Hence, different developmental stages are often found in the mesohyl of adult sponges (Fig. 2.4). During later embryonic development, the free-moving flagellated larval form (coeloblastula, amphiblastula, or parenchymula larva) is created, subsequently leaving the sponge and serving for the distribution of the species. Sperm and oocytes are, in most cases, formed within an individual at different times. Thus, most sponges are consecutive hermaphrodites, although a few gonochoric species are also known.

Gonochorism is a sexual system in which each organism is either male or female. In addition to sexual reproduction, all sponges are also capable of asexual reproduction through fragmentation, budding, or, in freshwater sponges, through the formation of gemmules. These resting stages differentiate into a new sponge after a diapause stage.

Resting Stages of Freshwater Species

Freshwater sponges are relatively inconspicuous in their body shape and color, but can cover large substrate areas in crust-like growth forms. One may find the most common native species, *Spongilla lacustris* (Fig. 2.6a), in practically every creek, stream, river, or lake. The body of these freshwater sponges is completely broken down in the fall. Before this, dormant stages, known as gemmulae, are formed. Gemmules are very elaborate, highly resistant resting stages. The formation of gemmules starts with food-filled archaeocytes, now called amoebocytes. When enough amoebocytes have formed a discrete mass, this archaeocyte core becomes enveloped by hardened spongin sheets, forming a highly protective shell that allows the organism to remain in a dormant state over the winter (Fig. 2.6b, c). In the following spring, the archaeocytes migrate through the micropyle of the gemmules and form new sponge bodies by differentiating all sponge cell types from the archaeocytes.

Diversity of Sponges

On our marine biology excursions, we always encounter sponges of different body shapes and colors in the intertidal zone. Some of the nearly 10,000 described Porifera species reach a size of up to 3 m. These giant sponges, primarily found in Caribbean reefs, have several cubic meters of water flowing through their bodies per minute. Species found in colder waters do not reach such a size. Sponges can attain enormous population densities. For example, they locally account for up to 75% of the benthic biomass in the Antarctic Ocean. Most sponges are marine inhabitants, from the intertidal down to the deep sea. Only about 200 species inhabit limnic biotopes.

The largest group, roughly two-thirds of all sponges, is the demosponges (Demospongiae), all members of which belong to the leucon type. The demosponges include the ordinary bath sponge (*Spongilla officinalis*) and all freshwater sponges, such as *Spongilla lacustris*, the most common freshwater sponge in central Europe (Fig. 2.6a). Closely related to them are the glass sponges, Hexactinellida. They are characterized by six-rayed needles of amorphous silicon dioxide. Glass sponges live exclusively in the deep sea. About 600 species have been described worldwide. An example of a glass sponge is Venus's flower basket sponge, *Euplectella aspergillum*, shown in Fig. 2.7. Pairs of shrimp of the species *Spongicola venustus* often live inside the body. The shrimps grow inside the sponge to a size that

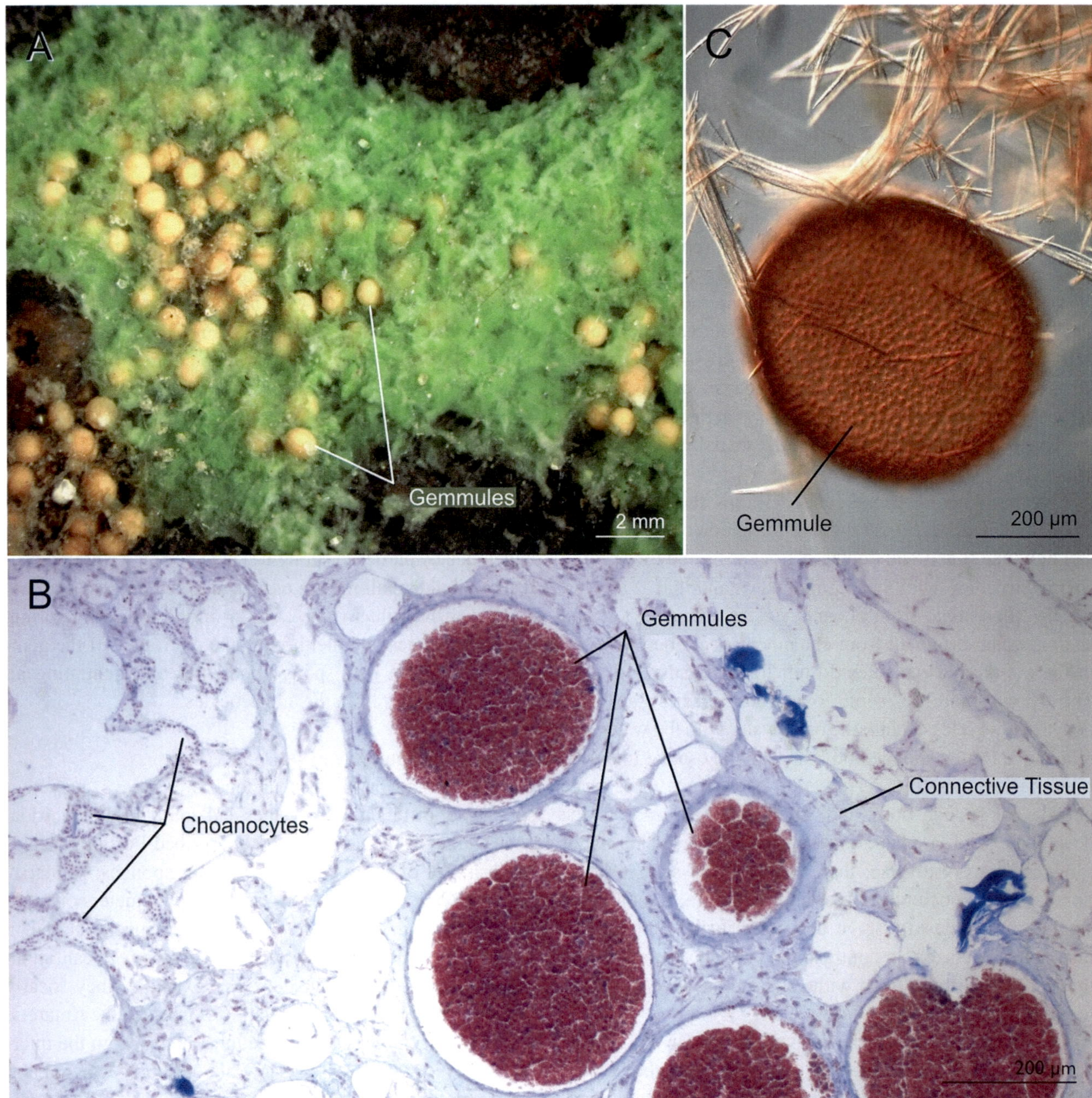

Fig. 2.6 *Spongilla lacustris*. (**a**) Live image of a growing colony, river Hase, Osnabrück. (**b**) Histological section through *Spongilla lacustris* with gemmules. Hematoxylin-Eosin. (**c**) Living gemmule with spicules. Specimens from the Biological Collection of the University of Osnabrück

prevents them from ever leaving it. As a unique adaptation to their way of life, the symbiotic shrimp no longer form a calcified carapace, because they are protected from predators through the spicules of their host sponge. In addition, they use the water stream produced by the sponge to filter food particles. Due to this symbiosis, these sponges are known as *Venus's flower baskets*.

The exclusively marine calcareous sponges (Calcarea) are characterized by sclerites made of calcium carbonate (calcite, double spar). They include, for example, *Sycon raphanus*, which is presented in this chapter (Figs. 2.2, 2.3, and 2.4). Only in this group, which includes about 1000 species, are sponges of the sycon, leucon, and ascon type found.

Sponges in Biotechnology and Bionics

To date, well over 10,000 different natural substances have been isolated from marine invertebrates. Sponges are particularly interesting as a source of novel chemical compounds. For sessile organisms, producing highly active, sometimes toxic compounds, is essential to protect themselves from predators or prevent other species from colonizing the same site. The pharmaceutical industry uses a diverse spectrum of natural substances, for example, to develop new drugs. Recently, it has been found that many natural substances isolated from marine invertebrates were produced by symbiotic microorganisms (bacteria, cyanobacteria, fungi, and algae). In some sponges, up to 50% of the biomass can be attributed to microorganisms. The glass sponge shown in Fig. 2.7, *Euplectella aspergillum*, also serves as a model for calculating novel architectures with unique structural statics. Inspired by sponges, the search is on for maximum strength with minimal weight and material effort. Our movie clip collection, which includes movies showing sponges, can be found at figshare: sn.pub/xr9qhi.

Fig. 2.7 *Euplectella aspergillum* (Venus's flower basket), a glass sponge. Specimen from the Biological Collection of the University of Osnabrück

▶ **Summary** The body of Cnidaria shows a radial symmetry and is composed of two adjacent epithelia connected by an extracellular matrix. Three of the four major groups show a life cycle including an alteration of generations from a sessile or sedentary polyp to a free-swimming medusa. Their exclusive feature is their cnidocytes, a highly specialized cell type that plays an important role in prey acquisition. A high regenerative capacity combined with unique adaptations of reproductive modes and only a few natural enemies have made cnidarians a widespread metazoan group. Apart from some freshwater species, most cnidarians live in marine habitats.

The aquatic cnidarians are an extraordinarily diverse and successful group of animals, with about 10,000 species. Most cnidarians live in marine habitats, either attached to the substrate or as free-living forms swimming and floating in the water column. Only a few species live in freshwater, such as the polyps of the genus *Hydra*, presented in this chapter. Cnidarians are divided into four subtaxa: sea anemones (Anthozoa), jellyfish (Scyphozoa), box jellyfish (Cubozoa), and hydrozoans (Hydrozoa). Only the latter three taxa exhibit a dimorphic life cycle that includes two entirely different body morphologies, the so-called morphs. These are a sessile, asexual reproducing polyp and a free-living (pelagic) gonochoristic medusa that reproduces sexually. Each body form is, therefore, closely associated with particular biological functions, such as reproduction and distribution of the species. However, in the sea anemones (Anthozoa), the sister group of all other cnidarians, only the sessile polyp form exists, which itself becomes sexually mature and produces eggs and sperm or the individual reproduces asexually (Fig. 3.1).

A specific feature of all Cnidarians is their radial symmetry. The body of a hydra or a medusa is built symmetrically around the oral-aboral axis (Fig. 3.1a, d). The tentacles, used for catching prey and transporting the food toward the mouth opening, are arranged radially around the mouth. Since all Cnidaria are sessile, sedentary, or pelagic, a radial body symmetry is advantageous for capturing food from all directions. The body of a polyp is divided into four regions: the pedal disc, the trunk or column, the tentacle part or oral disc, and the elevated hypostome region with the mouth opening (Fig. 3.1a, b). Polyps attach themselves with their pedal disc to different substrates and wait for passing prey, which they then grab with their tentacles. They belong to ambush predators or the so-called sit-and-wait predators, which recognize the prey by sensory cells in the tentacles. The trunk region harbors the gastric cavity or coelenteron, which, in hydrozoan (and anthozoan) polyps, extends into the tentacles (Fig. 3.2a). The cone-shaped hypostome region rises above a ring of tentacles up to the single body opening, which serves both as a mouth and an anus. The most common form of reproduction of the polyps is asexually by budding. In this process, often induced by the local accessibility of large amounts of food, the daughter hydras are formed in the transition area between the stem and trunk region. The growth and budding of daughter hydra occur within only a few days. Figure 3.1b shows an animal with two buds. Some hydra species, like the marine living *Protohydra*, also reproduce through transverse division (Fig. 3.1c).

Besides asexual reproduction, which allows for the attainment of high population densities in a very short time and thus is highly advantageous for populating a habitat as quickly as possible, hydrozoan and scyphozoan species also reproduce sexually, generating a new life form, the free-swimming medusae. Medusae are female or male and reproduce sexually by producing eggs or sperm. The sexual mode of reproduction is often induced only under suitable environmental conditions. Water temperature, day length, moon phase, or limited food supply are essential to inducing this process. Such a change between different reproductive modes and morphologically different morphs is called

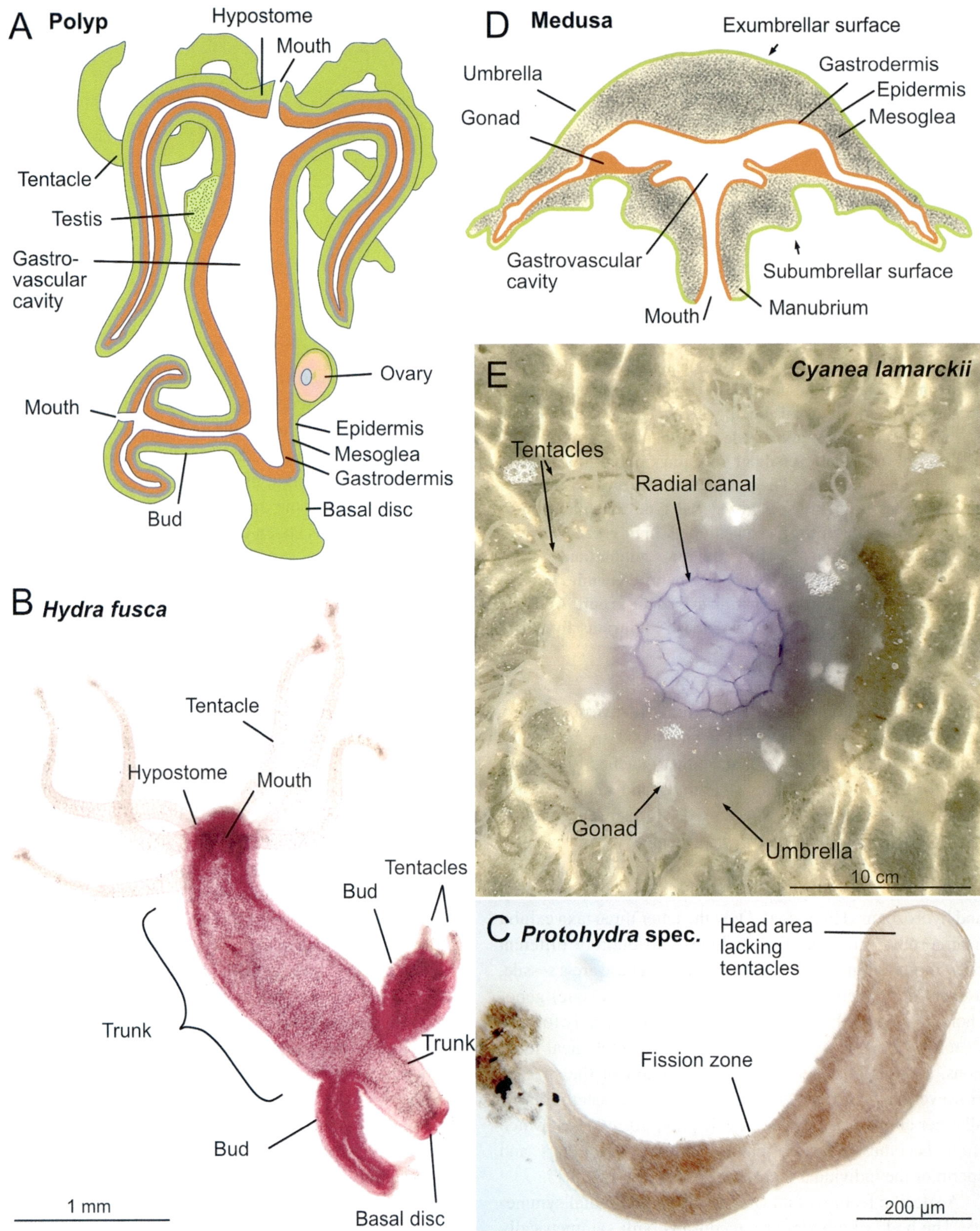

Fig. 3.1 Cnidaria, basic body form. (**a**) Morphology of a hydroid polyp, according to various authors. (**b**) *Hydra fusca* with two buds, stained. Specimen from the Biological Collection of the University of Osnabrück. (**c**) *Protohydra* sp., in division. Specimen from the Zoological Teaching Collection, Philipps University Marburg. (**d**) Anatomy of a medusa, according to various authors. (**e**) Medusa, *Cyanea lamarckii* (Blue jellyfish), live, North Sea, Borkum, Germany

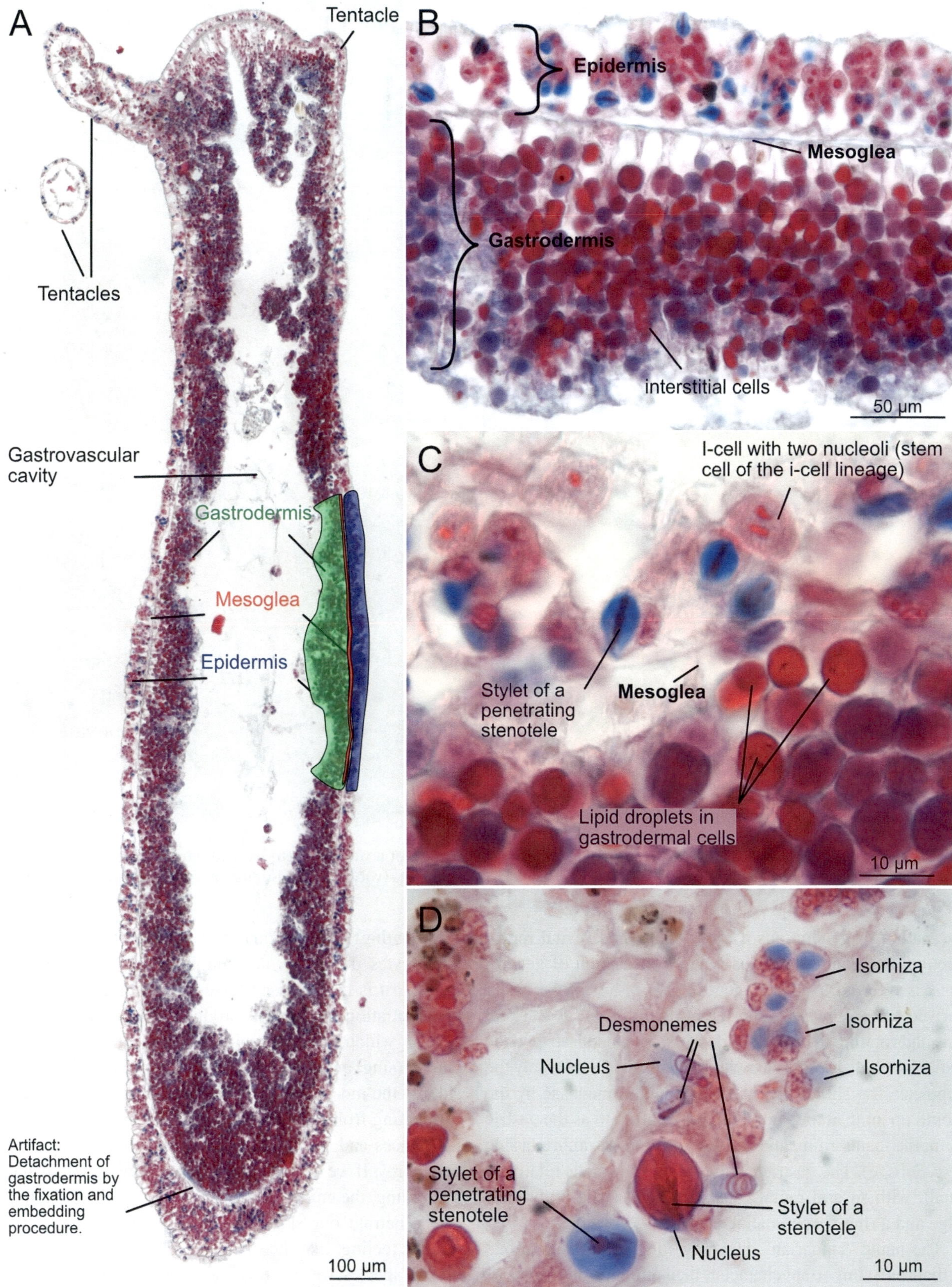

Fig. 3.2 *Hydra* sp., stained. (**a**) Complete longitudinal section through an animal with sectioned tentacles. (**b**) Epidermis, mesoglea, and gastrodermis. The mesoglea appears bluish in Azan-stained sections. (**c**) Nematocysts in the epidermis. Lipid droplets in cells of the gastrodermis. (**d**) The ectodermal battery cell of a tentacle with various types of nematocysts. Specimen from the Zoological Teaching Collection, Philipps University Marburg

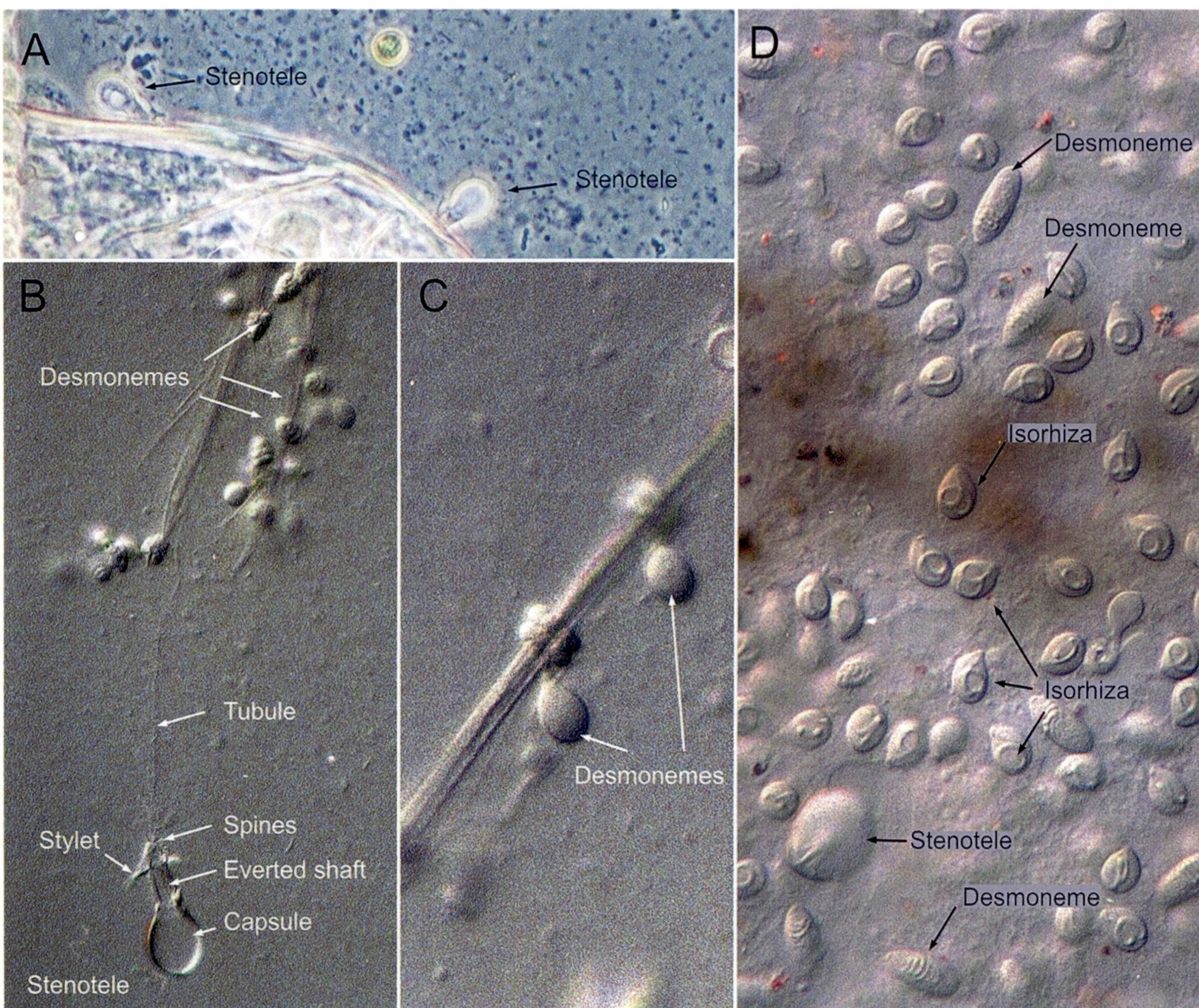

Fig. 3.3 *Hydra* sp., discharged cnidocytes. (**a**) Nematocytes stick in the cuticle of a water flea. (**b**) Isolated, fired stenotele and desmonemes. (**c**) Volvent (desmonemes) on the bristle of a water flea. (**d**) Stenoteles, desmonemes (volvents), and isorhiza. **a – d**: Phase contrast micrographs

alternation of generations. Here, asexual and sexual reproduction modes alternate, this type of alternation of generations is referred to as metagenesis. Medusae can be easily derived from polyps by shortening them in the aboral-oral axis and rotating them by 180 degrees. The medusae face the ground with their oral side; with the polyps, it is precisely the opposite. The aboral-oral axis is defined in medusae by the mouth opening at the end of a tube, referred to as the gastric stem, the manubrium, and the opposite convex umbrella side (Fig. 3.1d, e). The gastric cavity in the medusae is relatively small, with several radial canals running to the umbrella's edge and a ring-shaped radial canal.

The name "cnidarian", meaning stinging animals, refers to their most important character, their stinging cnidocysts (= cnidae or nematocysts), stinging organelles that are fired from specialized cells, the cnidocytes (= nematocytes). This cell type, the cnidocytes, exists exclusively in Cnidaria, thus representing an autapomorphy of this taxon (Fig. 3.3). Cnidocytes develop from cnidoblasts. In cell biology, the term blast is usually used for undifferentiated cells. During differentiation, cnidoblasts produce a giant intracellular vacuole in which, upon complex reorganization, the so-called cnidae (sing. cnida) are produced. Cnidae are currently considered the most specialized and complex secretory vesicles originating from the Golgi apparatus. Cnidae can be formed like lances and typically contain toxins that paralyze or kill small prey. If we accidentally touch a stinging jellyfish while swimming, the cnidocytes release their "cargo," and the cnidae penetrate our skin. The toxins cause a more or less intense feeling, like the injection of hot needles, depending on the species we meet. Some stinging cnidarians possess potent toxins that can trigger painful skin reactions or even an allergic shock and paralysis. Powerful and dangerous toxins have been detected, for example, in the Australian box jelly, also known under their vernacular name sea wasp (*Chironex fleckeri*, Cubozoa) or the Portuguese man o' war (*Physalia physalis*, Hydrozoa, Siphonophorae) found at Atlantic coasts and, in recent years, also at Mediterranean beaches. The sea wasp and the Portuguese man o' war are among the most venomous animals to humans. The painfully stinging yellow sea nettles or lion's mane jellyfish (*Cyanea capillata*) and the related blue jellyfish (*Cyanea larmarckii*) can be found in the North and Baltic Seas. The strongly nettled luminous mauve stinger or purple-striped jelly (*Pelagia noctiluca*) lives in the Mediterranean Sea and glows in the dark when touched. However, the most common jellyfish on European coasts is the moon jellyfish (*Aurelia aurita*), which is harmless to humans and occasionally washes up in huge numbers on bathing beaches.

Structure of the Body Wall

All Cnidaria have a body wall that consists of two true epithelia, which are connected to each other with their basal parts by an extracellular matrix (Figs. 3.1 and 3.2). The outer epithelium is called the epidermis in all Metazoa and originates embryonically from the ectoderm. The epithelium lining the gastric cavity, the coelenteron, is called the gastrodermis and originates embryonically from the endoderm. Epithelia are always formed by cells that have a defined apical-basal polarity. They are connected by a belt-like connection complex located near the apical membrane. In all invertebrates, the connective structures comprise an outer zonula adherens or adhering junction, followed by a septate junction, a sealing junction. In vertebrates, the septate junction is replaced by a sealing junction called the zonula occludens (= tight junction), which lies above the adhering junction. These connective structures have two primary functions: they ensure the lateral anchoring with the neighboring cells and, as a consequence, they separate the intercellular spaces from the external medium so that these cell–cell connections build the structural prerequisite for controlling the internal milieu.

Moreover, all epithelial cells secrete a collagen-containing extracellular matrix at their base, the basal lamina, in which the cells are anchored by an additional junctional type called hemidesmosome. The polarity of epithelial cells can often be seen at the ultrastructural level in the asymmetrical distribution of the cell organelles. In addition, apical and basal cell membranes have different molecular properties. Specific structures such as microvilli and cilia can only be formed by the apical cell membrane and are, therefore, only found in epithelial cells.

In Cnidaria, the extracellular matrix between epidermis and gastrodermis is called the mesoglea, for historical reasons only. At the time of naming, it was not clear that the extracellular matrix is a commonly widespread and exclusive property (autapomorphy) of all Metazoa. The mesoglea is quite thin in *Hydra* and most other Cnidaria. Only in scyphozoan medusae is it massively developed. The cells of the inner epithelium, the gastrodermis, and the outer epithelium, the epidermis, anchor themselves to the ECM by the abovementioned hemidesmosomes using specific transmembrane proteins, the Integrins. The Cadherin-Catenin protein complex primarily creates the lateral connections. In Anthozoa, Cubo- and Scyphomedusae, amoeboid cells move within the mesoglea (the ECM). This is one of the main reasons why some evolutionary biologists consider the mesoglea to be the ancestor of the mesodermal germ layer from which the musculature and other tissues develop in all bilaterian Metazoa.

Cnidaria possess a gastrovascular system. This means that the gastric cavity lined by the gastrodermis simultaneously serves for digestion and distribution of the ingested food (Fig. 3.2). There is no regionalization and specialization of the gastric cavity, as the indigestible food particles must leave the body the same way that they entered. The gastrodermis comprises different cell types: Cells that are specialized in the production and secretion of digestive enzymes for extracellular digestion in the gastral cavity. Such cells are called gland cells. Other cells are specialized in food uptake by endo- and phagocytosis. Through extracellular digestion, Cnidaria can ingest food objects larger than microscopic particles, for example, tiny crustaceans or larval stages from various species floating with the plankton. Extracellular digestion is, therefore, one of the essential evolutionary innovations that ultimately have established the extraordinary diversity and biological success of metazoans. In the gastral cavity, the digestive process is driven by a constant movement of the food pulp, an action in which the continuous ciliary activity of numerous cells of the gastrodermis plays a significant role. Undigested food particles are expelled through the mouth opening. This is also typical for a gastrovascular system, in which a single body opening simultaneously serves as mouth and anus. Both epithelia, the gastrodermis and the epidermis, contain numerous totipotent stem cells, ensuring the body's highly regenerative ability. In many cnidarians, these stem cells continuously replace cnidocytes, neurons, specific gland cells, sperm, and eggs.

Moreover, some species, for example, *Hydractinia echinata*, also replace epithelial cells. Each epithelial cell contains bundles of myofibrils and is, therefore, termed an epitheliomuscular cell. These cells are located in the epidermis and gastrodermis and are responsible for the contraction movements of the animal. In the epidermis, they are usually oriented in the aboral-oral axis, while in the gastrodermis, they are perpendicular, so that they form an antagonistic system of orthogonally arranged fibers for elongation and contraction, which *Hydra* and other solitary polyps use for locomotion.

Nervous System, Sensory Cells, and Sense Organs
Cnidaria possess a diffuse nervous system embedded in the epithelia, comprised by ganglion and sensory nerve cells. The ganglion cells form a neuronal network that extends over the entire body. There is no particular concentration of neurons, except for a condensation around the mouth opening and in the foot disc area. This means that signs of cephalization are absent in cnidarians. In addition to ganglion cells, sensory cells receive, process, and transmit external stimuli ultimately stimulating and coordinating the corresponding movement reactions.

The number of nerve cells increases with the size of the animal. Thus, the entire nervous system of adult large hydras consists of about 6000 interconnected neurons. This corresponds to about 10% of all cells in an animal. Polyps and medusae have functionally separate neuronal networks in the gastro- and epidermis. They coordinate the complex swimming movements or the processing of signals from sensory organs or cells. For instance, the moon jellyfish (*Aurelia aurita*, Fig. 3.4), native to European coasts, has two different neuronal networks. One is located in the epidermis and coordinates food intake with sensory cells on the surface of the umbrella. A second neuronal network accompanies the ring and radial canals and controls the swimming movements of the animal. In *Hydra vulgaris*, four independent neuronal networks were discovered just a few years ago. Three of these nets are found in the epithelia. They are selectively activated during body movements along the aboral-oral axis, stretching movements in response to light stimuli, and radial contractions. The fourth neuronal network is located around the hypostome region and is activated during nodding movements of the tentacle crown.

All Cnidaria possess various receptor cells, including mechano-, chemo-, and photoreceptor cells. These receptor cells occur within the epithelia or are part of more or less complex sensory organs. While polyps generally lack sense organs, medusae commonly possess eyes or statocysts. For example, the medusae of Scyphozoa and Cubozoa exclusively develop complex sensory bodies, called rhopalia, for sensing light and gravity. Rhopalia often appear as club-shaped, tentacle-like structures at the edge of the umbrella or between a pair of flaps, the lappets, of the Ephyra larvae, the juvenile stage of the Scyphomedusae (Fig. 3.4e, f). Rhopalia are sensory organs that have gastrodermal, mesogleal, and epidermal components. They harbor epidermal neurons, a pair of chemosensory pits, a balance organ, the statocyst, and eyes of various designs. Generally, two eyes are present in each rhopalium, one forming a pigment spot and one a pigment cup. So, rhopalia provide information about gravity, water movements, chemicals, and light. The rhopalia of the Cubomedusae, which include the very poisonous box jelly-

fish, harbor lens eyes that enable simple image vision. Cubomedusae are the only cnidarians to target potential prey directly through swimming movements. *Obelia*'s medusa (Fig. 3.5) has eight statocysts, which are evenly distributed at the base of the tentacles (Fig. 3.5c, d). The bubble-shaped statocysts each contain a statolith, whose position is perceived by receptor cells distributed in the wall of the statocyst, and thus by the animal. In fixed animals, statocysts and statoliths are usually not preserved. In contrast to *Aurelia* (moon jellyfish), *Obelia,* which we discuss later as an example of colony-forming cnidarian polyp, possess no eyes. With their sensory organelles, the stinging cells can also perceive stimuli and thus represent a combined sensor-effector cell.

Prey Capture, Cnidocytes, and Cnidocysts
Cnidaria catch their prey using their unique cnidae. These organelles are among the most complicated intracellular structures found in nature. According to current knowledge, they are built by the Golgi apparatus and are flask-like capsules, up to 100 μm or more in length, with a thick wall composed of collagen-like proteins. One side of the capsule is turned inward as a long, coiled, eversible tube; a hinged operculum may cover the prospective firing site. The pressure inside the capsule is immense (approx. 15 MPa), and when cnidocytes, cells with cnidae, are mechanically stimulated, the coiled tube of the capsule shoots out of the cell. Depending on the cnidocyte type, the ejected cnidae can wrap around the prey, bore into its integument, and release toxins, paralyzing or killing the prey (Fig. 3.2d). Cnidocytes (referred to as nematocytes in older literature) derive from stem cells and migrate upward from the body column to colonize the tentacles by occupying apicobasal tunnels. Specialized epitheliomuscular cells in the epidermis, the so-called battery cells, enclose and support multiple cnidocytes. Often, in the middle of a battery cell, one to two cnidocysts wait for the mechanical signal to fire the cnidae. About 70% of all cells of a *Hydra vulgaris* are cnidocytes. Cnidocytes are very effective "projectiles" with which all Cnidaria hunt for prey. Suppose prey animals like crustaceans or fish get between the tentacles that are densely populated with cnidocytes. In that case, they trigger the firing of adjacent cnidocytes through the mechanical stimulus and are ultimately paralyzed by their poison (Fig. 3.3). Once a cnidocyte has fired, it dies.

Cnidocytes are stimulated by mechanical stimuli to release their cnidae. They possess a modified cilium, the cnidocil, and 8–10 long microvilli arranged in a circle around the cnidocil. On the one hand, the cnida is under a very high osmotic internal pressure; on the other hand, the unique folding of the filament part contains an immense torsional force in itself. At the moment of stimulus recep-

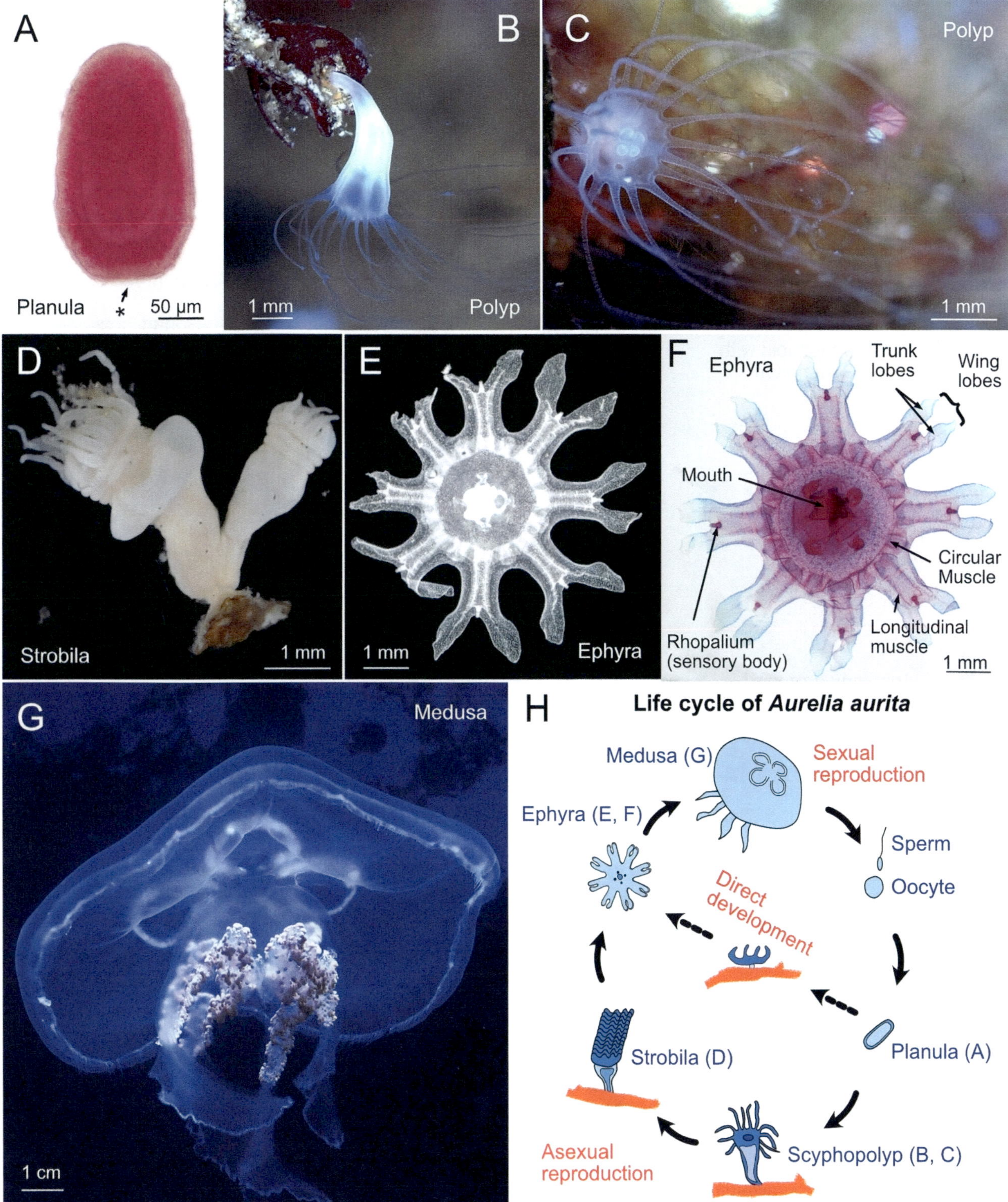

Fig. 3.4 *Aurelia aurita*, common moon jellyfish. (**a**) Planula larvae, developed from a fertilized egg, stained. (**b, c**) Scyphistoma polyps, live specimens. (**d**) Strobila, fixed, unstained. (**e**) Free swimming young medusa, called Ephyra, fixed, unstained. (**f**) Ephyra, stained. (**g**) *Aurelia aurita*, mature, live. (**h**) The life cycle of *Aurelia aurita*, according to various authors. Living specimens on display at the North Sea Aquarium Borkum, Germany. Unstained Strobila and an Ephyra from the BAH Heligoland, stained specimens were from the Biological Collection of the University of Osnabrück

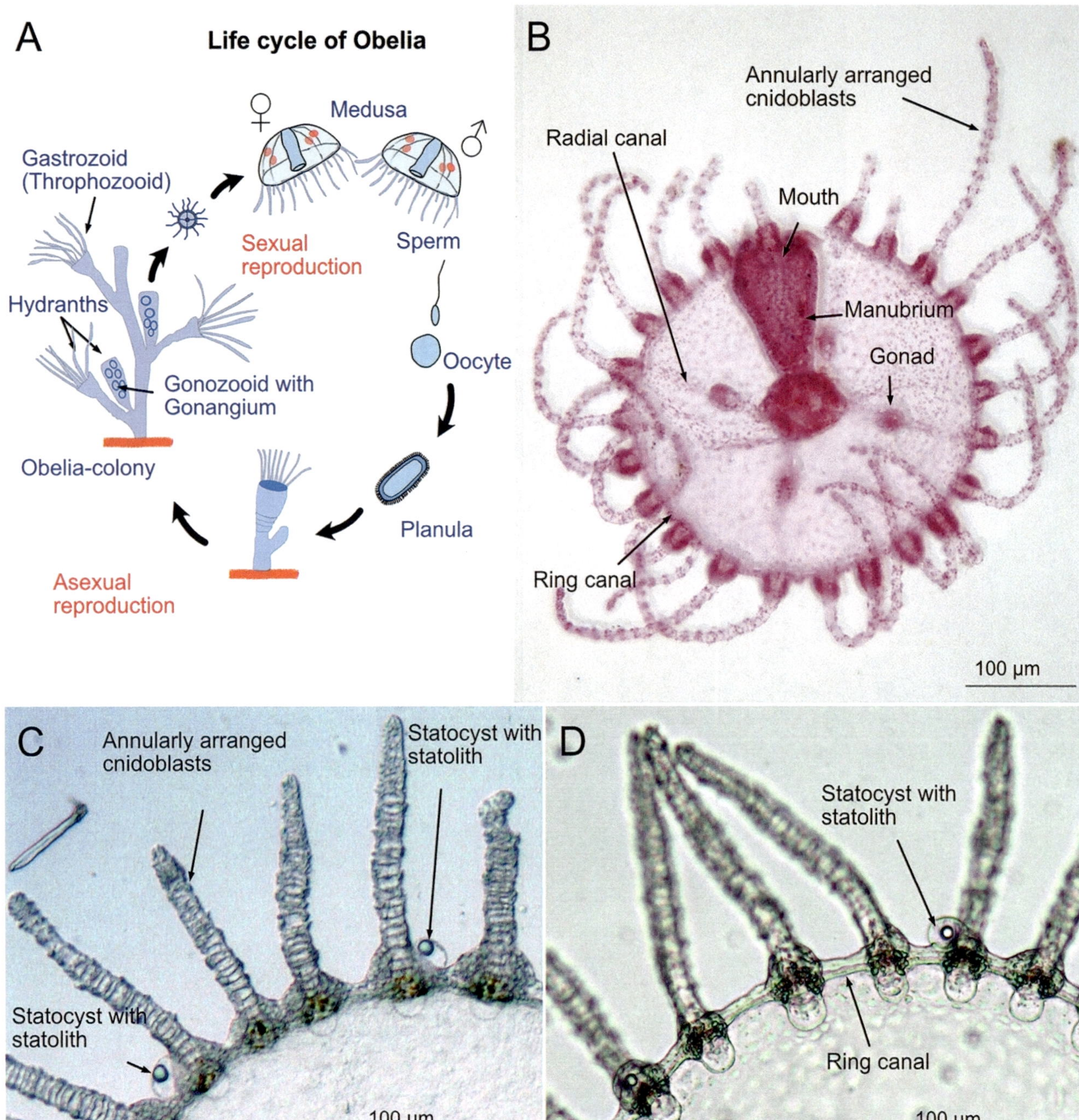

Fig. 3.5 *Obelia geniculata*. (**a**) Life cycle of *Obelia geniculata*, according to various authors. Hydranth = feeding polyp, gonozooid = sex polyp, trophozoid = gastrozoid. (**b**) Medusa, stained, the manubrium is pushed to the side. Specimen from the Biological Collection of the University of Osnabrück (**c**, **d**) *Obelia geniculata*, live, umbrella with tentacles and statoliths

tion at the cnidocil or the surrounding microvilli, the permeability of the cnidae membrane changes, and there is a rapid influx of water. This leads directly to the bursting of the operculum, and within nanoseconds, cnidae are discharged. About 30 kinds of cnidae have been identified and described in the literature. However, *Hydra* has three fundamentally different types, each with specific functions.

• Cnidocysts (also called stenoteles or penetrants) possess a double-walled capsule. When discharged, their stylet apparatus pierces the epidermis of many prey animals and even the chitin cuticle of crustaceans (Fig. 3.3). In doing so, the cnidocyst tube penetrates through the cuticle and the epidermis into the prey. The paralyzing poison is released from the terminal opening of the cnidocyst tube. Cnidocysts occur in all groups of cnidarians.

- Spirocysts (also called desmonemes or volvents) have a single thin-walled capsule filled with adhesive material, including muco- or glycoproteins. The undischarged tubule is coiled like a spring. After release, the tubule wraps around a bristle or other thin appendages of the prey (Figs. 3.2d and 3.3c, d).
- Ptychocysts (also called isorhiza) release a sticky tubule but do not have a lancet-like structure, and the tubule lacks a terminal opening. Some authors have also called them glutinants, due to their strictly adhesive nature. They secrete a short thread, which often contains short bristles and adhesive substances, with which the prey can be held. *Hydra* also uses ptychocysts for movement on a solid substrate like a looper caterpillar.

The morphology of cnidocytes is also used for systematic classification and to identify cnidarian species. The toxins transmitted by the cnidocytes act mainly as neurotoxins, for example, by inhibiting the Na^+-influx and thus preventing the formation of action potentials. In cardiac muscle cells, the uncontrolled release of Ca^{2+} ions from the ER into the cytoplasm is induced, which, in extreme cases, can lead to cardiac arrest. Histamine and prostaglandins in high concentration cause pain.

Are Hydras Immortal?

If a *Hydra* is divided into several equally sized pieces, a complete individual with foot, stem, trunk, hypostome, and tentacles arises from each piece. This extraordinary regenerative ability probably led the famous Swedish naturalist Carl von Linné (1707–1778) to name this genus *Hydra*. In Greek mythology, the *Hydra* is a multiheaded monster that, when it loses a head, grows two new heads. Moreover, the head in the middle is immortal, and its breath is lethal. Such regenerative ability has led scientists to describe *Hydra* as potentially immortal. This fascinating property of *Hydra* is primarily due to interstitial stem cells (I-cells) found in both epithelia. Two epithelial-specific stem cell lines renew the epithelial muscle cells in the gastro- and epidermis. A third I-cell line is responsible for continuously replacing cnidocytes, neurons, some gland cells, and germ cells. The stem cells retain their ability to regenerate forever, and thus ensure continuous maintenance of tissue functions. Visible aging does not occur in *Hydra*. Thus, I-cells differ from human stem cells, which, over time, lose their ability to regenerate, and aging can no longer be stopped. Most I-cells are located in the trunk area of both epithelia in *Hydra*. The descendants of the I-cell going into differentiation reach the head and tentacle region or the stalk area by being "pushed forward" by newly emerging cells from the trunk area. Especially in the tentacles, cnidocytes and other cell types are continuously lost through prey capture and must be replaced.

Reproduction and Alternation of Generations

A fascinating feature of the three cnidarian groups Hydrozoa, Scyphozoa, and Cubozoa is the change between two very different life forms, the polyp and the medusa. We see this, for example, in the moon jellyfish *Aurelia aurita* (Fig. 3.4) or *Obelia* (Fig. 3.5). The free-swimming medusae are almost always gonochoric. Sexes are monomorphic; distinguishing males and females is usually impossible without dissecting and staining the gonads. Only a few species are hermaphroditic. With the change between the two life forms, also called morphs, there is also a change in the mode of reproduction. The medusae consistently reproduce sexually, whereas the polyp reproduces asexually. Suppose the change between sexual and asexual reproduction, as in the cnidarians, is tied to a change between morphs. In that case, it is called a metagenetic alternation of generations, or simply metagenesis. As always, the exception proves the rule. There is a general evolutionary trend to reduce the medusa generation in hydrozoans, which likely led to their loss in many cases. However, the rule is also here: no rule without an exception. Within Hydrozoa, a *Hydra* does not develop medusae, but the polyp becomes sexually mature and develops testes and ovaries.

As an example of a jellyfish (Scyphozoa) with a metagenetic life cycle, *Aurelia aurita* from the North and Baltic Seas is shown here (Fig. 3.4). Sexual reproduction starts with the gonochoric medusae, whose gonads form within

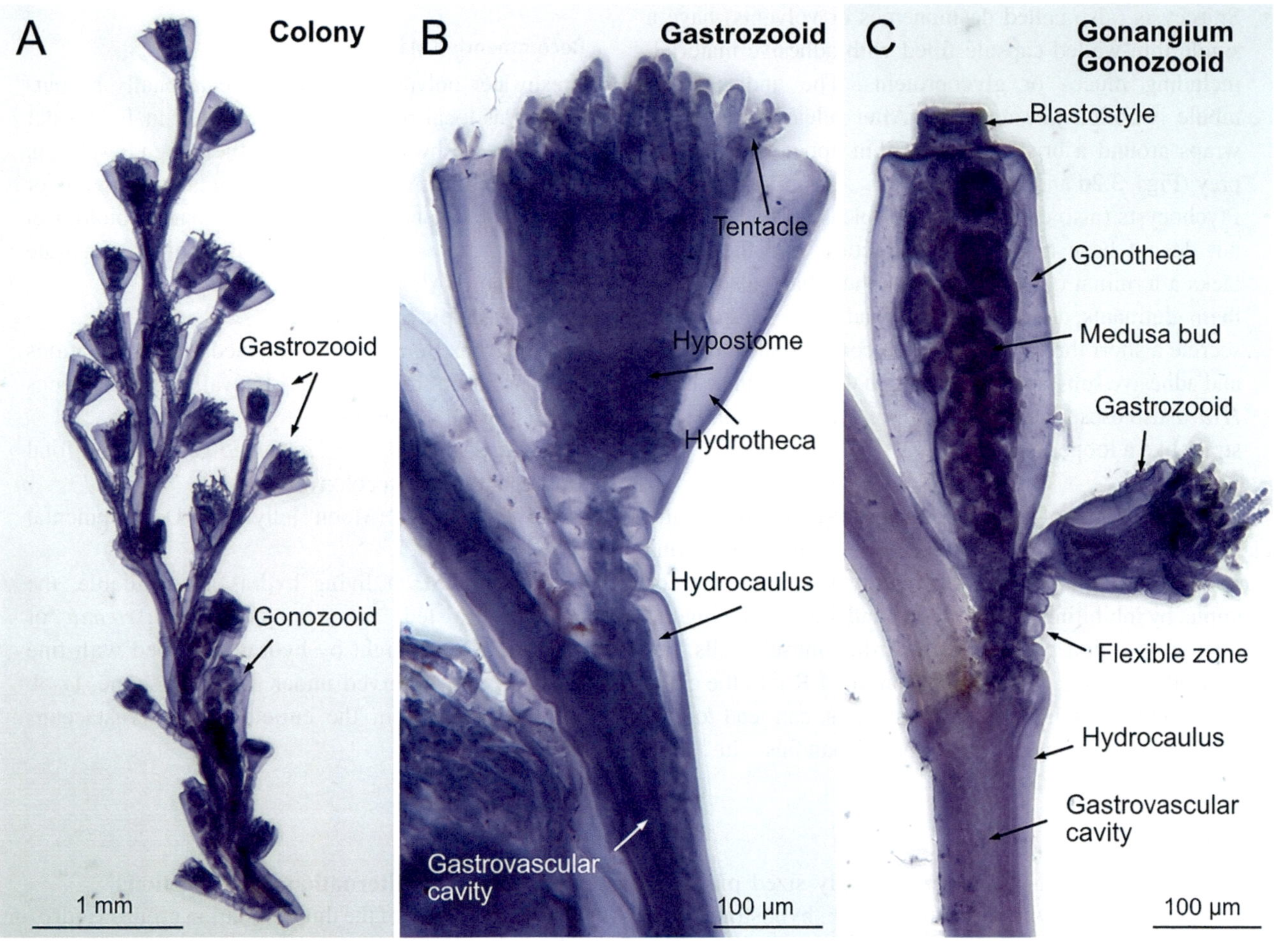

Fig. 3.6 *Obelia geniculata*. (**a**) Colony of polyps, stained whole mount, with hydranths in the hydrothecae (chitinous protective cover) and gonozooid in the gonotheca. (**b**) Detail with hydranth. (**c**) Detail with hydranth, gonozooid, and gonangium. The blastostyle is the central tissue where the medusa buds form. The gonotheca surrounds the blastostyle and the medusa buds. Specimens from the Biological Collection of the University of Osnabrück

the gastrodermis (Fig. 3.4g, h). Thereafter, the medusae release their eggs and sperm, first into the gastric cavity. From there, the gametes reach the open seawater, where the eggs attract sperm chemotactically and in a species-specific manner. Once eggs are fertilized, ciliated planula larvae are formed, drifting freely in the plankton (Fig. 3.4a). They attach themselves to suitable substrates with their anterior end, and their former posterior end develops into sessile polyps called scyphistostoma (Fig. 3.4b, c). The scyphistostoma differentiates into the strobila stage (Fig. 3.4d). Below the tentacle crown, these asexually form a series of disc-shaped medusae through constriction of body sections resembling a pile of discs. This process is called strobilation. The disc-shaped young medusae, called ephyrae, are the juvenile forms of the adult medusae. Ephyrae actively swim around with their eight radial arms (Fig. 3.4e, f) and, as the adult medusae, contribute to the distribution of the

species. Movies showing *Aurelia* medusea and polyps can be found at figshare: sn.pub/xr9qh.

Colony-Forming Cnidaria

In metagenetic species, the polyps consistently reproduce asexually. This may occur by fission, for example, in *Protohydra* (Fig. 3.1), or by budding, as in *Hydra* (Fig. 3.1). Some species form large colonies with many individuals. Colonies are formed if the young animals produced by budding remain attached to the mother animal. An example of a colony-forming hydrozoan species is *Obelia geniculata* (= *Laomedea geniculata*), which also has a complex life cycle with polyps and medusae (Fig. 3.6). An *Obelia* colony consists of many interconnected individuals that differ in structure and function, the so-called polymorphs (Fig. 3.6a–c). What kind of polymorphs do we find in an *Obelia* colony, and what are their functions? In one type, the polyp bodies are distally club-shaped, expanded toward the hydranth, and continue proximally into stalk-like sections (hydrocaulus) or are interconnected via creeping stolons (hydrorhiza). Hydranths, which serve for the capture and digestion of food, are referred to as trophozoites or gastrozoites, in other words, these are feeding polyps.

The tentacles are equipped with stinging cells. Within one colony, some species have, in addition to trophozoites, dactylozooids (defense polyps), which are tentacle-shaped and densely covered with stinging cells but have no mouth opening and, therefore, cannot ingest food by themselves. Another type of polyp in an *Obelia* colony is the gonozooids. Gonozooids are the reproductive polyps that form free-moving medusae through budding. The medusae are ultimately released from a gonotheca. Such a colony of polyps grows continuously through the progressive budding of new polyps (Figs. 3.5a and 3.6). Since the gastric cavities of the polyps in a colony are interconnected, and cilia generate a constant fluid flow, the food supply for all polyps is ensured. The medusae released by the gonozooids represent the sexual generation and develop gonads. Eggs and sperm of a species find each other chemotactically. From the fertilized egg cell develops the ciliated, free-moving planula larva. After settling on a suitable substrate, it differentiates into a sessile founder polyp, from which a new colony of polyps arises (Fig. 3.6a). In some colony-forming species, there are reductions of the medusae, which are no longer released, remain on the colony, and are referred to as medusoids. Only about 700 of the nearly 3200 species of Hydrozoa still form free-swimming medusae.

Reproduction in *Hydra*

Freshwater polyps are exotic among cnidarians. They do not form medusa, and neither do free-swimming planula larvae exist. The polyp forms female and male gonads and has thus become a consecutive hermaphrodite (Fig. 3.7). The absence of larvae, as well as the absence of medusae, can be interpreted as a secondary adaptation to the freshwater habitat. Free-swimming planulae or medusae drifting downstream with water currents would be a great disadvantage in small freshwater habitats. Therefore, they very likely disappeared in freshwater cnidarians during evolution.

So, how does sexual reproduction proceed in freshwater polyps? The gonads always form within the epidermis. Usually, the testes appear in the upper area of the trunk and the ovaries in the lower area (Fig. 3.7a, b).

The gametes are built by thousands of interstitial cells (I-cells); these first differentiate into spermatogonia, which undergo spermatogenesis to form sperm. The somatic part of the testes, the outer testes wall, is formed by ectodermal cells (Fig. 3.7a, b). The accumulation of many spermatogonia in a confined space leads to the bulging of the ectoderm. Therefore, the testes are easily recognizable from the outside (Fig. 3.7h). Through rupture (bursting) of the epidermis, the mature sperm are released into the surrounding water.

The formation of an ovary also begins with the accumulation of I-cells in the epidermis. However, only a single cell becomes the oocyte (Fig. 3.7c, d). All others differentiate into nurse cells. The egg cell phagocytizes nurse cells and thus continuously grows in size and mass. Eventually, the ectoderm ruptures and retracts around the edge of the egg cell. The resulting tissue ring at the base of the egg cell can be seen in Fig. 3.7e, f. When an oocyte is exposed to the surrounding water, fertilization, followed by meiosis, may occur. The nurse cells, taken up by the developing oocyte during oogenesis, remain throughout embryogenesis as spherical cells with pyknotic nuclei, which means degenerating cell nuclei in the oocyte (Fig. 3.7f). Even during the subsequent cleavages, as well as in the resulting blastula and the later gastrula, all embryo cells still have nurse cells in their cytoplasm. The egg cell is surrounded by a robust, spiky shell, which provides mechanical protection. Further development can last several weeks and ends with the hatching of a young tentacle-less *Hydra*, which immediately attaches its foot to a substrate and develops tentacles within a few hours.

In addition to sexual reproduction, *Hydra* also have the capacity for asexual reproduction. It is actually much more common, primarily when favorable living conditions exist. In that case, a *Hydra* will populate a habitat in a comparatively short time, with many individuals formed by budding. Buds are built exclusively in the transition zone between the trunk and the stem-foot region (Fig. 3.1b) by a unilateral protrusion of tissue from the gastric region. This body tissue (epidermis, mesoglea (extracellular matrix = ECM), gastrodermis) retains its proliferation, typical cell composition, and local differentiation ability. Through pattern formation processes, a new tentacle region is differentiated after about 2 days. A day later, the tissue bridge at the future foot region is strongly constricted. The young polyp is released from the mother animal, without a wound, on the fourth day. This happens through a rapid, neuronal-mediated contraction of the sphincter. Other hydra species, like *Protohydra*, whose life cycle and systematic position are still largely unclear, can also reproduce by transverse fission (Fig. 3.1c). Our movie clip collection, which includes movies showing living cnidarians, can be found at figshare: sn.pub/xr9qhi.

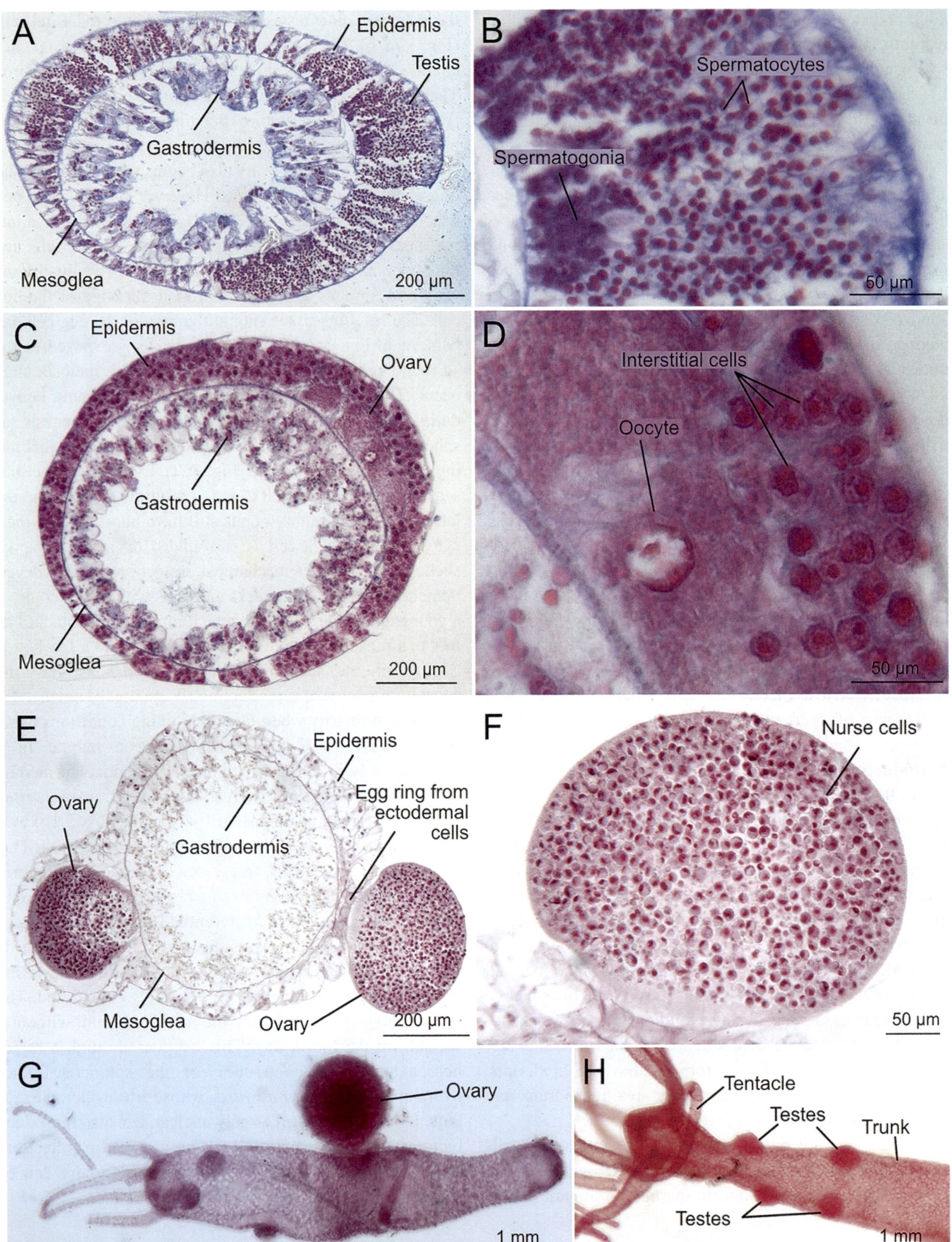

Fig. 3.7 *Hydra* sp. (**a**) Cross-section through the trunk area with testis. (**b**) Detail from (**a**). (**c**) Cross-section through the trunk area with ovary and egg cells. (**d**) Detail from (**b**). **a–d**: Azan staining. Specimen by Johannes Lieder, Ludwigsburg. (**e**) Cross-section through the trunk area of *Hydra fusca* with two ovaries. (**f**) Detail from **e**. The young ovaries are filled with I-cells, and the oocyte is invisible. **e, f**: Specimen of the Zoological Teaching Collection, Philipps University Marburg. (**g**) Total preparation of *Hydra viridis*, stained. The animal has a well-developed ovary and three ovaries in early stages of development. (**h**) Total preparation of *Hydra viridis*, with four developing testes stained. **g, h**: Specimen from the Biological Collection of the University of Osnabrück

▶ **Summary** **Platyhelminthes show three distinct body axes, dorsal-ventral, anterior-posterior, and left-right, with a visible bilateral symmetry of the body. The space between the epidermis and digestive tract is filled with mesodermal parenchyma and muscle cells, into which all organs are embedded. These animals do not possess a secondary body cavity, a coelom. About one-seventh of the 37,000 described Platyhelminth species are free-living predators. All the rest are parasites. Most of them live as endoparasites inside their hosts, for example, in the liver's bile ducts or the intestine in vertebrates. Other species colonize as ectoparasites, inhabiting the gills, the oral and pharyngeal cavities, or the rectum of their hosts. All parasites show complex life cycles.**

The parasitic lifestyle makes these animals utterly dependent on their hosts, a fact that is reflected in particular adaptations of their morphology, metabolic physiology, reproductive biology, and the anatomy of their organ systems. However, some of these adaptations were already present as the so-called preadaptation in their free-living ancestors, such as the hermaphroditic organization, extraordinarily complex sexual organs, and internal fertilization. Platyhelminthes were formerly divided into four taxa: flatworms (Turbellaria), flukes (Trematoda), monogeneans (Monogenea), and tapeworms (Cestoda). This classification can be picked up in a beginner's course, as it is still used in many older textbooks. However, it should be pointed out that this subdivision does not reflect our current knowledge about the phylogeny of Platyhelminthes. Recent morphological and molecular analyses suggest that Platyhelminthes are divided into the limnic Catenulida and the Rhabditophora (with special epidermal gland cells), which comprise the remaining flatworms, including all parasites. An essential evolutionary key event within Rhabditophora is the appearance of ectolecithal eggs,

composed of several yolk cells and a single fertilized egg cell (Neoophora). The parasitic species very likely go back to only one evolutionary event. Among others, they have in common that the primarily ciliated epidermis is shed upon penetrating the first host and replaced by a new body covering called neodermis. The neodermis is syncytial, comprising only a few cells, each containing hundreds or thousands of nuclei. The neodermis is newly formed from mesodermal stem cells and protects the parasite perfectly against the defense mechanisms of the hosts. Because of their new body covering, this group is named Neodermata ("bearing new skin").

Flukes (Trematoda)

About 18,000 fluke species (Trematoda) are known, and they are all parasites. The trematodes include the two species discussed herein, the common liver fluke (*Fasciola hepatica*) (Figs. 4.1, 4.2, 4.3, 4.4, 4.5, 4.6, 4.7 and 4.8) and the lancet liver fluke (*Dicrocoelium dendriticum*) (Figs. 4.9 and 4.10).

We distinguish between two types of hosts in trematode life cycles. There are the intermediate hosts, in which only the asexually reproducing generations of the parasites occur. And we have the definitive hosts, where the usually significantly larger sexually reproducing generations live. In the case of trematodes, the first intermediate host is almost always a mollusk, usually a snail. The larvae (cercariae) of the sexual generation leave the intermediate host and are then taken up by a definitive host with contaminated water or food. This is usually a vertebrate species. Only when the cercariae arrive in a suitable host do they develop into sexually mature hermaphroditic flukes. The mature flukes live as endoparasites in the intestine, lymph vessels, or other organs. *Fasciola hepatica* colonizes the bile ducts in the liver of its host (Fig. 4.1a–d). All known trematodes undergo a complex generation and host alternation, showing numerous adaptations to the parasitic lifestyle. For example, they develop dedicated suckers, which

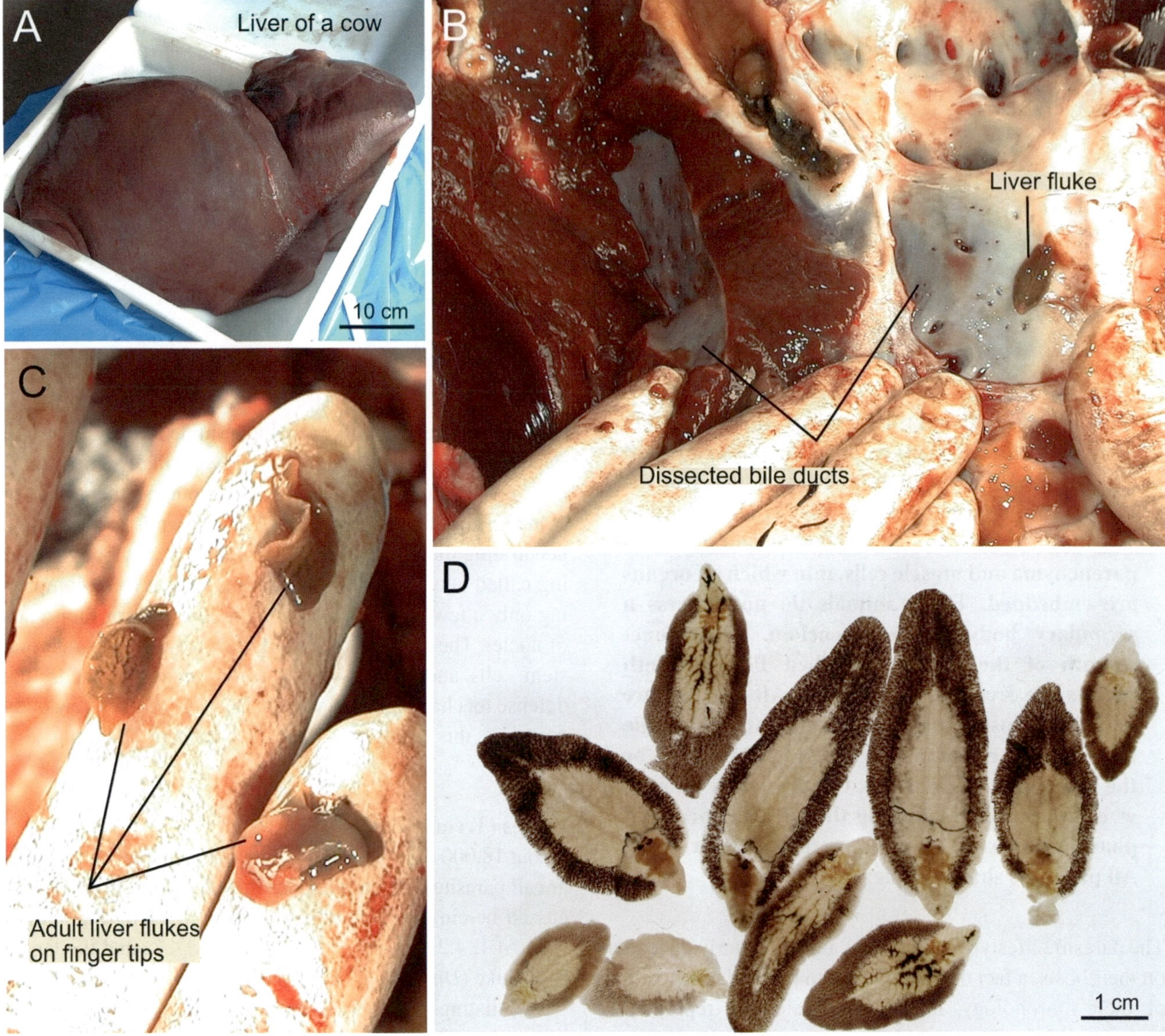

Fig. 4.1 Common liver fluke (*Fasciola hepatica*). (**a**) Liver of a mature cow. (**b**) Dissected cow liver. Common liver flukes live in the bile ducts, which run through the liver. (**c**) Flukes removed from bile ducts. (**d**) Fixed liver flukes, brightfield illumination, unstained

are used to anchor themselves to the walls of the bile ducts (Figs. 4.1, 4.2, and 4.3, 4.6). Moreover, they have a unique body surface optimized for directly exchanging nutrients or gasses, the neodermis.

Almost every vertebrate, including ourselves, can become infected with various trematodes and potentially develop symptoms of disease. For example, a severe infestation with *Fasciola hepatica* causes edema and inflammation of the liver, which can potentially result in permanent liver damage. Malfunctions of the liver are caused by the migration of the fluke through the organ and by the continuous release of toxic excretory products. The disease is referred to as fasciolosis. Often, the parasitic species are not addicted to one single definitive host species. The common liver flukes are primarily found in herbivorous mammals such as sheep or cattle, but can also infect other mammals, including humans, as a definitive host, as mentioned above. In addition, the common liver fluke has numerous close relatives, each preferring a specific definitive host. For example, *Fasciola jacksoni* colonizes elephants, *Fasciola gigantica* water buffaloes, cows, and humans, while *Fasciola nyanzae* prefers hippos. Some species are only locally present, *Fasciola hepatica*; however, like its definitive host, our domestic cattle, is cosmopolitan,

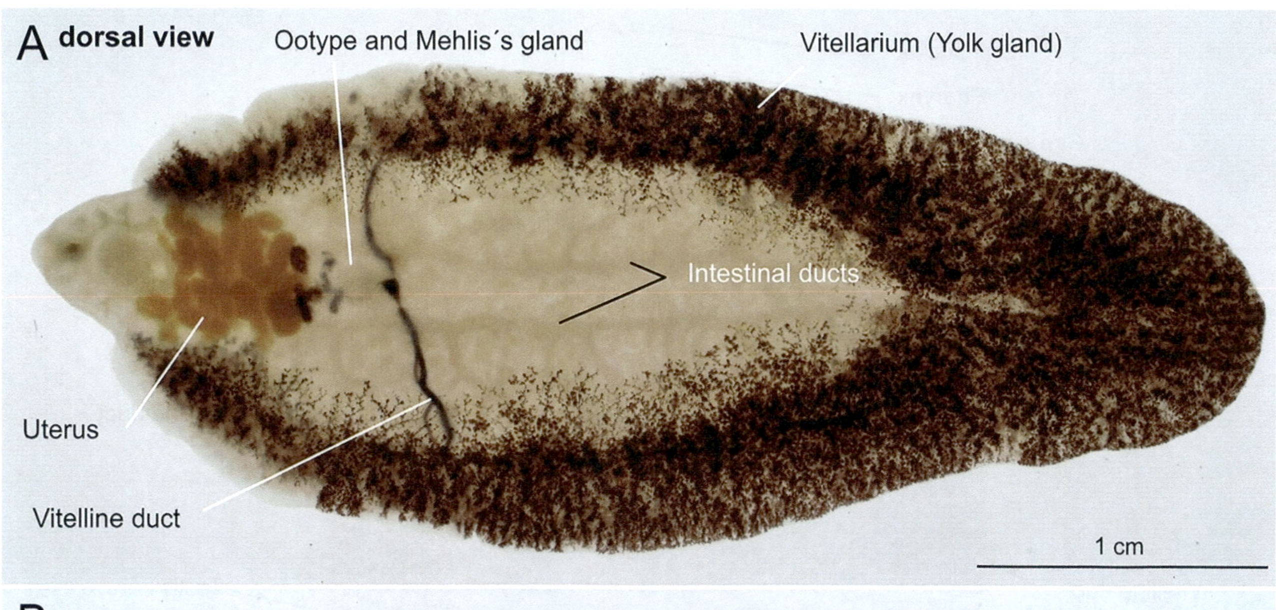

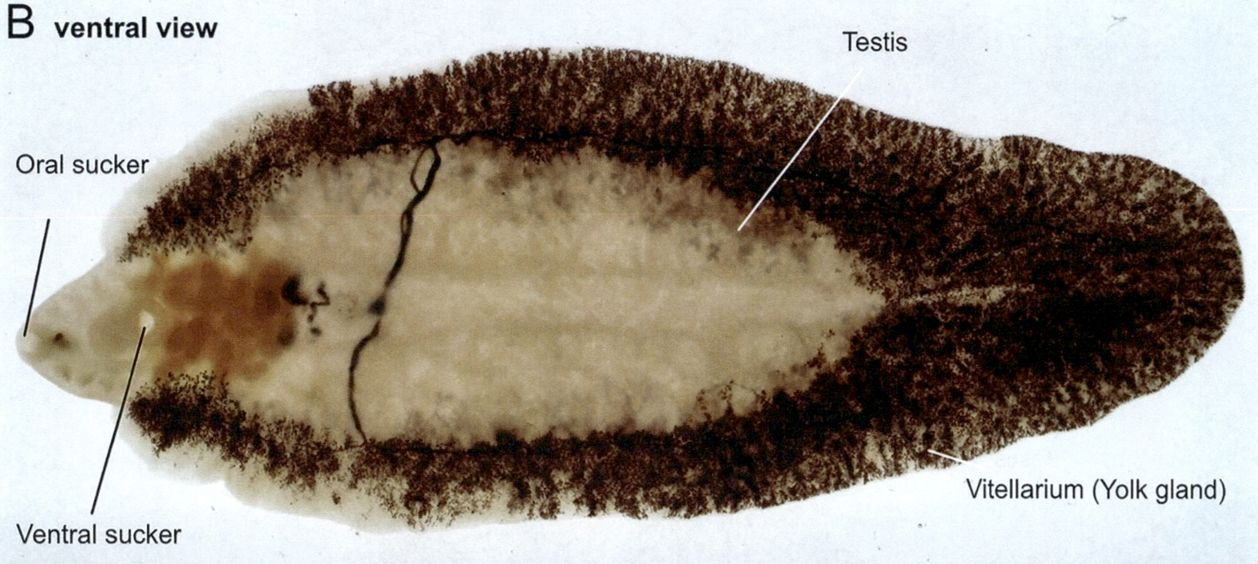

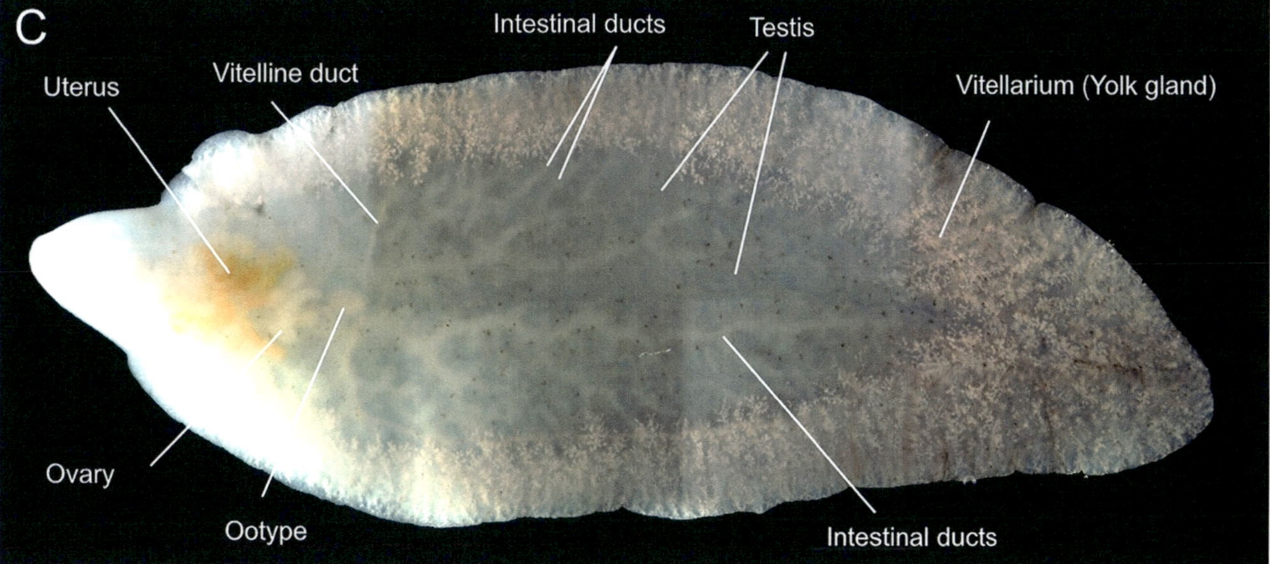

Fig. 4.2 Common liver fluke (*Fasciola hepatica*). (**a**) Fixed, brightfield illumination, dorsal view. (**b**) Fixed, brightfield illumination, ventral view. (**c**) Fixed, incident light, dorsal view. All animals were unstained

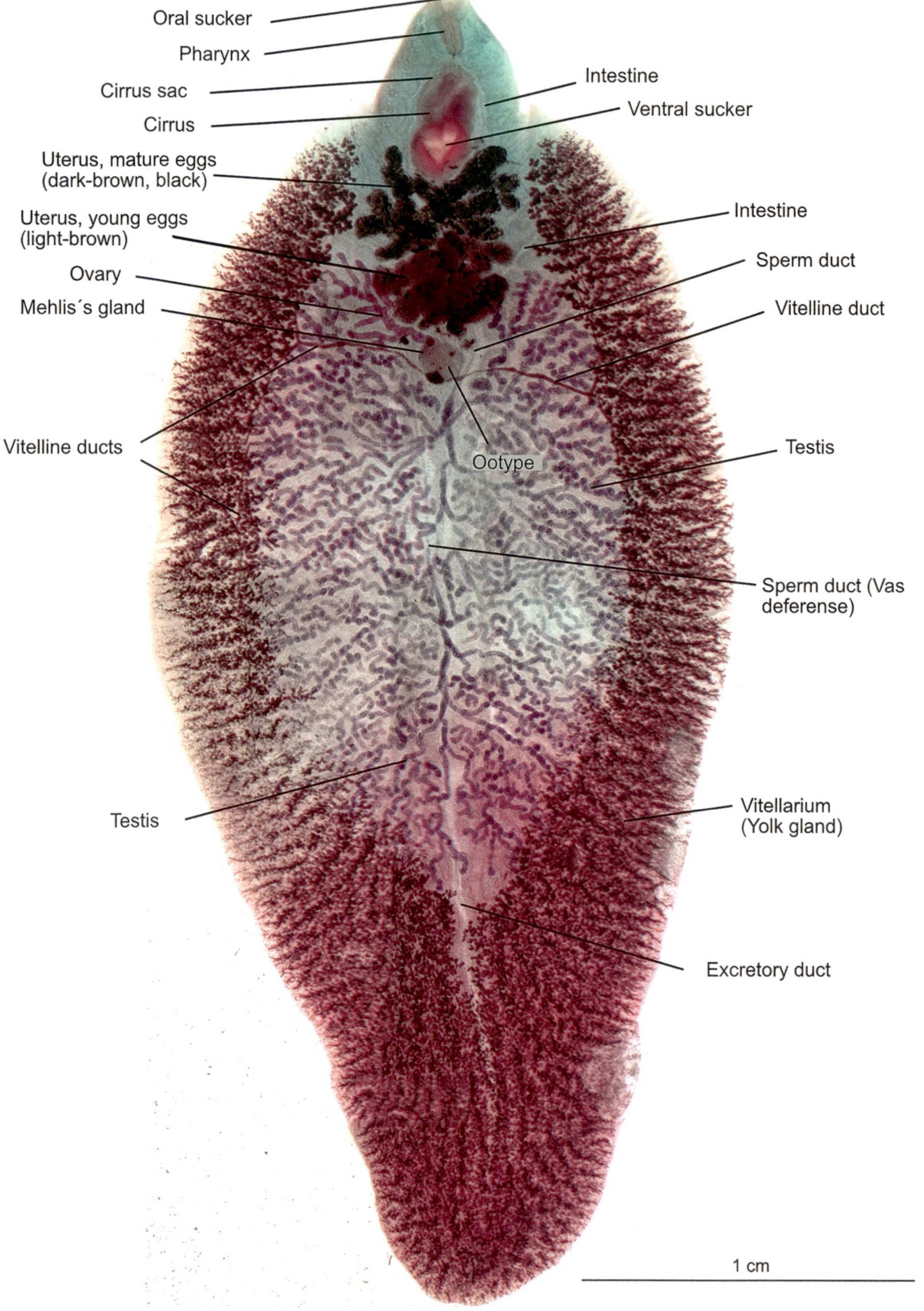

Fig. 4.3 Common liver fluke (*Fasciola hepatica*) and its reproductive organs. Azan-stained specimen. Organs of the reproductive system, the intestine, and the suckers are visible (brightfield illumination). Specimen from the Biological Collection of the University of Osnabrück

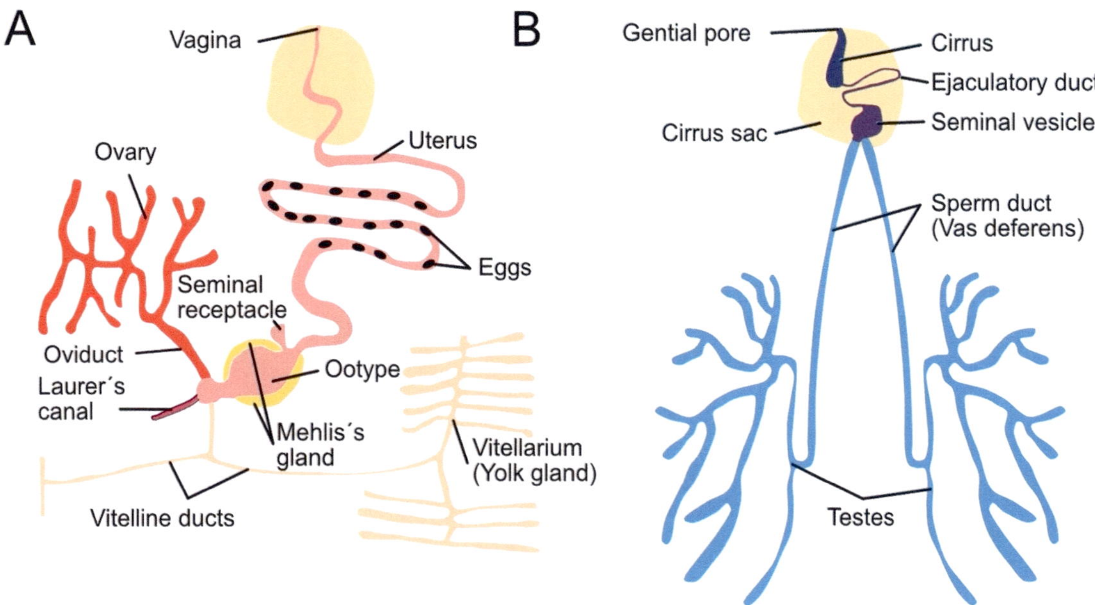

Fig. 4.4 The female (**a**) and male (**b**) sexual organs of *Fasciola hepatica*

with local infection hotspots such as on the Bolivian high plateau Altiplano.

Tapeworms (Cestoda)

The second major group of parasitic Platyhelminthes is named Cercomeromorpha, which comprises the tapeworms (Cestoda) and hookworms (Monogenea). They have a primary ciliated larva in common. The larvae possess a posterior appendage with crescent-shaped hooks for holding on. The approximately 5000 known cestode species live exclusively as endoparasites in the intestines of their final hosts. Their life cycle is significantly simpler than that of the trematodes (Fig. 4.13). A host change is almost always included, while an alternation of generations is lacking. Tapeworms are probably the most highly adapted endoparasites living on Earth. In all developmental stages, the digestive tract is missing. The absorption of nutrients takes place solely through the neodermis. Thus, the entire body surface serves as an organ for food uptake. Sexually mature animals are strongly flattened, and in tapeworms, serial multiplication of the sexual apparatus is found, accompanied by band-shaped body dimensions, which accounts for their name (Figs. 4.13 and 4.14). A beef tapeworm (*Taenia saginata*) can reach up to 12 m, with only a 7 mm body width (Fig. 4.13). Among the longest species is the fish or broad tapeworm (*Diphyllobothrium latum*), which has a length of up to 20 m and whose final hosts are various mammals that feed on fish, including domestic dogs. Infected people may suffer from multiple complications, including intestinal obstruction and gallbladder disease caused by the migration of proglottids, the repetitive body segments of tapeworms (see below).

The approximately 8000 species of hookworms (Monogenea) live ectoparasitically on the gills or integument of their hosts, mostly amphibians or fish. These parasites are often found in body cavities that have direct contact with the outside, for example, in the mouth and throat, in the bladder, the cloaca, or the rectum. Monogeneans show no alternation of generations and only rarely a host change. Humans are not among their hosts.

Free-Living Predators

Among the free-living predators, two groups with larger species stand out: the freshwater planarians (Tricladida, with a three-branched intestine) and the marine Polycladida (with a multibranched intestine). The latter are often very brightly colored and are easily mistaken for marine nudibranchs or sea slugs. Movies showing marine polycladida can be found at figshare: sn.pub/xr9qhi. It is only to these relatively large forms of 10–25 mm and the parasitic species that the name flatworm applies. The other free-living species usually have a round body cross-section. Tricladida also include our native free-living turbellarians, planarians, which inhabit various biotopes in fresh water. The term turbellarian refers to the locomotion movement of all free-living Platyhelminthes, which is achieved by the cilia of the epidermis. The musculature of the body wall is mainly used for changing the body shape, except in the larger species. Triclads are often found in creeks or the shore zone of ponds and lakes. There, they hunt on the undersides of leaves or stones. With a body size of 10–25 mm, *Dugesia gonocephala*, one of the most common species in Central Europe, is one of the larger species within the free-living

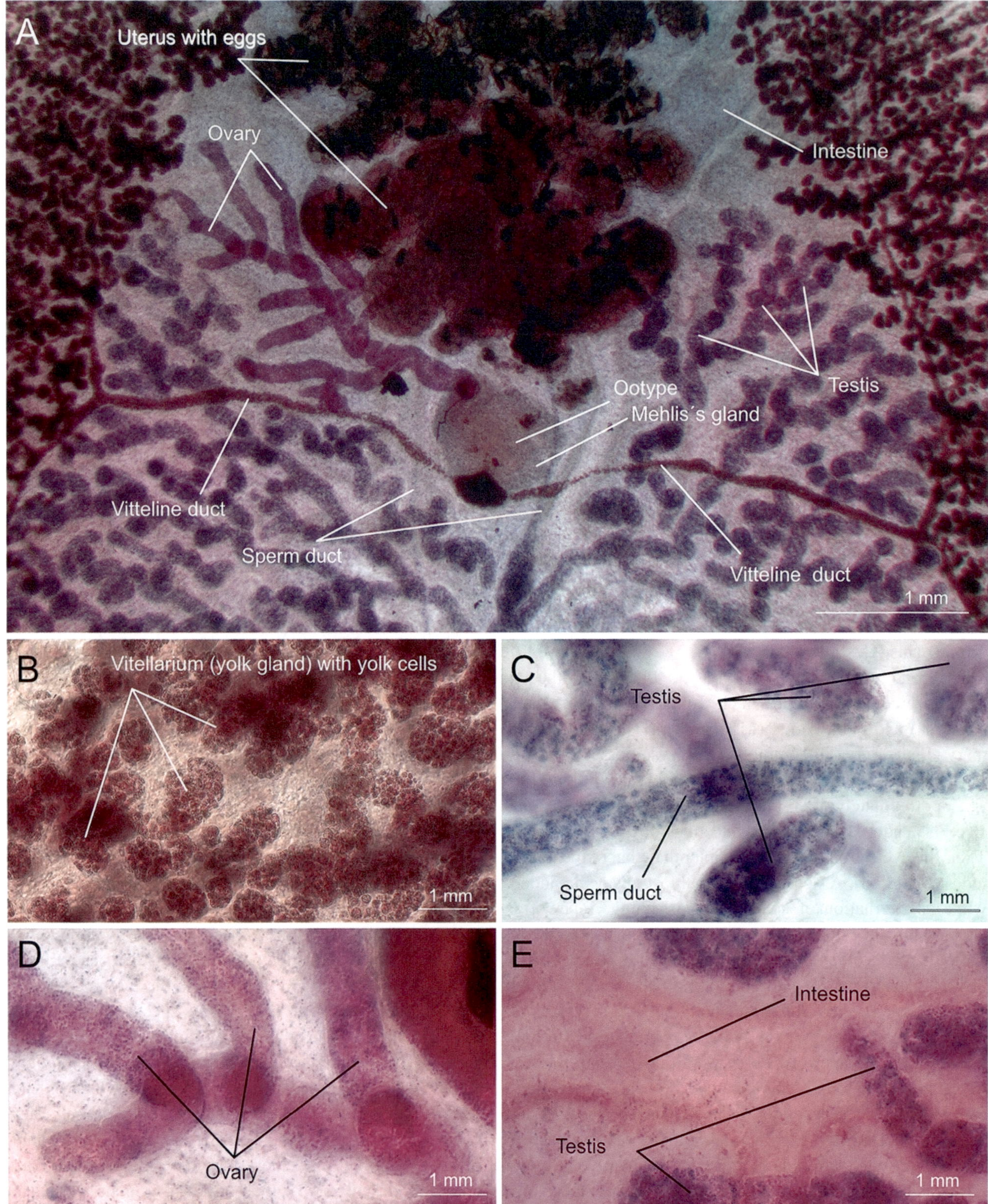

Fig. 4.5 Common liver fluke (*Fasciola hepatica*). Selected organs at higher magnification (brightfield illumination). (**a**) In the ootype, a single egg cell and about 50–60 yolk cells form the composite eggs of the liver fluke. (**b**) Yolk stock with yolk cells. (**c**) Testis with spermatogonia. (**d**) Ovary with immature egg cells. (**e**) Intestinal diverticulum and testis. Specimen from the Biological Collection of the University of Osnabrück

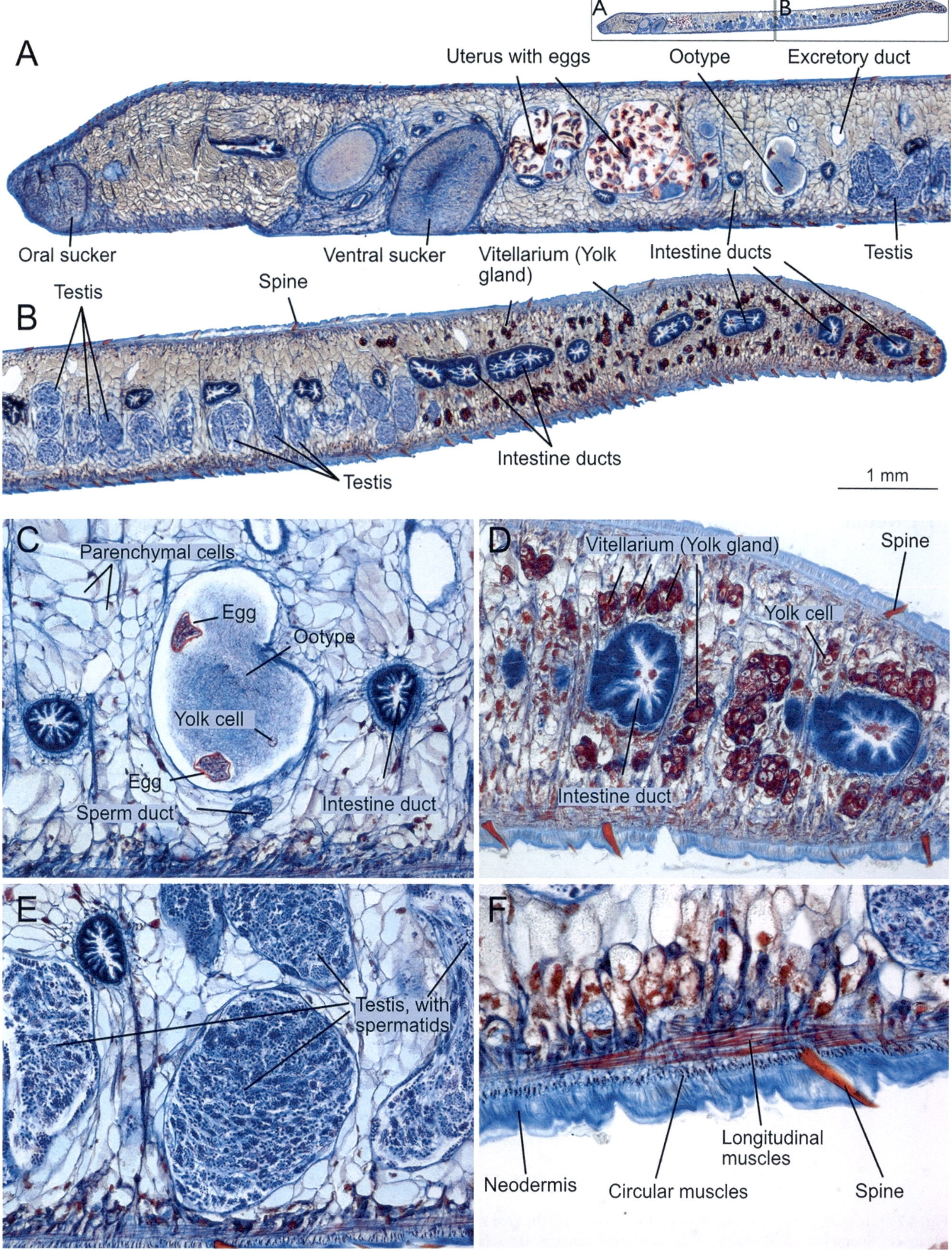

Fig. 4.6 Common liver fluke (*Fasciola hepatica*). Histology. Azan staining. (**a**, **b**) **a** and **b** show the anterior (**a**) and posterior (**b**) parts of a median longitudinal section at the level of the mouth and ventral sucker. (**c**) The ootype with two composite eggs and a yolk cell. (**d**) In the periphery of the animal, sections of the intestine (intestinal diverticula) and the yolk ducts can be seen. (**e**) Sections of the testis show spermatogonia, spermatids, and various stages of spermatogenesis. (**f**) The neodermis contains spines. Specimens by Johannes Lieder, Ludwigsburg

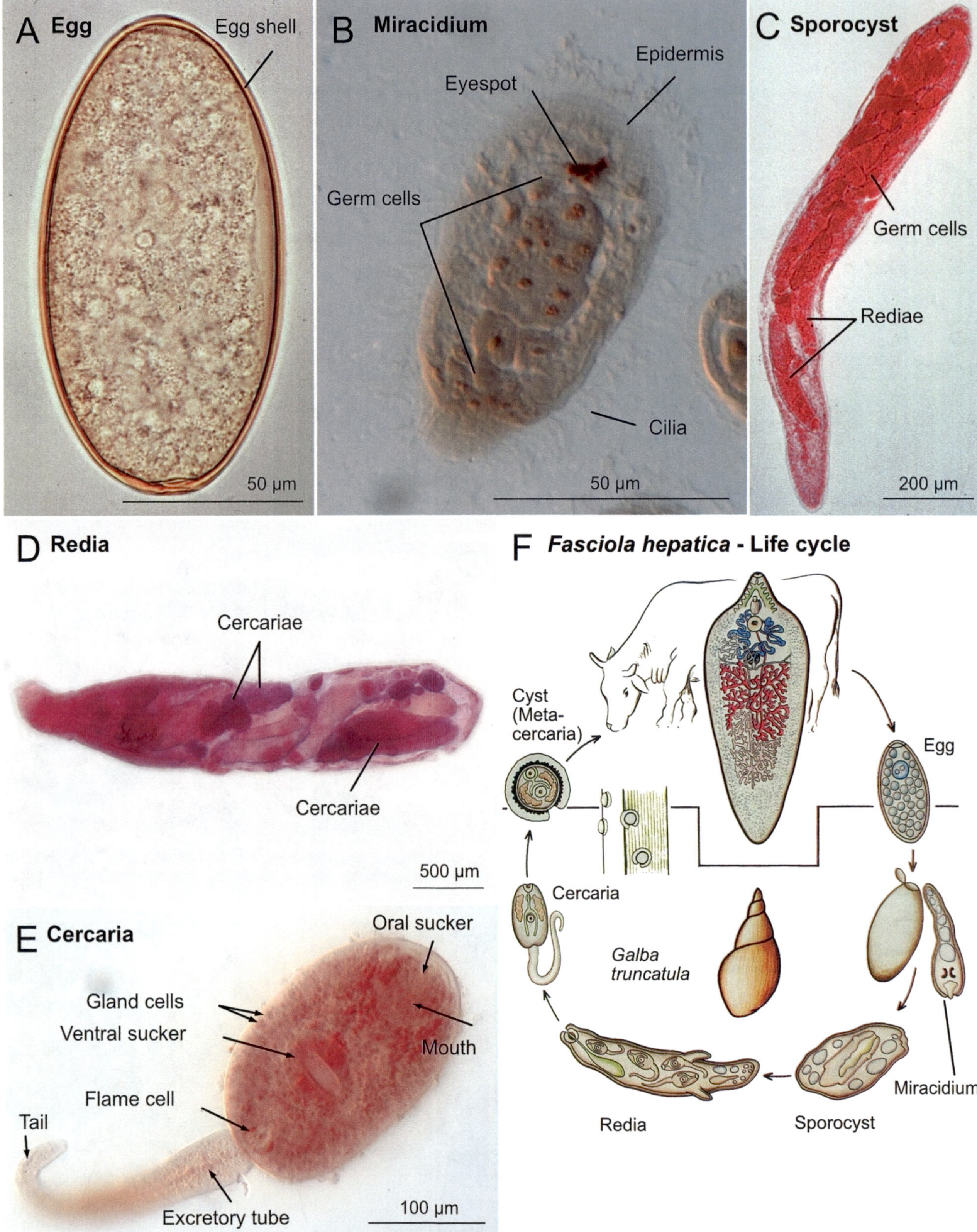

Fig. 4.7 Life cycle of *Fasciola hepatica*. (**a**) Composite egg. (**b**) Free-swimming miracidium larva with ciliated primary epidermis, light-sensing organ. (**c**) Sporocyst. (**d**) Redia. (**e**) Free-swimming cercaria. (**f**) Schematic representation of the life cycle. Specimens from the Biological Collection of the University of Osnabrück, the Zoological Teaching Collection of the Philipps University Marburg, and the Zoological Teaching Collection of the Heinrich Heine University Düsseldorf

Life cycle of a digenetic trematode, for example, *Fasciola hepatica*			
Generation	**Stage**	**Reproductive mode**	**Host**
1	Hermaphroditic Fluke	Sexual	Mammals
2	Egg	-	-
2	Miracidium (larvae)	-	Free living (water)
2	Sporocyst (without intestine)	Asexual (stem cells)	Mollusk
3	Redia (with intestine)	Asexual (stem cells)	Mollusk
1	Cercaria (larvae)	-	Free living
1	Metacercaria (permanent stage)	-	Often encapsulated

Fig. 4.8 Alteration of generations and hosts of *Fasciola hepatica* as an example of the digenetic Trematoda

Recommended Material

Live specimens, fixed specimens, unstained and stained, and histological sections of trematodes are well suited for studying body organization. Histological sections of the common liver fluke (*Fasciola hepatica*), the Chinese liver fluke (*Clonorchis sinensis*), or the lancet liver fluke (*Dicrocoelium dendriticum*) are usually available in most zoological collections of universities for teaching purposes. Otherwise, teaching material can be purchased from biology supply companies. Live common liver flukes (*Fasciola hepatica*) are occasionally available at nearby cattle slaughterhouses. Ask for the liver of infected animals. Scalpels and larger types of tweezers are needed to extract the liver flukes. The flukes can be fixed in 4% formaldehyde.

Representatives of Tricladida (planarians) can be collected in ponds or creeks and then observed alive in the laboratory. Marine forms are washed out of the sediment after being immobilized with magnesium chloride and can be observed alive, most easily during marine biology excursions. Cestodes should be demonstrated as fixed total specimens. Histological sections of the cattle or pig tapeworm are often already available in a university's zoological collection or can be obtained through specialist biology suppliers.

interstitial spaces between the sediment particles using their ciliated epidermis. Being good swimmers, they hunt there for animals damaged by the waves.

Anatomy and Histology of Trematoda: The Common Liver Fluke (*Fasciola hepatica*) and the Lancet Liver Fluke (*Dicrocoelium dendriticum*)

Platyhelminthes entirely lack a coelom. A coelom, the secondary body cavity, is lined with epithelial cells from the mesoderm. In Platyhelminthes, the space between the epidermis and gastrodermis is filled with muscle or parenchymal cells, which also derive from the mesoderm. All parenchymal cells are collectively referred to as connective tissue. The circular, diagonal, and longitudinally arranged muscle fibers are located beneath the epidermis, enabling the animals to move actively. All muscles are directly embedded in the parenchyme. Accordingly, Platyhelminthes also do not have a vascular system, and their excretory organs are protonephridia. Since a body cavity is missing, this level of organization is referred to as acoelomate. In living animals or whole mounts (Figs. 4.1, 4.3, 4.9, and 4.10), the bilaterally symmetrical body organization and all three body axes— anterior-posterior, dorsal-ventral, and left-right— can be easily recognized. Whereas almost all organs are bilaterally organized, the ovaries become secondarily unpaired and are situated on the left side. The suckers are located ventrally and anteriorly, and thus the dorsal and posterior sides are opposite to each other.

Digestive System

Liver flukes use their oral and ventral suckers to attach to the walls of the bile ducts that traverse the liver (Fig. 4.1b). The oral sucker is located at the tapered anterior end of the animal. The ventral sucker is located more posteriorly on the ventral side.

Platyhelminthes. However, most free-living Platyhelminthes species are marine predators that live interstitially in marine sands. They are tiny animals, usually less than 10 mm long. For example, the members of Otoplanidae live in the surf zone of sandy beaches and move through the water-filled

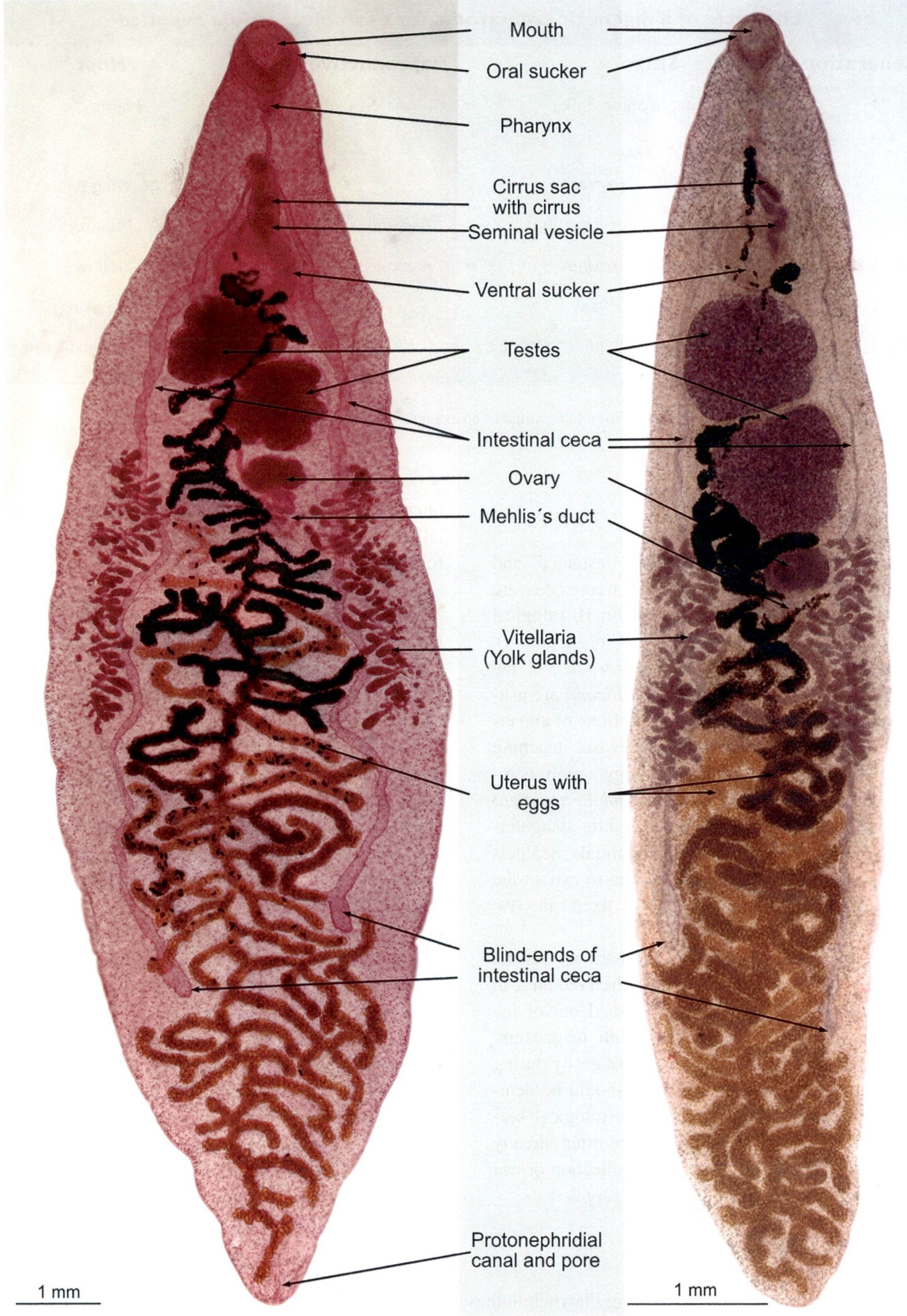

Fig. 4.9 Lancet liver fluke (*Dicrocoelium dendriticum*) and its organs, two Azan-stained total preparations from the Zoological Teaching Collection, Philipps University Marburg

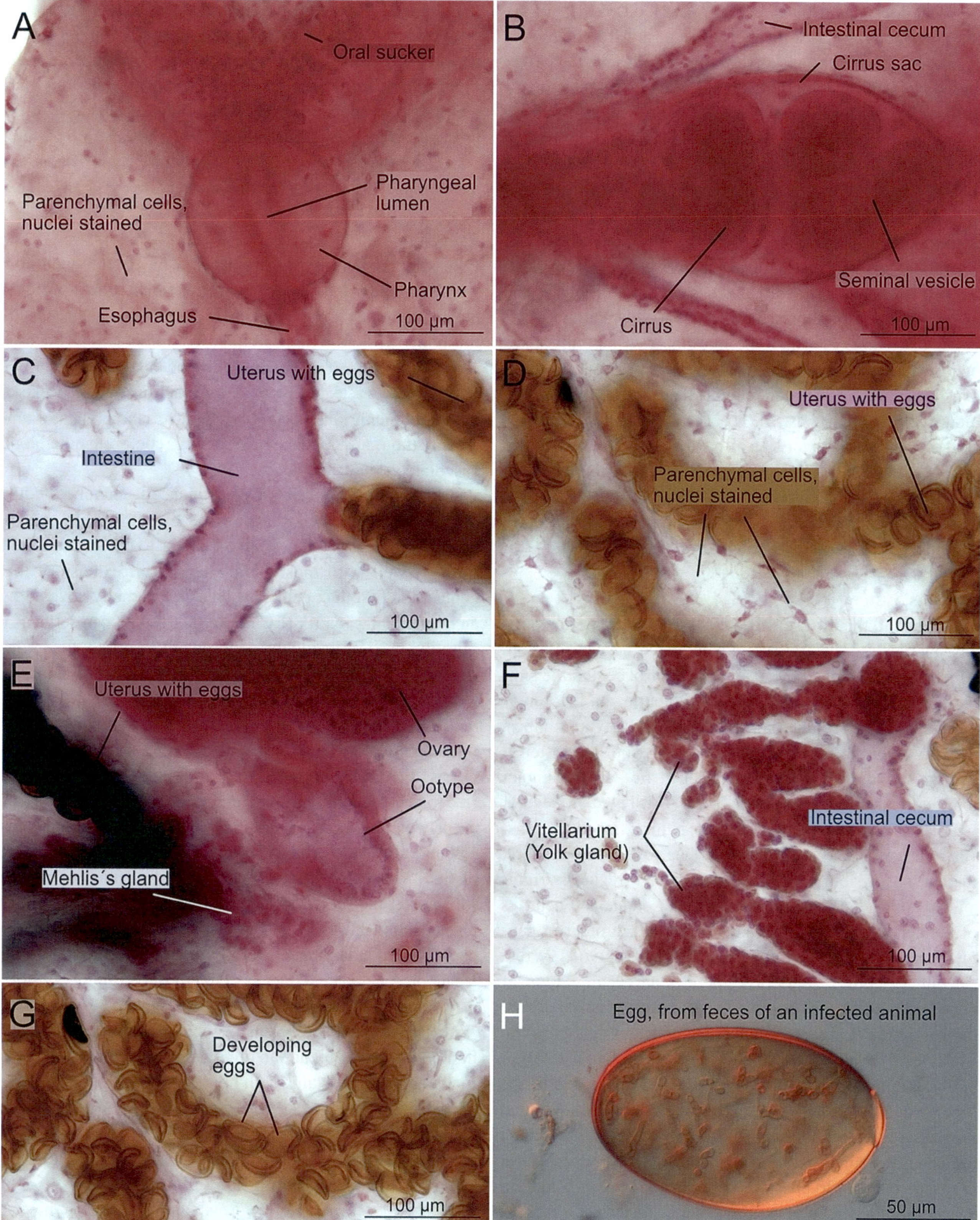

Fig. 4.10 Organs of *Dicrocoelium dendriticum* at higher magnification (brightfield illumination). (**a**) The head region with oral sucker. (**b**) The retractable cirrus ("penis") in the cirrus sac is located anterior to the ventral sucker. The male genital opening is usually not visible. (**c**) One intestinal branch. (**d**) Cell nuclei of the parenchyma. (**e**) Section with yolk tubules, intestinal branch, and uterus. The initially light yellowish eggs turn black during the hardening of the eggshell. (**f**) Branches of the yolk ducts. (**g**) Maturing eggs in the uterus. (**h**) Egg from the feces of an infected animal. Specimens from the Zoological Teaching Collection, Philipps University Marburg

What does a liver fluke feed on? The food consists of the fluid taken up from the bile ducts of its host. After the food is ingested, it enters the intestine through a muscular pharynx and a short esophagus. In the simplest case, the intestine of trematodes consists of an unpaired anterior section with a powerful muscular suction pharynx and two blind-ending ceca, which can extend along the animal's entire body. In the case of the common liver fluke, the intestinal ceca are greatly expanded by numerous lateral branches (diverticula). Digestion is extracellular in the ceca and their diverticula. Therefore, we can speak of a gastrovascular system here. Such a branched gastrovascular system occurs only in the larger Trematoda. It represents a secondary adaptation to their larger body size, in which the intestine also partially takes over functions of the missing blood vascular system. Since an anus is lacking, food residues are again given off through the mouth. While, in the past, the absence of an anus was considered a primary character state, it is now assumed that, in the ancestors of Platyhelminthes, an anus was initially present in the ground pattern. During evolution, the anus was lost. Trematodes are facultative anaerobes; both liver fluke species gain most of their energy through glycolysis.

Nervous System

In contrast to the diffuse net-like nervous system of cnidarians, Platyhelminthes possess a centralized nervous system, which is characterized by an anterior control center, the simple brain, which receives the sensory input and a few longitudinally running conducting nerve cords. In Platyhelminthes, the two main nerve cords run in two anterior and posterior ventral strands. They are accompanied by several thinner longitudinal nerve cords, which are distributed in regular intervals around the body. These are, in turn, linked by transverse nerve cords (commissures). Due to this architecture, the nervous system is called an orthogonal (regularly right-angled connected) nervous system. The somata and neurites of the nerve cells are not separated from each other in ganglia, but are distributed throughout the entire nervous system (medullary cords). Anteriorly, there is an accumulation of nerve cell somata and neurites with synapses (nerve nodes) in the simple brain. In conjunction with the endoparasitic lifestyle of *Fasciola hepatica*, the nervous system and, especially, the sensory structures are reduced compared to the free-living relatives. For example, sense organs like eyes and statocysts are often missing. In histological sections, which are usually stained with Azan for practical purposes, nerve cords are difficult to observe. The advanced cephalization in Platyhelminthes is best seen in living planarians or polyclads, as they have two (or more) large eye spots that roughly mark the location of the brain, which, in certain species, is visible as a less pigmented area. These light sensory organs include dark-shading pigment cells and rhabdomeric photoreceptor cells. The pigment cells formed the pigment cup, and can be seen as dark spots (the eye spots). However, the

rhabdomeric photoreceptor cells, the second cell type in the eye spots, can only be seen in histological sections. Due to the arrangement of pigment and photoreceptor cells, only light from specific directions is perceived, and thus they facilitate directional vision. Therefore, one can speak of the presence of real eyes in flatworms. In contrast to the endoparasitic stages, simple pigment cup ocelli occur in Trematoda's primary larvae, the miracidia (Fig. 4.7b).

Excretion

Like all Platyhelminthes, trematodes have protonephridia, which account for two critical functions. First, protonephridia maintain ion homeostasis and water balance inside the body. This function is called osmoregulation. Second, protonephridia are responsible for the disposition of toxic nitrogenous waste products. This function is called excretion. Protonephridia are highly characteristic organs of animals without a secondary body cavity (coelom) and a blood vascular system.

Protonephridia consist of a terminal cell, a number of duct cells, and a pore cell forming the excretory duct, and a nephridiopore opening to the outside. The terminal cell forms a weir with mostly rod-shaped cell extensions that contact the first duct cell. This weir is externally covered by an extracellular matrix (ECM). In this way, it can serve as a filtration barrier: Inside the weir, flagella, built by the terminal cell, generate a constant fluid flow down the nephridioduct, thereby generating a lowered pressure within the tubule lumen. The resulting negative pressure forces body fluids to flow into the protonephridial duct. On its way, it has to pass the ECM, allowing only molecules up to a specific size to get through. Thus, the interstitial tissue fluid is filtered at the weir spaces of the terminal cell and the adjacent canal cell and forms the primary urine. In the subsequent sections of the protonephridial duct system, reabsorption of certain substances and secretion of substances that are too large for ultrafiltration take place. The primary urine thus gradually becomes secondary urine and is ultimately excreted. The filtering protonephridial duct system has only a single, posteriorly located collecting excretory duct and pore (Fig. 4.3). With the exception of Xenacoelomorpha, representing the first branch in the bilaterian phylogenetic tree, this process is fundamentally the same in all bilaterians with different types of nephridia, which is why they are also referred to as Nephrozoa (see Fig. 1.1).

The Neodermis, an Adaptation to Parasitic Life

All parasitic flukes, hookworms, and tapeworms have a highly specialized outer body covering that allows them to live inside their host. This is the neodermis, an unusual form of integument unique to Metazoa. The neodermis is always unciliated and serves for the direct exchange of substances with the environment. This includes the uptake and release of substances from and to their environment (excretion and

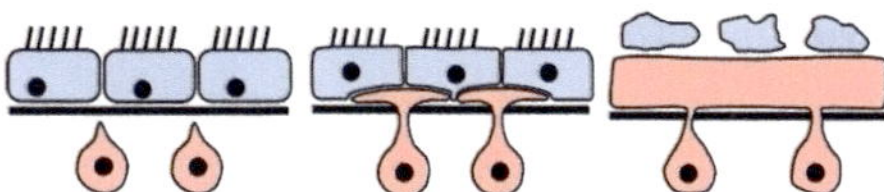

Fig. 4.11 Formation of the neodermis. Mesodermal neoblasts (red) push cell extensions under the old epidermis (blue) and displace it. Neodermis cells fuse and form a syncytium. After Burda, Hilken, and Zrzavy (2008) modified

osmoregulation), such as that done by the bile ducts or intestine of the host animal. In tapeworms, it takes up all nutrients, leading to the complete loss of the parasite's intestine.

The second primary function of the neodermis is the defense against their hosts' immune response or digestive enzymes. A system that can effectively allow them to evade the immune response of their hosts cannot be achieved by the primary epidermis, and thus the neodermis has evolved as an adaptation to endoparasitic life. The neodermis is a pharmacological target for certain antihelminthics. Triclabendazole, for example, damages the neodermis of juvenile and adult liver flukes, which ultimately leads to the death of the parasites.

The neodermis is built from cells of mesodermal origin and replaces the original ciliated cellular epidermis (Fig. 4.11). This is achieved by already present larval mesoderm tissue, which pushes itself under the epidermis through extensions and detaches it. The individual cells of the neodermis subsequently fuse to form a uniform tissue without cell boundaries, known as syncytium, in which the cell nuclei lie further inside the body, below the ECM.

The neodermis differentiates apically specific subcellular structures, e.g., many microvillus-like extensions. Moreover, the outermost part of the neodermis is denser than the inner part but does not form a hard layer. Both subcellular structures help to evade the defense reactions of the host, while simultaneously favoring nutrient uptake. The densely arranged microvilli or labyrinthine infoldings increase the surface area, which can be used to exchange substances. Due to the syncytial organization of the neodermis, there are no points of apical cell–cell contact, whereby the inner body is much better separated from the outside than it would be by a typical epithelium, with its numerous cell–cell contact zones. Thus, the syncytial epithelium regulates what the parasite takes up and releases more effectively. In summary, the formation of an exchange-friendly and protective neodermis most likely represents the essential adaptation of Neodermata to exclusively parasitic life.

While all adult parasite stages rely on the physiology provided by the neodermis, the first larval form shows a ciliated ectodermal epidermis required for swimming. The neodermis is formed during often complicated alternation of generations and hosts, which we describe in more detail in the next section (Figs 4.7 and 4.8). For example, a ciliated epidermis can be found in the primary larvae of flukes, the miracidia (Fig. 4.7b). The epidermis is shed when the larva penetrates its first host and is replaced by the neodermis, which newly differentiates as described above. By the way, the migration of stem cells into the epidermis for regeneration can also be observed in free-living flatworms.

Neodermis, Epidermis, and Other Body Coverings: A Short Explanation

As we have seen, the neodermis of parasitic Platyhelminthes is unique. The outer body covering of all invertebrate Metazoa with one exception (arrow worms, Chaetognatha) consists of a single-layered epithelium, which emerges from the ectoderm and is referred to as the epidermis. In contrast, the neodermis originates from mesodermal cells. Epithelia always consist of cells with an apical-basal polarity. These epithelial cells are interconnected by specific cell–cell junctions that ensure the mechanical stability of the cell layer. Moreover, they control substance exchange and prevent uncontrolled diffusion into the interstitium. The polarity of the epithelia is reflected in the properties of the respective cell membrane domain and the distribution of the cell organelles. Although an extracellular matrix is secreted apically and basally, only the basal ECM is responsible for the mechanical anchoring of the epithelial cells and for substance exchange and transport with the body's interior. However, the apical ECM is referred to as cuticle and is a protective anchoring layer. The starting point of cuticle formation is the glycocalyx, a thin extracellular layer composed of glycoproteins and glycolipids that form outside the cells. The distinction between a glycocalix and a cuticle is gradual. The extracellular matrices on both sides of epithelial cells can contain various amounts of collagen, the most common structural protein of extracellular matrices.

In Ecdysozoa, represented here by nematodes and arthropods, the cuticle contains chitin, a highly polymerized nitrogenous polysaccharide. A unique feature is found in the tunicates (Tunicata). Here, the cuticle contains cellulose, a substance known only from plant cell walls. Vertebrates, Craniota, have a multilayered epidermis, whose innermost layer produces new cells for the outside throughout life, and thus replaces the outer cells. In all amniotes (Amniota), these cells secrete structural proteins like filaggrin, keratin, lipids, and other protective substances. Finally, these cells die and form a waterproof protective layer, which functionally replaces a cuticle in terrestrial arthropods. This process is called keratinization.

Reproduction and Development

Like all flatworms, liver flukes are hermaphrodites. Each individual develops both female and male reproductive organs. However, the male germ cells (sperm) mature first, followed by the female germ cells (eggs). Thus, the common liver fluke is a protandric consecutive hermaphrodite.

In the unstained animal, eggs of different ages and degrees of maturity can almost always be seen. Young eggs appear whitish or yellowish and are located in the oviduct and uterus (Fig. 4.2). The composite eggs of Trematoda each contain 50 to 60 yolk cells and one fertilized oocyte. These are called ectolecithal eggs, representing the composite eggs of an even larger clade, called Neoophora (see above) (Figs. 4.5 and 4.6). When the eggshell hardens, the eggs change their color, and appear black. They are released through the uterus and the genital pore into the host's bile ducts. The animals usually mate reciprocally, but self-fertilization is not excluded. During mating, the male copulatory organ, the penis or cirrus, is everted through the male genital pore and inserted into the female genital pore. The transferred sperm are initially stored in the receptaculum seminis until they are released into the ootype for fertilization of the oocytes (Figs. 4.5 and 4.6).

It should be noted that asexual reproduction is quite common among certain freshwater flatworms, such as planarians. It generally occurs by transverse fission. However, flukes only reproduce sexually, and asexual reproduction is restricted to sporocysts and rediae, with intermediate generations occurring during the complex life cycle of trematodes. We will discuss this separately.

The Gonads of Flukes and Egg Formation

The paired testes extend into the entire periphery of the animal (Fig. 4.6). The mature sperm are transported via the vasa efferentia to the seminal vesicle, where the sperm are stored before being passed to the cirrus for transfer to another animal. The ovary is divided into two functional units: the germanium, where oocytes develop meiotically, and the vitellarium, a massive network of blind-ending ducts, where the yolk cells, which later nourish the embryo, develop. The originally paired germarium (ovary) is unpaired in trematodes (Figs. 4.2a–c and 4.3). The oocytes from the germarium (ovary), the sperm from the receptaculum seminis (received from a mating partner and stored here), and the yolk cells from the vitellarium come together in the ootype (Fig. 4.5). The oocyte must have already been fertilized by sperm from a sexual partner before the shell is formed around the oocyte and yolk cells to give rise to the compound egg. The Mehlis's gland produces the material needed to form the prospective eggshell. Therefore, it is also called the shell gland. Thus, several ducts of different organs of the sexual apparatus open into the ootype: the oviduct coming from the ovary, the united vitelline duct, which splits to the lateral vitellaria, as well as the receptaculum seminis (sperm pouch, storage of foreign sperm) and the Mehlis's gland. Additionally, the Laurer's canal, a connection to the dorsal outside of the animal, starts here. It is believed that overaged functionless sperm are dispelled via the Laurer's canal. The uterus branches off of the ootype, leading the composite eggs to the genital pore. In the case of the common liver fluke, as with almost all parasitic flatworms, the reproductive organs take up a large part of the body (Figs. 4.2 and 4.3). Perhaps the essential function of adult liver flukes,

which is true for all parasitic Platyhelminthes, is to reproduce sexually, provide offspring, and thus colonize new hosts. However, the probability of finding the proper host at the appropriate time is very low, so an enormous number of eggs are produced to ensure successful reproduction. Parasitic Platyhelminthes achieve this with sometimes very complex alternation of generations and host changes that many species undergo. For instance, the tapeworms of carnivorous hosts can only reach new host animals when they ingest the infectious stages with their prey. Thus, a herbivore (rodent, mouse, rat) must next be infested by a carnivore (fox, cat, dog) so the reproductive order does not break off. Only a massive number of fertilized eggs, usually hundreds of thousands per individual, is sufficient for a few offspring to find their way into a new final host. Once such a host is found and successfully colonized, the parasite lives protected and nourished from within it like a "maggot in bacon," but at the price that a large part of its energy must be invested in reproduction.

The Life Cycle of Trematodes

The parasitic life makes the liver fluke and other species of this taxon utterly dependent on their host. If the host dies, the life of the parasite also ends. Therefore, it is essential for the success of a species to produce high numbers of juveniles in time, so as to successfully colonize new proper hosts. Host–parasite relationships of flatworms are well balanced so that the vertebrate hosts are damaged but do not perish from moderate infestation (Figs. 4.7 and 4.8).

In the case of the common liver fluke (*Fasciola hepatica*), the fertilized eggs are discharged with the host's feces and will survive for several months (Fig. 4.7a). Ciliated miracidia larvae develop within 1.5–3 weeks in water or on a damp meadow. Miracidia possess organs, a cellular epidermis, and eyes (Fig. 4.7b). This is the only stage in the life cycle of the trematodes possessing an ectodermal primary epidermis with cilia. Miracidia actively search for their intermediate host, the pond snail *Galba truncatula* (Fig. 4.7f). They penetrate the body wall of the snail, shed off the ciliated epidermis of the miracidium, and develop into sporocysts. Sporocysts are very simple organized adult stages that now display the highly specialized neodermis as an outer covering (Fig. 4.7c). In the sporocyst, all internal organs are regressed. The sporocyst feeds solely through the neodermis. In the sporocyst, the next generation, the rediae, named after the Italian parasitologist and naturalist Francesco Redi (1626–1697), emerges from diploid stem cells (Fig. 4.7d). Rediae leave the sporocyst, but not the host. Rediae, in contrast to sporocysts, have organs and actively move to the digestive gland of their host snail. The rediae harbor stem cells from which the larvae of the next generation of liver flukes, the cercariae, derive. Cercariae are true larvae, with a tail to enable them to swim around. They leave the snail and populate the surrounding water (Fig. 4.7e). The infective stage of the common liver fluke is the metacercariae, the permanent forms that wait on plants for a suitable final host. As a rule, the cercariae attach themselves to plants, crawl to the upper parts of the plant, shed their tail, and form an encapsulated resting stage (metacercaria). In this way, they will survive dry weather periods for a long time. For example, the metacercariae of the common liver fluke are ingested by cattle with grass, and thus enter the intestine of their final host. The cyst shell is digested, and the metacercariae are released into the intestine, piercing the intestinal wall and reaching the liver via the body cavity. They then migrate through the liver tissue into the bile ducts, where they grow into adult, sexually mature animals. The only ones who are infectious for humans and mammals, as with all trematodes, are the metacercariae. Therefore, students cannot become infected in a zoology course when examining the contaminated liver of a cow or a sheep. Today, due to high hygiene standards, human infestation by the common liver fluke is rare in Central Europe. However, about two to three million people worldwide are infected with *Fasciola hepatica*. The excretory products and the destruction of the liver parenchyma can lead to jaundice (icterus), anemia, and calcification of the bile ducts, and thus to severe health restrictions.

Some other parasitic trematode species can infest humans as intermediate or definitive hosts and cause severe zoonoses. Zoonoses represent all diseases caused by animals. Of clinical relevance are, in particular, the Chinese liver fluke (*Clonorchis sinensis = Opisthorchis sinensis*), the lung fluke (*Paragonimus* species), and the blood fluke (*Schistosoma* spp.), to name just a few. *Paragonimus westermani* (Japanese lung fluke or Oriental lung fluke) infests mammals that feed on crabs. By eating raw infected crustaceans, such as lobsters, the infectious metacercariae enter the human body and cause lung paragonimiasis with cough, sputum expectoration, hemoptysis, and lung anomalies. Worldwide, about 22 million people are infected yearly. The Chinese liver fluke (*Opisthorchis sinensis*) spreads through the consumption of raw freshwater fish and is relatively widespread in the Asian region, with about 20 million infected people. *Schistosoma* (blood flukes) are the only known dioecious species among trematodes. The female lives permanently in the male belly fold, called the gynacophoric canal, in a kind of permanent copulation (Fig. 4.12). *Schistosoma mansoni* is the most important pathogen of schistosomiasis worldwide. Infected patients suffer from allergic reactions with edema, cough, and fever. In mass infestations, the course can even be fatal. Schistosomes remain in the host for up to 20 years, resulting in chronic symptoms. Unlike the trematodes described so far, an infection with *Schistosoma* does not occur through the persistent stages (metacercariae), but directly by the cercariae. Cercariae bore through the skin when one is bathing or working in contaminated water. From there, they enter the

blood vascular system and settle in the mesenteric vessels of the intestine or the portal vein system. The clinically relevant *Schistosoma* species are limited to the warmer climate regions of Asia, Africa, and South America. The number of infected people worldwide is estimated to be about 300 million. Other species of the Schistosomatidae (*Ornithobilharzia, Trichobilharzia, Bilhariella* species), which generally infect waterfowl as final hosts, have also been known to cause a so-called swimmer's itch or cercarial dermatitis in humans in Central Europe. Cercariae of certain schistosomatid species can actively penetrate the skin, causing the skin irritation mentioned. Readers have perhaps encountered signs at natural bathing lakes during the summer, warning of a mass occurrence of cercariae in the water. The cercariae do not develop further after penetrating the skin in humans, the typical inflammations being caused by their death. However, cercariae of fluke species typically present in Central Europe will not enter the host's internal organs and die after a while.

Tapeworms (Cestoda)

The tapeworms are only briefly mentioned here due to their veterinary and clinical relevance (Figs. 4.13 and 4.16). According to their even stronger adaptation to the parasitic lifestyle than trematodes, they are less suitable for representing the ground pattern of Platyhelminthes. The most common species in Europe are the beef tapeworm (*Taenia saginata*) (Fig. 4.13a), the pork tapeworm (*Taenia solium*), the dog tapeworm (*Echinococcus granulosus*, Fig. 4.16), and the fox tapeworm (*Echinococcus multilocularis*).

The ribbon-like body of the tapeworms is always divided into several easily distinguishable sections. The anterior head section, the scolex, is often equipped with suckers and a hooked crown. A tapeworm uses both structures to anchor itself in the intestinal wall of its host (Fig. 4.14). The scolex is followed by a growth zone (neck) from which a varying number of body segments (proglottids) emerge (Fig. 4.14). Mitotic divisions of stem cells produce these; thus, the youngest proglottids follow the neck region. Since the active uptake of nutrients occurs directly through the neodermis, tapeworms do not have an intestine. In every proglottid, complete female and male reproductive organs are formed, similar in structure and arrangement to those which we already know from the liver fluke (Fig. 4.14). Since the male gonads mature first, followed by the female gonads later (Fig. 4.14b–d), tapeworms are, like the flukes, also protandric hermaphrodites with reciprocal fertilization. The posterior body region of the tapeworm consists of mature proglottids, each of which can contain up to 100,000 eggs. This is the point at which the mature proglottids separate from the rest of the body and are released outside, mainly with the host's feces. There, the proglottids burst and release their eggs. Still in the eggshell, the six-hook larva (oncosphaera) develops. An intermediate host, such as a cow in the

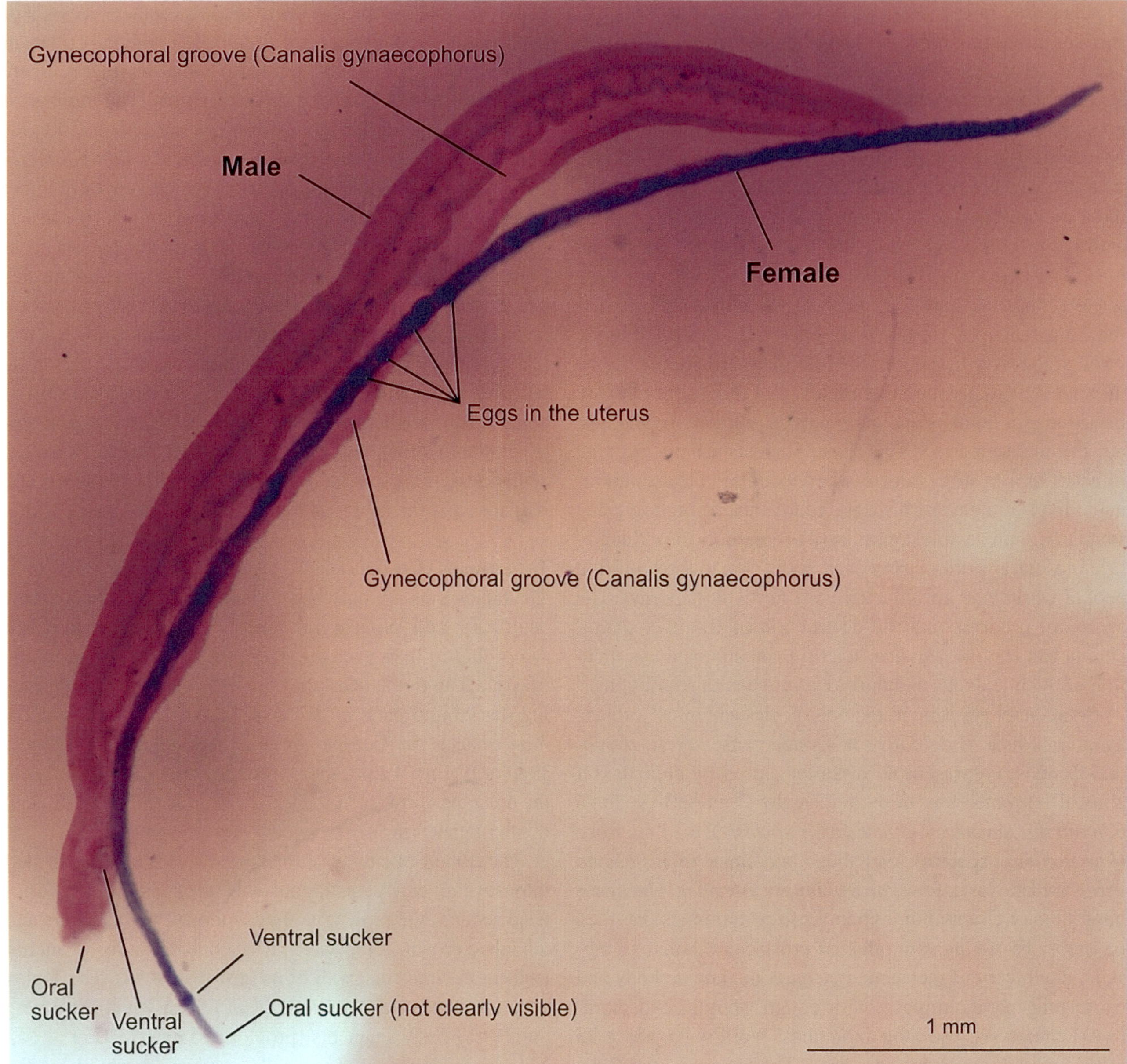

Fig. 4.12 Blood fluke (*Schistosoma* sp.). The male is stained reddish, the female bluish. The animals remain in permanent copulation for a lifetime. The ventral fold, with which the male covers the female, is called *Canalis gynaecophorus*. Specimen of the Zoological Teaching Collection, Philipps University Marburg

meadow, ingests oncospheres. The eggshells are digested in the host animal's intestine and the oncospheres hatch. Then, they penetrate the intestinal epithelium, shed the ciliated epidermis, which is replaced by the neodermis, finally reach the blood vessels, and are distributed throughout the body. They are transported to striated musculature, where the so-called bladder worm (cysticercus, pl. cysticerci) develops. This is a thin-walled, small fluid-filled sac-like stage resembling a bladder (hence why it is called a bladder worm), into which the head of the tapeworm (scolex) invaginates from the wall. The 7–9 × 5 mm large cysticerci of, for example, the beef

tapeworm, represent the juvenile stage of the tapeworm. Thus, various organ systems of the host can be infested. Final hosts usually ingest the parasite by consuming raw, contaminated muscle tissue. As soon as the cysticerci are in the final host's intestine, they evert the scolex and attach themselves to the intestinal wall with their suckers, developing into the adult tapeworm as described above (Fig. 4.15).

Infestation with adult beef tapeworms can be treated relatively easily and without consequences to health. This is not the case when people get infected with dog or fox tapeworm (*Echinococcus multilocularis* and *Echinococcus*

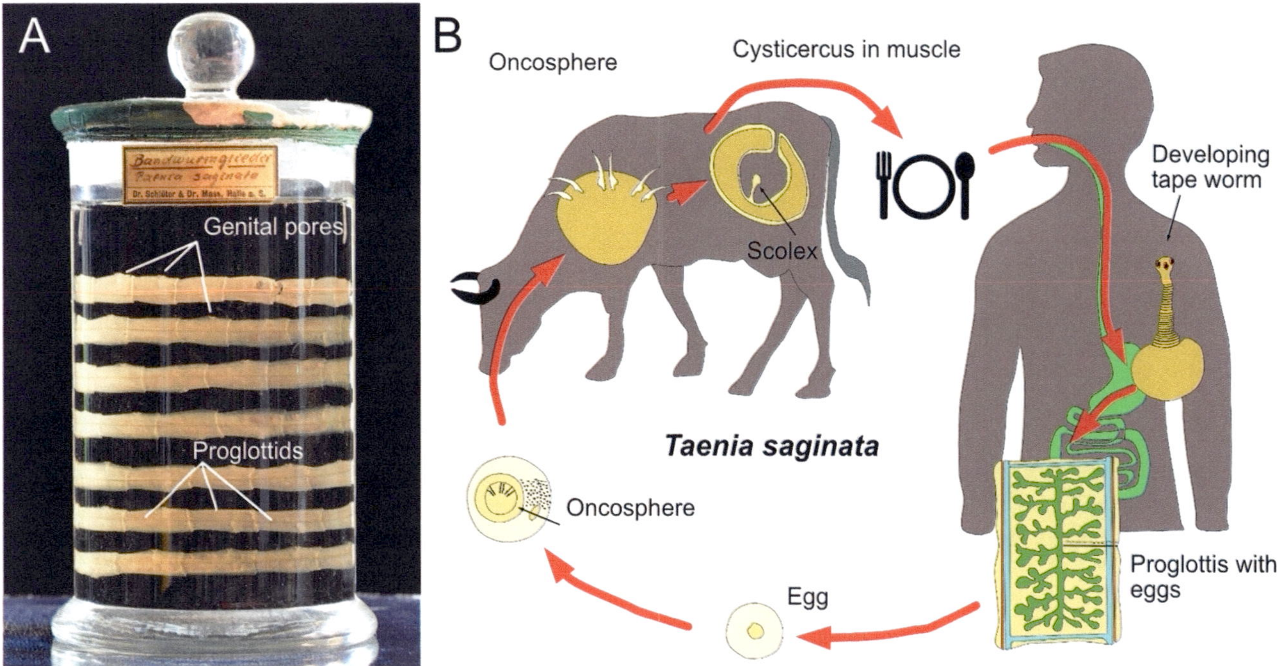

Fig. 4.13 Beef tapeworm (*Taenia saginata*). (**a**) Wet preparation of a beef tapeworm (*Taenia saginata*), private collection. (**b**) Schematic representation of the life cycle of *Taenia saginata* (according to various authors)

granulosus) (Fig. 4.16). These two species are among the most dangerous tapeworms for humans. Humans inadvertently become intermediate hosts through too close contact with infected dogs or cats. Up to soccer ball-sized, ulcer-like cysts develop, infesting essential organs such as the lungs, brain, and liver, and cannot be removed either chemotherapeutically or surgically. Thus, such infections are almost always fatal for the affected people. Humans usually do not become the definitive host, as the cysticerci of *Echinococcus multilocularis* are generally found in small mammals like mice, so the definitive hosts are predators like foxes, wolves, or dogs (Table 4.1). Our movie clip collection, which includes movies showing some living platyhelminthes, can be found at figshare: sn.pub/xr9qhi.

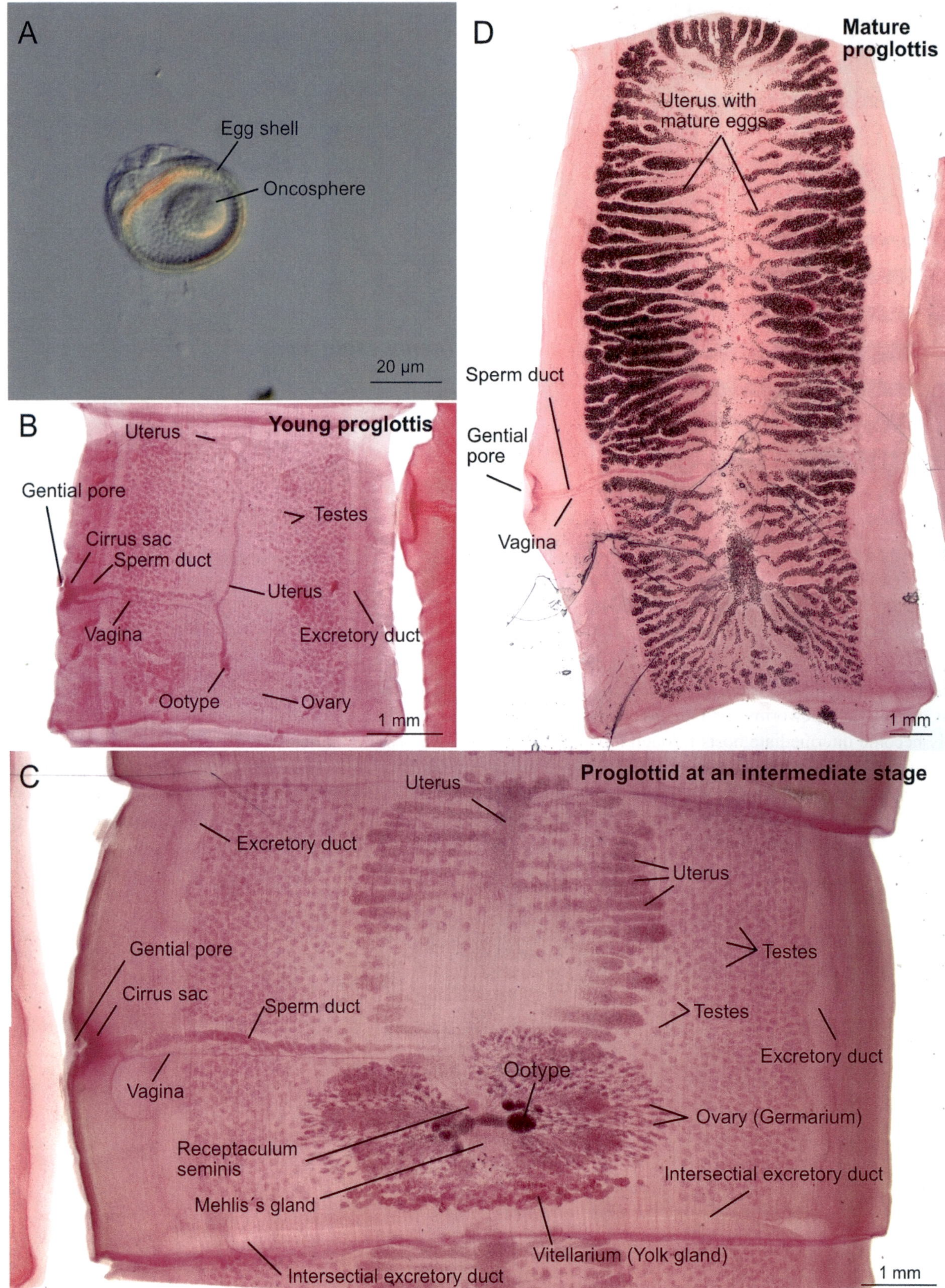

Fig. 4.14 Beef tapeworm (*Taenia saginata*). (**a**) Egg with oncosphere larva. (**b**) Young proglottid with well-developed testicular vesicles. (**c**) Proglottids of medium maturity with developing uterus. (**d**) Mature proglottid with numerous eggs. Whole mounts from the Zoological Teaching Collection, Philipps University Marburg

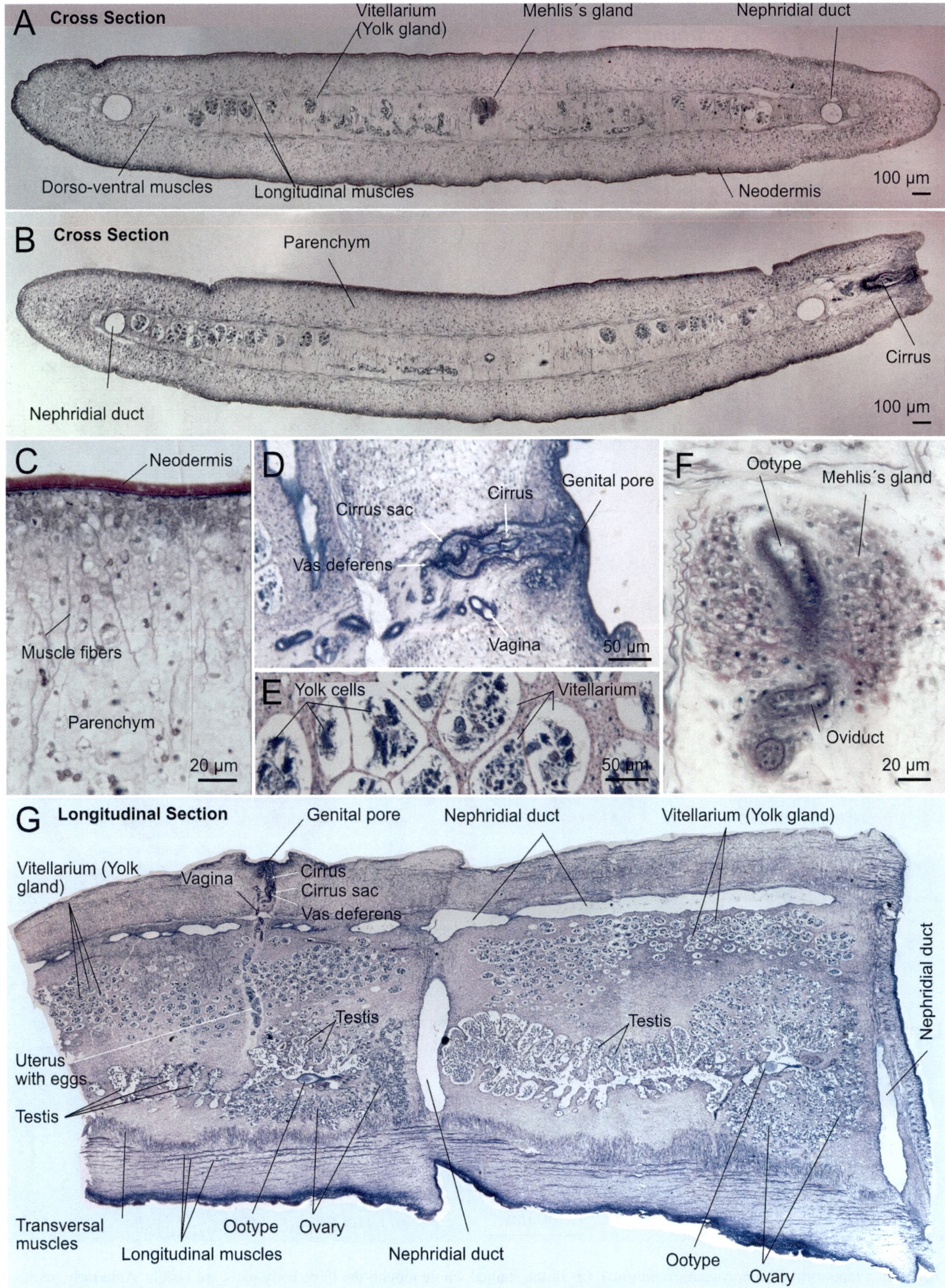

Fig. 4.15 Beef tapeworm (*Taenia saginata*) (**a**) Cross-sections through a mature proglottid with vitellarium and Mehlis's gland. (**b**) Cross-sections through a mature proglottid with cirrus. (**c**) Enlargement, showing the neodermis and the parenchym with muscle fibers. (**d**) Enlargement of the genital tract. (**e**) Yolk cells in the evitellarium. (**f**) The ootype, ooduct and Mehlis's gland at higher magnification. (**g**) Longitudinal section through two proglottids, overview. Histological sections from the Zoological Teaching Collection of the Heinrich Heine University Düsseldorf, Azan staining

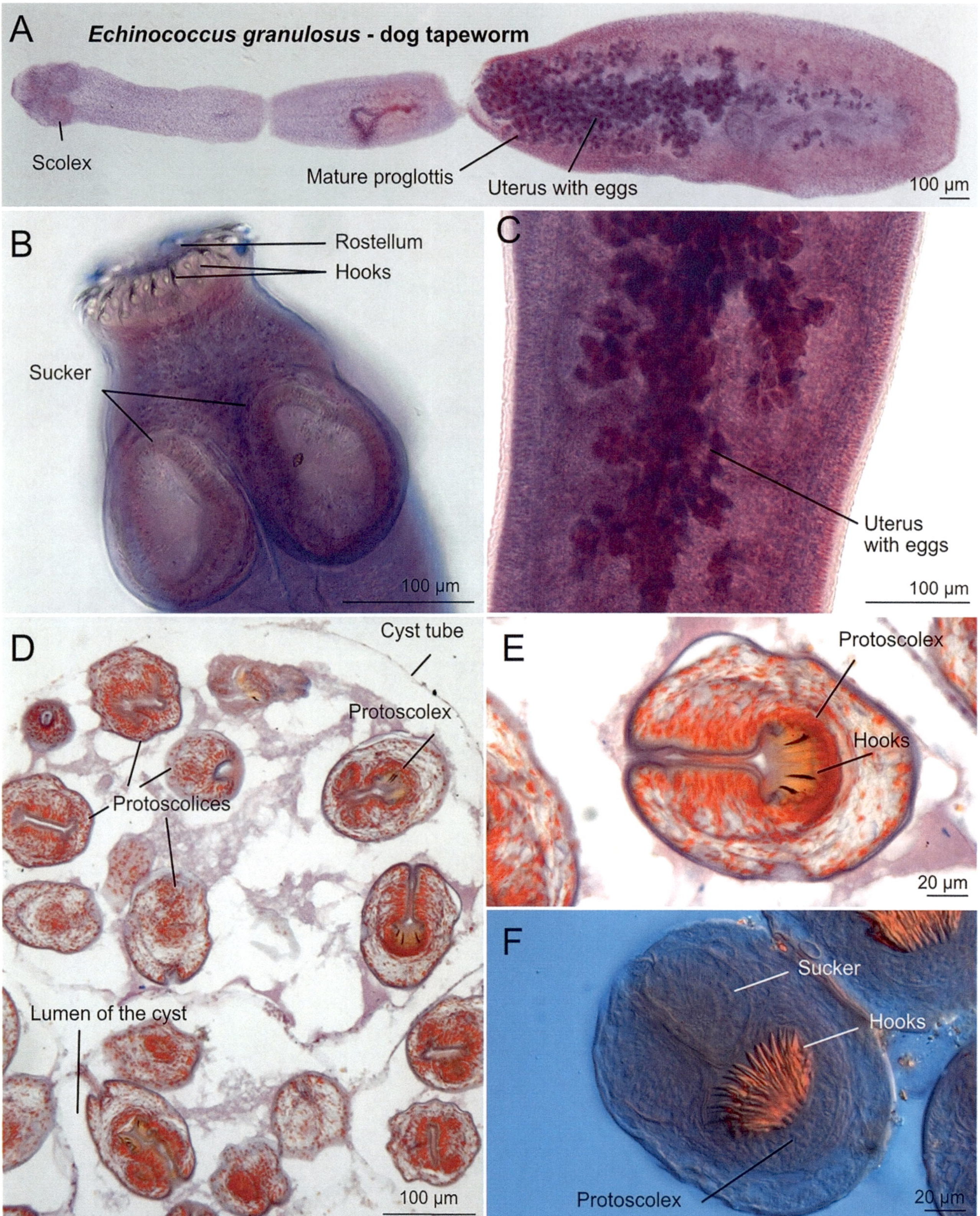

Fig. 4.16 Dog tapeworm (*Echinococcus granulosus*). (**a**) In the stained whole mount, the three body parts are visible. Anteriorly, scolex with rostellum and hook crown, followed by (**b**) a sexually mature proglottid. (**c**) The posterior proglottid contains mature eggs and partly mature larvae, the oncospheres. Specimen from the Biological Collection of the University of Osnabrück. (**d**) Cross-sections through a stained cyst with still invaginated protoscolices. (**e**). Cross-section through an invaginated protoscolex. (**f**) Hook crown of an invaginated protoscolex, unstained, Nomarski interference contrast. Specimens from the Zoological Teaching Collection, Philipps University Marburg

Table 4.1 Tapeworms (Cestoda)

	Size, age, etc.	Definitive host	Intermediate host	Symptoms
Beef tapeworm (*Taenia saginata*)	6–10 m, can live up to 20 years, 100,000 eggs per proglottid, 5–10 proglottids are released per day	Human Carnivores	Cow	Hunger, weight loss, abdominal discomfort (Taeniasis)
Pork tapeworm (*Taenia solium*)	4–6 m, can live up to 20 years, 80,000 eggs per proglottids, 5–10 proglottids are released per day	Human Carnivores (in case of infection with cysticerci)	Pig, human (in case of infection with oncosphere)	Like beef tapeworm, in case of infection with oncosphere larvae, cysts form in the brain, eye, etc. (Taeniasis, Cysticercosis)
Dog tapeworm (*Echinococcus granulosus*)	3–6 mm	Carnivorous vertebrates, dog, fox, badger, or cat	Herbivorous ruminants: cattle, sheep, horses or goats, Human	In case of infection with oncosphere larvae, cysts form in the brain, liver, and other organs (cystic echinococcosis)
Fox tapeworm (*Echinococcus multilocularis*)	3–6 mm	Mostly in foxes (in southern Germany, up to 30% of foxes are carriers), also found in dogs and cats	Human, rodents, mostly mice	In case of infection with oncosphere larvae, cysts form in the brain, liver, and other organs (alveolar echinococcosis)

Annelida (Segmented Worms)

▶ **Summary** **Segmented worms (Annelida) are a relatively large group of primarily free-living aquatic and terrestrial forms, displaying a fantastically large diversity in their morphology, physiology, and ecology. They play a critical ecological role, especially in the seas' benthic and terrestrial detritivore communities. Annelids are characterized by a body structure consisting of numerous identical sections, the segments. A collagenous cuticle covers their epidermis, and they possess a coelom and a closed blood vascular system.**

Annelida (segmented worms) is a relatively large and extraordinarily diverse group of animals with about 20,000 known species. Annelids are found in marine, limnic, and terrestrial habitats, where they are essential members of the respective food webs and ecosystems. Aquatic species are typically endobenthic but are less commonly found in open waters or on the substrate. Thus, these animals lead a more or less hidden life and remain largely unknown to most people. However, a few species occur in large numbers and dense aggregations, such as the lugworm (*Arenicola marina*) and the so-called sand corals, the *Sabellaria* species, which can form extensive reefs from sand tubes. Unfortunately, the reefs of *Sabellaria alveolata* native to the North Sea have been destroyed in the Wadden Sea area by trawling. Most annelids live in the sea, where they occur from the intertidal zone down to the deep sea. The marine annelids are collectively united as bristle worms or polychaetes and comprise about 12,000 species. All other annelids belong to the predominantly terrestrial Clitellata (girdle worms). Clitellata include the oligochaetes and the parasitic or predatory leeches (Hirudinea).

Annelids represent very different lifestyles and feeding habits. Their reproductive biology is also highly diverse. All these characteristics are reflected in a large morphological and systematic diversity. Current phylogenomic analyses have clarified the relationships within the group and placed Annelida within Spiralia as a subtaxon of Lophotrochozoa. In Spiralia, the fertilized eggs undergo a series of cell divisions, in which the daughter cells are arranged in a spiral pattern. The first mitotic cell divisions in embryogenesis are generally called cleavages, because they show a modified cell cycle pattern that usually lack certain replication checkpoints. A common feature of many Lophotrochozoa is the so-called trochophore larva. The trochophore develops a characteristic midbody ciliary band, called the prototroch. This band divides the larva into an upper and a lower part, called episphere and hyposphere. Although belonging to the plankton, these cilia allow for vertical movements in the water column. Annelids and mollusks develop from such trochophore larvae.

Most annelids belong to one of the two high-ranked sister groups, Errantia and Sedentaria. These names derive from their different lifestyles. Representatives of Errantia (from the Latin *errantus*, wandering) are marine mobile species that are predominantly herbivorous and carnivorous, but there are also microphagous species. Members of Sedentaria (from the Latin *sedere*, to sit) are usually less mobile, often tube-dwelling, and frequently micro- or detritivorous. The tubes are permanent or temporary and made of consolidated sediment particles or secreted substances such as chitin or mucus. This clade includes, in addition to many marine forms, the predominantly terrestrial Clitellata (girdle worms). We exemplarily selected one species from each of these two major groups, within which numerous typical structures for annelids are observable: The sandworm or king ragworm, *Alitta virens* (= *Nereis virens*), as representative of Errantia, and the common earthworm, *Lumbricus terrestris*, from the group Sedentaria. As the name implies, earthworms are terrestrial. They burrow in soils and only occasionally come to the surface, a fact that applies to most Clitellata except for leeches. In addition to the terrestrial forms, some species within Clitellata live in freshwater or the sea. Many morphological, physiological, and developmental characteristics of

© The Author(s), under exclusive license to Springer-Verlag GmbH, DE, part of Springer Nature 2025
A. Paululat, G. Purschke, *Metazoa – Morphology and Evolution of Animals*, https://doi.org/10.1007/978-3-662-69904-1_5

earthworms can be explained by the unique lifestyle, biology, and reproduction of Clitellata, rather than being considered primitive or typical of Annelida.

Annelida, like Arthropoda and Chordata, are segmented Metazoa. As we know today, segmentation has evolved several times independently in metazoans. As their name (Latin *annulus*, small ring) indicates, the segmentation of annelids is easily visible externally through their ringed appearance. Segments are repeating body sections (modules) that are primarily identical to each other (homonomous segmentation) (Fig. 5.1). However, during evolution, specializations of individual body sections arose, resulting in the formation of the so-called tagmata. Tagmata are groups of multiple segments converted into coherently functional morphological units. Through segmentation, very different organizational forms have been realized during evolution, and among the most successful metazoan groups, we find three segmented taxa: the annelids, the arthropods, and the chordates. There are small non-segmental sections called prostomium and pygidium at the front and rear ends of the annelid body. The prostomium at the front contains the brain and carries the most important sensory organs, for example, tactile organs (palps), eyes, chemosensory organs (nuchal organs), and sometimes mechanosensory antennae. The pygidium at the posterior end of the animal carries the anal opening and, in certain annelid species, also tactile sensory appendages, the anal cirri. The trunk segments lie between the prostomium and the pygidium. The first segment contains the ventrally located mouth opening and is called the peristomium. The number of segments varies greatly in annelids, ranging from less than 10 to over 1000. Most annelids grow after embryonic development by forming new segments in a pre-anal budding zone in front of the pygidium. Thus, in most annelid species, new segments are formed throughout their lifetime. Accordingly, the number of segments in adult individuals varies around an average. In contrast, the number of segments is constant in many species with a low number of segments, mostly up to about 20, and in all leeches (Hirudinea).

Each segment contains a pair of large fluid-filled coelomic cavities lined by a mesodermal epithelium called coelothelium. In the median, the coelothelium forms a mesentery that separates the left and right coelomic cavities and surrounds and suspends the intestine. Above and below the intestine, the mesentery forms the two main longitudinal blood vessels of the closed circulatory system. The coelomic cavities of adjacent segments are separated by transverse septa, also called dissepiments, which likewise are formed by a coelothelial bilayer. Thus, dissepiments and mesenteries consist of two thin epithelia facing each other on their basal sides and having an extracellular matrix (ECM) in between.

The excretory organs of all annelids are segmentally arranged paired metanephridia. Each metanephridium begins with an open ciliated funnel in the anterior coelomic cavity leading to a dissepiment. It continues with the nephridial canal into the following segment, where it opens to the exterior.

The body wall of annelids consists of a collagenous cuticle, epidermis, and ECM, as well as circular and longitudinal muscle fibers. The longitudinal musculature is typically much stronger than the circular. Most annelids move primarily using their longitudinal muscles and the parapodia. Each segment is equipped with paired lateral outgrowths, the parapodia, which bear chitinous chaetae. Parapodia can be entirely or partially reduced, for example, in earthworms where only the two bundles of chaetae remain.

The central nervous system comprises a brain located dorsally in the prostomium connected to the ventral nerve cord via paired circumesophageal connectives. The ventral nerve cord consists of a ventral chain of paired segmental ganglia and longitudinal nerve cords. In each segment, ring-shaped segmental nerves emanate from the ventral nerve cord, forming the peripheral nervous system.

The digestive tract of the annelids is designed as a straight tube, with an ectodermal foregut, hindgut, and endodermal midgut. The foregut (pharynx) is often muscular and can, as in our course object, *Alitta*, carry tooth-like structures and jaws that serve to capture prey.

The reproductive biology of annelids is highly diverse. Many marine annelids have a biphasic life cycle comprising an acoelomate planktonic larva (trochophore) and a coelomate adult. The larval stage is often suppressed, and we find a direct development. This is also the case for the common earthworm, an example of the taxon Clitellata. The loss of free-swimming trochophore larvae was very likely a prerequisite for the conquest of land by Clitellata. Annelids are primarily gonochoristic, but simultaneous or consecutive hermaphrodites have repeatedly evolved with clitellates as the best known example of simultaneous hermaphrodites. The formation of gametes may occur in all or many body segments or may be limited to only a few segments. In many large marine species, the gametes are released directly into the water and fertilized there. The zygotes then develop within the water column into trochophore larvae. In other forms, however, special gonads and genital organs are developed, often correlated with direct sperm transfer and internal fertilization. Various forms of parental care may be observed, particularly in the latter case. Especially within Errantia, sexually mature individuals of certain species leave the sediment to seek their sexual partners in open water. Many of these species show the phenomenon of epitoky, a form of adult metamorphosis in which large parts of the body are rebuilt, and the animals transform into a free-swimming, mating-ready epitoke. These species often reproduce only once in the entire life cycle. Shortly after the release of their gametes, the animals die. For example, in spring, individuals of *Alitta virens* can often be found in large numbers in the surf zone of our coasts. These are usually males who are dying after having released sperm.

Many annelid species are characterized by a remarkable regenerative capacity, with which they can respond to naturally occurring injuries or even the loss of body parts. The most common events are transverse cuts through the body, such as loss of the head region or the posterior end, caused by predators. The greatest danger for certain tube dwellers is the loss of their tentacle crown. The compartmentation of the body into several repetitive units separated by septa plays a crucial role in regeneration. Upon an injury, the musculature of the nearest intact segment contracts and closes the injured body, thereby limiting the loss of body fluids. The molecular mechanisms of regeneration in annelids are not yet fully understood. Mostly, regeneration starts with the dedifferentiation of cells, which transform into stem cells. These begin to proliferate and generate new tissue that differentiates into missing segments. During regeneration, the nervous system plays a crucial role. However, the regenerative capacity is extraordinarily different. Some species can only replace lost appendages or do not regenerate at all, while others can form the whole body anew from just one undamaged segment. Adult Nereididae like *Alitta virens* can only replace a lost posterior end. After wound closure, a new budding zone and a new pygidium are initially created. Additional segments are then successively formed by the new preanal growth zone. Many clitellates, including the common earthworm, regenerate anterior and posterior body segments. However, anterior regeneration requires a minimal number of intact segments after injury. Our movie clip collection, which includes movies showing living annelids, can be found at figshare: sn.pub/xr9qhi.

5.1 The Ragworms (*Hediste* sp. and *Alitta* sp.)

General Anatomy

The elongated body of Nereididae is divided into the prostomium, peristomium, numerous trunk segments, and the pygidium (Figs. 5.1a and 5.2b–e). A pair of large, biarticulated palps emerge from the prostomium, equipped with various receptor cells (Figs. 5.1a, b and 5.2c–e). When moving on the substrate, the region in front of the prostomium is continuously scanned and checked with the palps. Furthermore, the prostomium carries two small antennae and, like many errant annelids, two pairs of eyes (Fig. 5.1b–d). The chemoreceptive nuchal organs, characteristic of annelids, are located at the posterior edge of the prostomium and are usually only visible under the microscope (Fig. 5.1c, d) or in histological sections. They are presumably crucial in finding prey, food, or mating partners. The following body part, the peristomium, carries four pairs of finger-shaped sensory appendages, the tentacular cirri. They support and complement the sensory appendages of the prostomium. In the subsequent segments, the paired, pincer-shaped jaws can be seen through the body wall in small juvenile and transparent individuals (Fig. 5.1a, b). If the animals have everted their pharynx during fixation, the jaws can also be seen from the outside (Fig. 5.2c–e).

The segments following the peristomium each carry a pair of biramous parapodia (Figs. 5.1, 5.2, and 5.3). Biramous means comprising two branches. The dorsal branch of a parapodium is called notopodium; the ventral branch is called neuropodium. Parapodia are lateral outgrowths of the annelid body, leg-like extensions that can be actively moved. Usually, in histological sections, only parts of a given para-

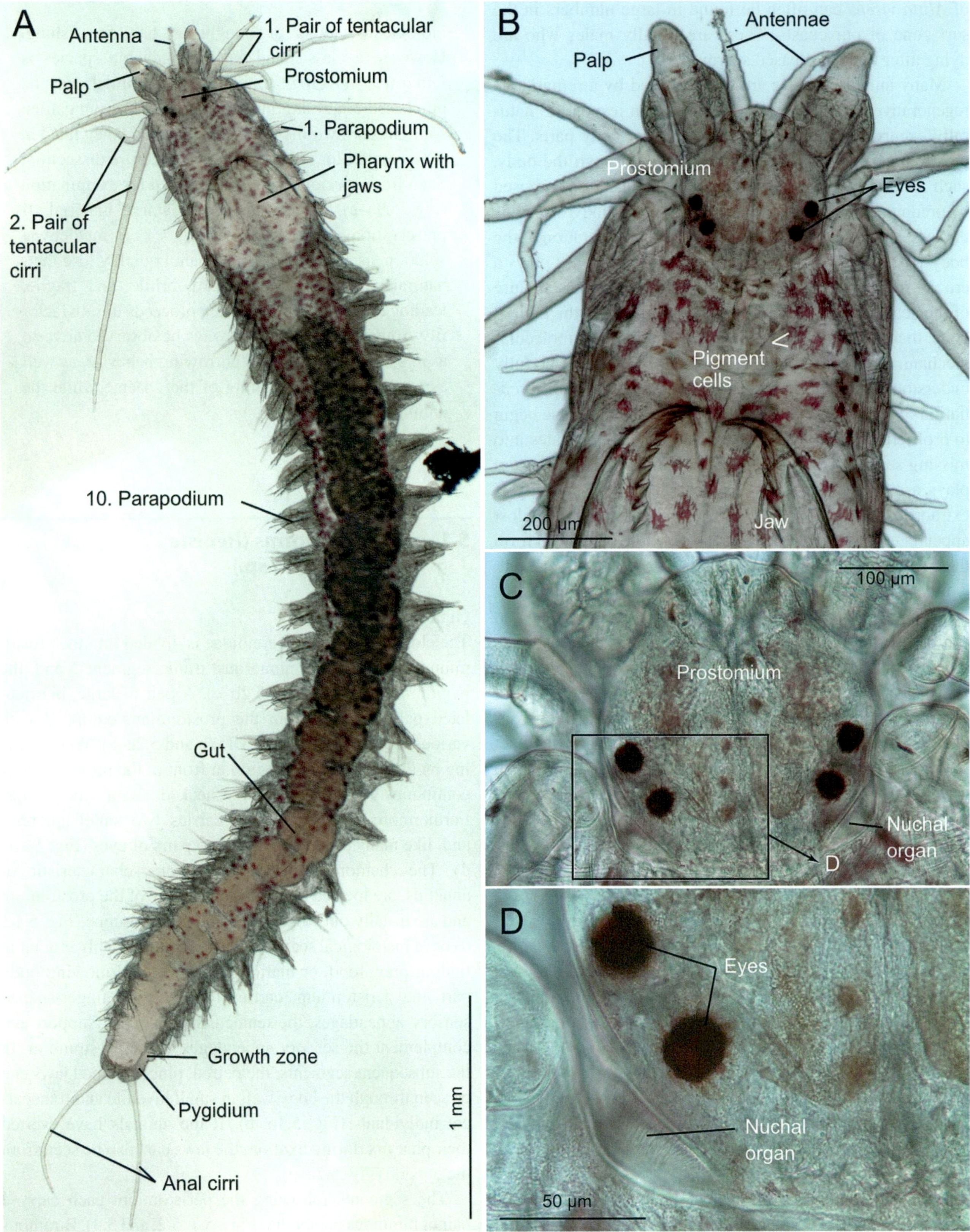

Fig. 5.1 *Alitta* sp. Habitus and external morphology. Juvenile, living animal, brightfield illumination. (**a**) Complete animal. (**b**) Front end with sensory organs and jaw pincers. (**c**, **d**) Close-up of the left part of the prostomium with eyes and nuchal organs

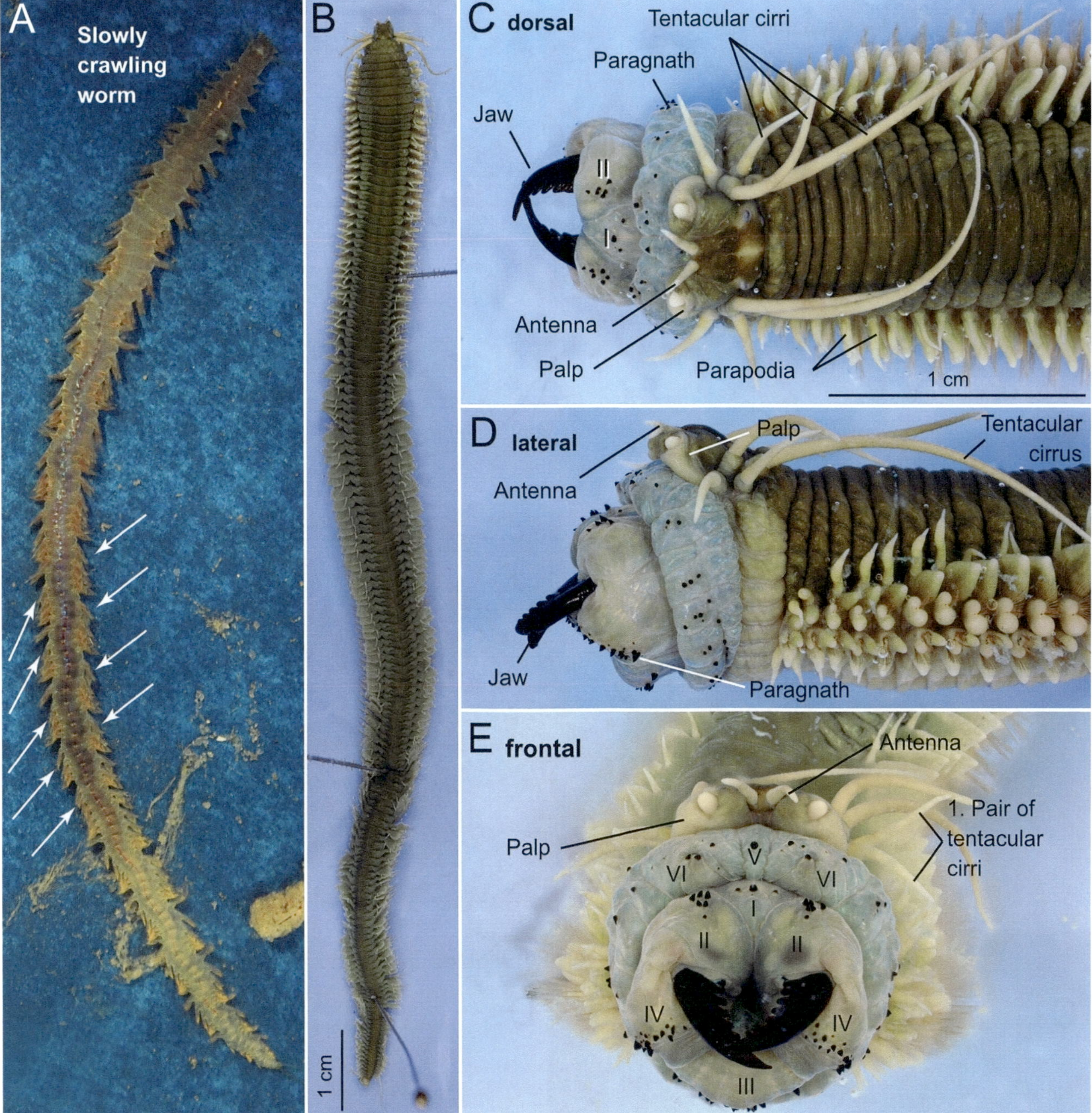

Fig. 5.2 External morphology. (**a**) *Hediste diversicolor*. Slowly moving living animal, arrows point to parapodia after backward movement (power stroke; about every fifth parapod is in the same movement mode). Longitudinal musculature supports the parapodia and forces the body into a slight sinoid movement. The dorsal blood vessel is visible as a dark line in the median. (**b–e**) *Alitta virens*, fixed specimen, ♂. (**b**) Entire animal. (**c–e**) Front end with sensory organs (antennae, palps, eyes, tentacular cirri) and everted pharynx with jaws and paragnaths; Roman numerals denote the individual pharyngeal sectors in which the paragnaths are situated in a species-specific pattern. **c**, **d**, and **e** each have the same scale

pod are visible, but several parapodia are usually cut at different levels (Fig. 5.3a, b).

The typical structure of a parapod can be studied by cutting out a trunk segment with a sharp razor blade or a scalpel, placing it in a drop of water on a microscopic slide, and examining it under a microscope under a cover slip (Fig. 5.3c,

d). In *Alitta virens*, the notopodium, the dorsal branch, consists of several tongue-shaped lobes, the dorsalmost of which is the largest, and a fan-shaped chaetal bundle. Inside the parapod, an exceptionally thick supporting chaeta, the acicula, is recognizable. Dorsally, a finger-shaped sensory appendage emerges, the dorsal cirrus. The dorsal lobe of the

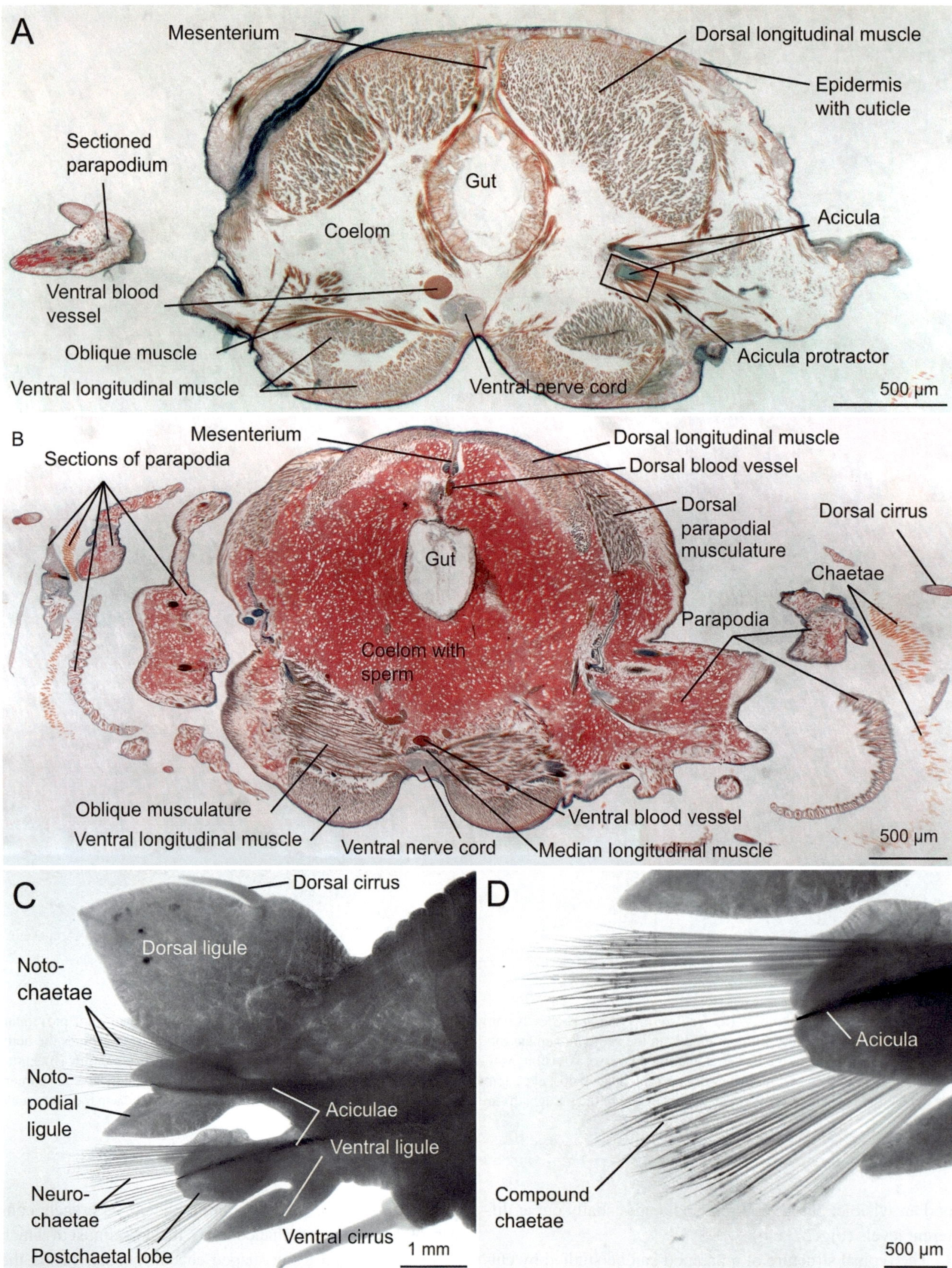

Fig. 5.3 (**a**, **b**) *Alitta* sp., cross-sections through the trunk region of immature (**a**) and sexually mature (**b**) individuals. This species undergoes an adult metamorphosis from the atoke to the epitoke stage upon reaching sexual maturity. Azan staining. (**c**, **d**) *Alitta virens*. Biramous parapodium consisting of noto- and neuropodium. (**c**) Whole parapodium. (**d**) Neuropodium with acicula and compound chaetae. **a**, **b** Specimens from the Biological Collection of the University of Osnabrück

parapod is particularly well supplied with blood vessels. In marine annelids, respiration occurs across the entire body surface or with the help of branchiae. In Nereididae, the large dorsalmost lobes in each parapod also serve as branchiae. The ventral parapodial branch, the neuropodium, is similarly constructed, with only the lobe-shaped sections being smaller. When crawling on the substrate, the neuropod is a flexible foot used to walk on the ground. When the animal crawls in the tubes or swims, both parapodial branches, noto- and neuropodium, are used for locomotion. If one lets living animals run in a large petri dish or aquarium, the use of the parapodia can be easily observed. A movie showing parapodia in motion can be found at figshare: sn.pub/xr9qhi. The longitudinal musculature assists the parapodia, which sets the body in sinusoidal undulations parallel to the sagittal plane (Fig. 5.2a). This support of movement increases when the animal is running faster, up to the point when it finally turns to undulating swimming, in which the parapodia are used as paddles.

In epitoke individuals, for example, in *Platynereis dumerilii*, the posterior parapodia are remodeled at sexual maturity. They differentiate into numerous paddle-shaped structures with swimming chaetae, which make swimming much easier (Fig. 5.3a, b). In males, the transformed parapodia also carry numerous chemoreceptors, facilitating the search for mature females. Uric acid released into the water stimulates the males to release sperm. Since the trunk muscles are less critical for swimming than for moving on the ground, they are largely reduced in epitokes, while the parapodia musculature enlarges. Since sexually mature epitoke animals no longer feed, the intestine is regressed. At the same time, the coelomic cavities become filled with gametes (Fig. 5.3b). In addition to these functions, the animals can perform respiratory movements with their longitudinal muscles, in which the body undulates at 90° to the sagittal plane.

Toward the body's posterior end, the size of the segments and parapodia decreases continuously (Figs. 5.1a and 5.2a, b). Approximately the last three to five segments show an incomplete differentiation. The body ends with the rather inconspicuous pygidium, with its two long chemo- and mechanosensory anal cirri. All these appendages bear a large number of different receptor cells. Thus, the animals perceive numerous stimuli from their environment. Due to the distribution of the sensory organs along the body's longitudinal axis, gradients of chemical stimuli can easily be detected.

The internal organization is best studied in histological sections. Both cross- and longitudinal sections are suitable (Figs. 5.3, 5.4, 5.5, 5.6, 5.7, and 5.8). The brain, nervous system, and sensory organs are usually easier to observe in longitudinal sections close to the sagittal plane. In cross-sections, the musculature of the body wall, which is organized in four longitudinal muscle bundles, is particularly noticeable (Fig. 5.3a, b). These occupy the largest part of the body cavity and allow the animal to crawl on the substrate or swim freely, supported by the paddling movements of the parapodia. Dorsally, in the median, we find the mesentery with the dorsal blood vessel; the ventral nerve cord lies ventrally, and directly above it is the main ventral blood vessel. A circular musculature is weakly developed (Fig. 5.5a, b). Other muscle groups are the diagonal and parapodial muscles running obliquely through the body cavity. Usually, several parapodia are visible at the same time, as these are mainly directed backward. Thus, a cross-section through an animal cuts several parapodia, but only partially. In the middle of the body is the alimentary canal. The body cavity appears more or less empty, apart from the chaetae and their musculature (Fig. 5.3a), or, in sexually mature animals, it is densely filled with gametes (Fig. 5.3b). The body cavity forms a hydrostatic skeleton. The longitudinal muscles determine the movement of the trunk; all four muscle groups can be moved independently of each other, so that the possible movements have significantly more degrees of freedom than, for example, in nematodes, where the longitudinal muscles only form two functional groups.

Body Wall

The body wall is formed by the cuticle and the single-layered epidermis (Fig. 5.3a, b, 5.4a–c, and 5.5a–d). The cuticle and the ECM underlying the epidermis appear as blue structures in Azan stainings. The cuticle also continues into the ectodermal foregut and the pharynx (Fig. 5.4a, b). The epidermis of Nereididae mainly consists of non-ciliated supporting cells and various gland cells. The so-called spinning glands are situated in the tongues of the parapodia, whose somata reach deep into the body cavity and send long cell processes to the surface (Fig. 5.5c–e). Along this path, the composition of the secretion changes, and the color in the histological preparation changes accordingly. The secretions of the spinning glands serve, in some species, to form a tube in which they reside. In other species, the secreted products stabilize the tubes burrowed into the sediment or form a mucus net, where fine particles and plankton are trapped. This is then eaten from time to time and digested as a whole. Another form of epidermal formation is the chitinous chaetae (Fig. 5.5e–g), which are moved by muscle fibers (Fig. 5.5e, g).

Nervous System and Sensory Organs

The nervous system of nereidids includes a relatively large brain in the prostomium (Figs. 5.4b, c and 5.6a–c). In sections close to the sagittal plane, the large, forward-facing mushroom bodies are visible (Figs. 5.4b and 5.6a, c). These are association centers that receive their sensory inputs from the palps. The mushroom bodies consist of a stalk-like neuropil and the small, grape-like arranged somata of the globuli cells. Numerous nerves extend from the brain to the sensory organs of the head region. Particularly noticeable is the innervation

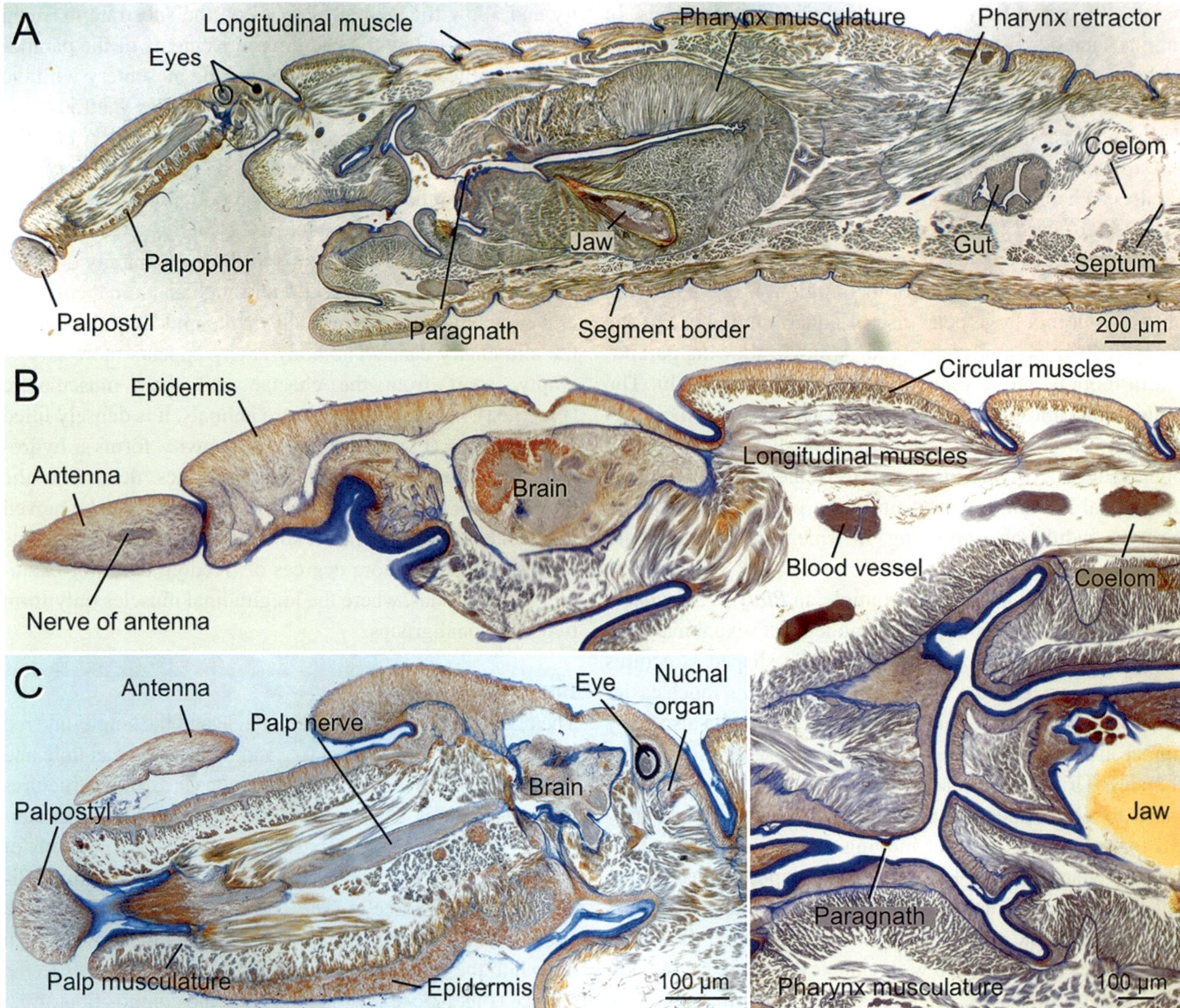

Fig. 5.4 *Alitta* sp., longitudinal sections through the anterior end. Azan staining. (**a**) Parasagittal section, prostomium, and anterior trunk region with pharynx and pharynx retractors. (**b**) Sagittal section with antenna, brain, pharynx, jaws, and the dorsal blood vessel. In the brain, the forward-directed mushroom bodies are recognizable as red structures. (**c**) Palp with main nerve, lateral view of the brain, posterior eye, and nuchal organ in the fold between pro- and peristomium. Specimens from the Biological Collection of the University of Osnabrück

of the palps (Figs. 5.4c and 5.6b), whose nerves are connected to the brain via several nerve roots. In addition to antennae and palps, the head's other sensory organs are the two pairs of eyes and the nuchal organs (Figs. 5.4c and 5.6b, d, e). The eyes have a vitreous body and a retina, on which three layers can be distinguished (Fig. 5.6d): The innermost layer contains the light-perceiving parts of the photoreceptor cells and extensions of the pigment cells, which pass through this layer and form the vitreous body. The middle layer is characterized mainly by the dark layer of the shading pigment, and in the third layer, we find the somata of the pigment and photoreceptor cells. The eye muscles are located beneath the ECM. The chemosensory nuchal organs consist of ciliated supporting and receptor cells as well as a retractor muscle (Fig. 5.6e).

The circumesophageal connectives, which connect the brain with the ventral nerve cord, also originate from the brain; each splits into two roots (Fig. 5.6b). The ventral nerve cord bulges into the body cavity (Fig. 5.7a) and is recognizable as part of the epidermis thanks to a continuous ECM, stained in blue (Fig. 5.7c). The ventral nerve cord between the ganglia consists of three longitudinally running connectives, each containing a giant fiber, particularly fast-conducting nerve fibers (Fig. 5.7b). Such giant fibers are present in many annelids in different numbers. The most prominent giant fibers reach a diameter of 1 mm (!). In addi-

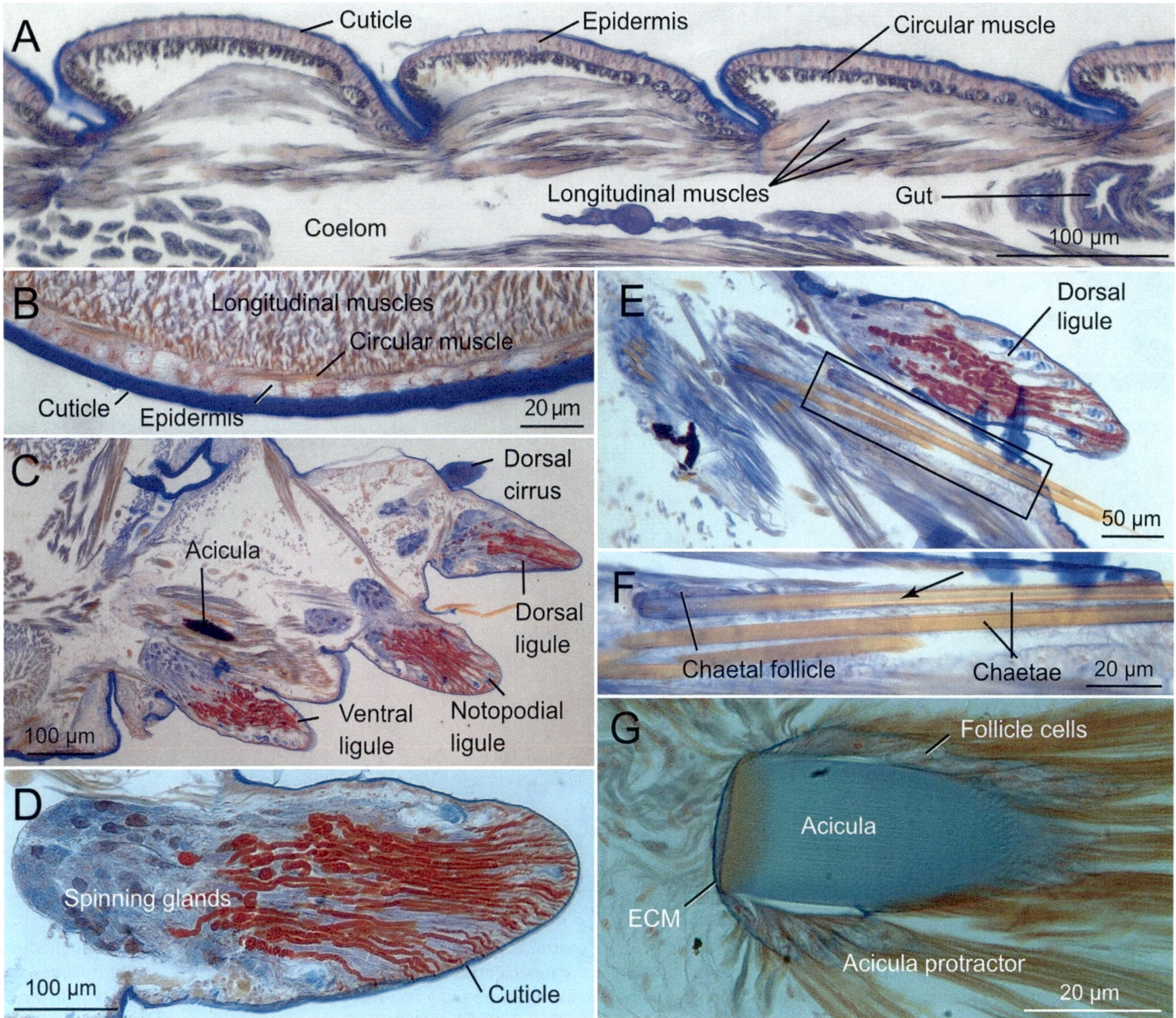

Fig. 5.5 *Alitta* sp., body wall musculature and chaetae, Azan staining. (**a**) Longitudinal section with three segments, dorsal. (**b**) Cross-section, ventral. (**c**) Parapodium with lobe-shaped extensions; note the groups of gland cells, the so-called spinning glands. (**d**) Enlargement of the ventral parapodial tongue, with a group of gland cells; somata appear light blue; nuclei dark blue-violet; cell processes with mature, red-stained secretory granules extend to the periphery. (**e**) Notopodium with a bundle of chaetae. (**f**) Enlargement of the framed region from E with chaetal follicle cut longitudinally; arrow points to the hollow core of the chaeta. (**g**) The base of the acicula with follicle cells and protractors; longitudinal striation of the chaeta reflects the fine chitin channels which make up the chaeta. Specimens from the Biological Collection of the University of Osnabrück

tion to the neurons, many glial cells surround the nerve cells. In the ganglia, the longitudinal nerves are interconnected by commissures, and segmental nerves project toward the periphery (Fig. 5.7a). The somata of the corresponding neurons are located laterally and ventrally in the ganglia, whereas the dorsal side lacks somata (Fig. 5.7c).

Digestive System

Highly characteristic of the intestinal canal of Nereididae is the extraordinarily muscular pharynx. The two powerful epidermal jaw pincers are embedded into the musculature (Fig. 5.4a–b). The pharynx is extended so the two jaw pincers can freely point forward and grasp food (Fig. 5.2c–e). A system of various muscle groups connects the pharynx with the body wall. The retractor muscles are particularly strongly developed (Fig. 5.4a). The intestine has a single-layered ciliated epithelium with fine circular and longitudinal muscle fibers (Figs 5.3a, b and 5.8c, d).

Body Cavity

The body cavity is a coelom; it contains the coelomic fluid and different types of cells, collectively called coelomocytes.

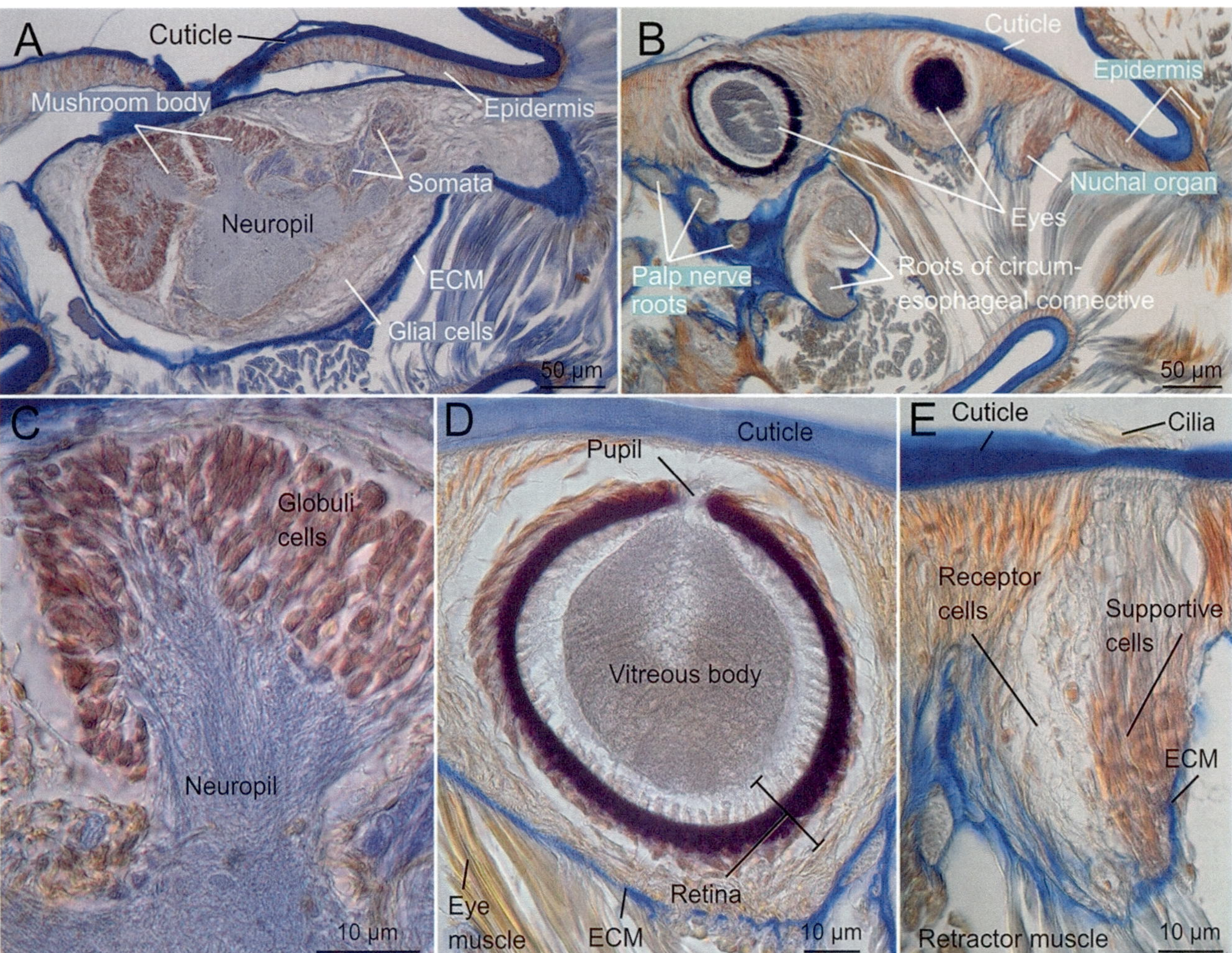

Fig. 5.6 *Alitta* sp., brain and sensory organs. Azan staining. (**a**) Brain; central neuropil, peripheral somata, and mushroom bodies; longitudinal section (cf. Fig. 5.4a–c). (**b**) Peripheral longitudinal section with palp nerve roots, dorsal and ventral root of the pharyngeal connective; eyes and nuchal organ embedded in the epidermis. (**c**) Mushroom bodies with stalk-like neuropil and grape-like positioned globuli cells. (**d**) Eye with three-layered retina: inner lighter layer with rhabdomeres, middle layer with shading pigment, outer layer with somata of photoreceptor cells and pigment cells. (**e**) Nuchal organ; cilia of the supporting cells penetrate the cuticle; receptor cells are recognizable by the lighter cytoplasm. Specimens from the Biological Collection of the University of Osnabrück

At sexual maturity, the coelom also contains the gametes (Figs. 5.3b and 5.8a–f). Coelomocytes have diverse tasks; they play a crucial role in defense (phagocytosis of microorganisms and viruses, immune reactions, uptake of foreign bodies), wound closure, and yolk synthesis. The coelom is traversed by the dissepiments, mesenteries, and blood vessels (Figs. 5.3a, b, 5.4a, b, 5.7a, b, and 5.8c). A movie showing blood streaming in the dorsal vessel of a polychaete can be found at figshare: sn.pub/xr9qhi. The endothelium, or coelothel, the epithelial cell layer that lines all coelomic cavities, consists of flat, plate-shaped cells attached to the ECM. These cells are so thin that often only the ECM is recognizable on histological sections and at high magnification (e.g., Figs 5.7a–c and 5.8c). To identify these cells, searching for the cell nuclei is recommended. There these cells are thicker, and with some effort, their nuclei are visible as red dots on the band-shaped ECM (Figs. 5.7b and 5.8c).

The nephridia are typical metanephridia. They are relatively small structures located lateral to the ventral longitudinal muscles at the bases of the parapodia (Fig. 5.8a, b). While the ciliated funnel (nephrostome) is hard to find and often not observable, the nephridial canal, which forms several loops, is easier to detect. The lumen of the canal is ciliated and distally formed, creating a straight section running to the epidermis, where it opens to the outside ventrally (Fig. 5.8b).

Reproduction

Nereididae are gonochoric. Thus, the sexes are separate. Their reproductive organs are simple. At the onset of sexual

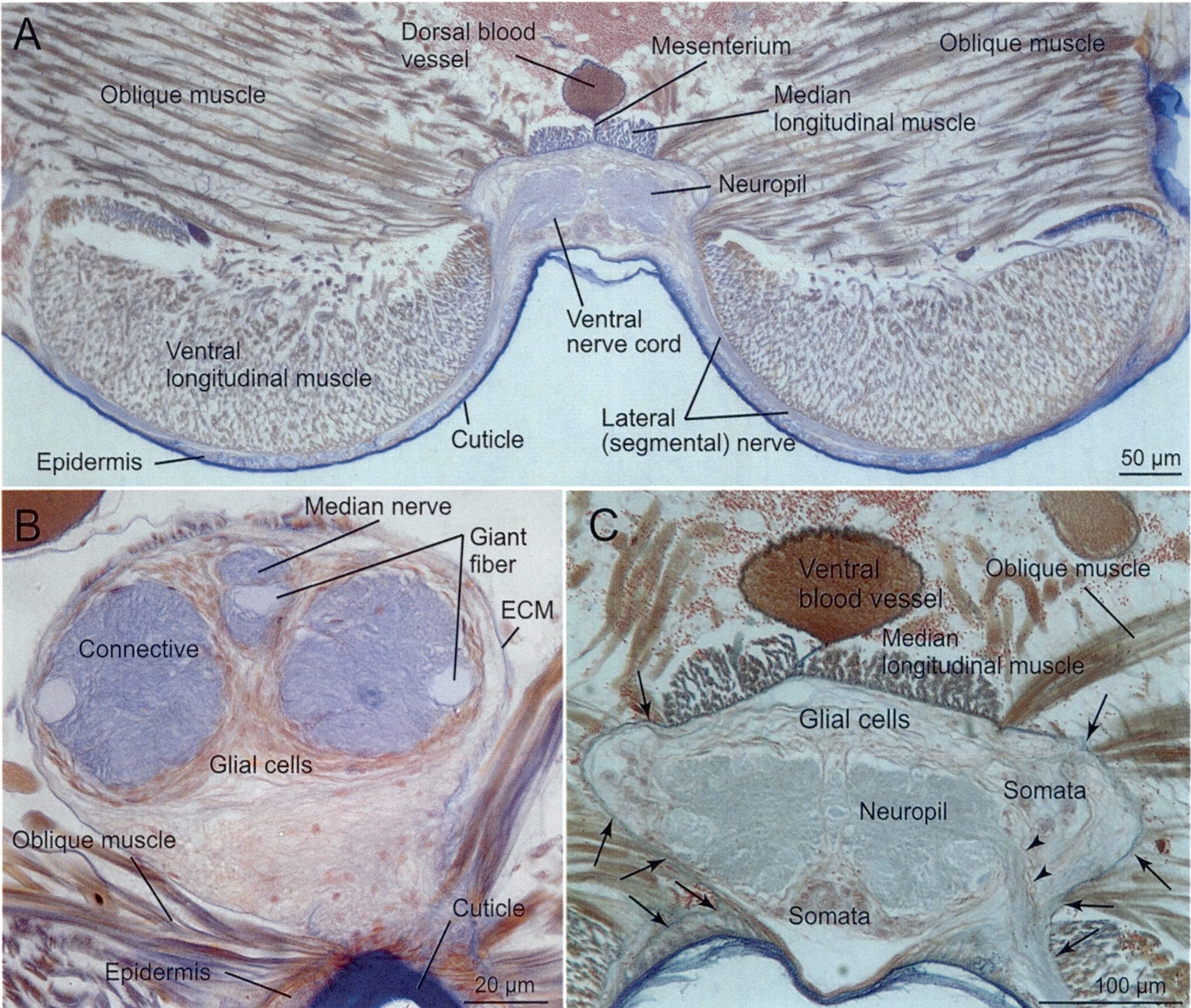

Fig. 5.7 *Alitta* sp., ventral nerve cord. Azan staining. (**a**) Overview; ganglion with branching lateral or segmental nerves. Strong longitudinal muscles; the parapodial muscles traverse the body cavity. Above the ventral nerve cord, the median longitudinal muscle, the mesentery, and the ventral blood vessel are visible. (**b**) The area between the ganglia (connectives) and the ventral nerve cord consists of three longitudinal nerves, surrounded by glial cells, each with a giant fiber. (**c**) Ganglion with uniform central neuropil and peripheral somata; basiepithelial position of the nervous system recognizable by the ECM continuous with the epidermis (arrows). Arrowheads point to the branching lateral nerve. Specimens from the Biological Collection of the University of Osnabrück

maturity, groups of developing oogonia or spermatogonia appear in the coelomic cavities (Fig. 5.8e). From these inconspicuous gonads, cells are released into the coelom, where further development into oocytes or sperm occurs. Gonads are not limited to specific body segments but include the entire trunk. The mature gametes are collected within the body cavity and then released into the open water via the nephridia. Due to this form of reproduction, which is then followed by the pelagic larval phase in the plankton, the number of gametes is enormous, so that sexu-

ally mature individuals always have coelomic cavities that are densely filled with gametes (Fig. 5.3b). The different species of the family show different mating behaviors. In species with epitoky, such as *Platynereis dumerilii*, the individuals enter the open water column and release their gametes under swarm formation (or pairwise in culture). In other species, either both sexes remain on the ground (*Hediste diversicolor*) or only one does (*Alitta virens*); here, either the males seek out the females (*Alitta virens*) or vice versa (*Hediste diversicolor*).

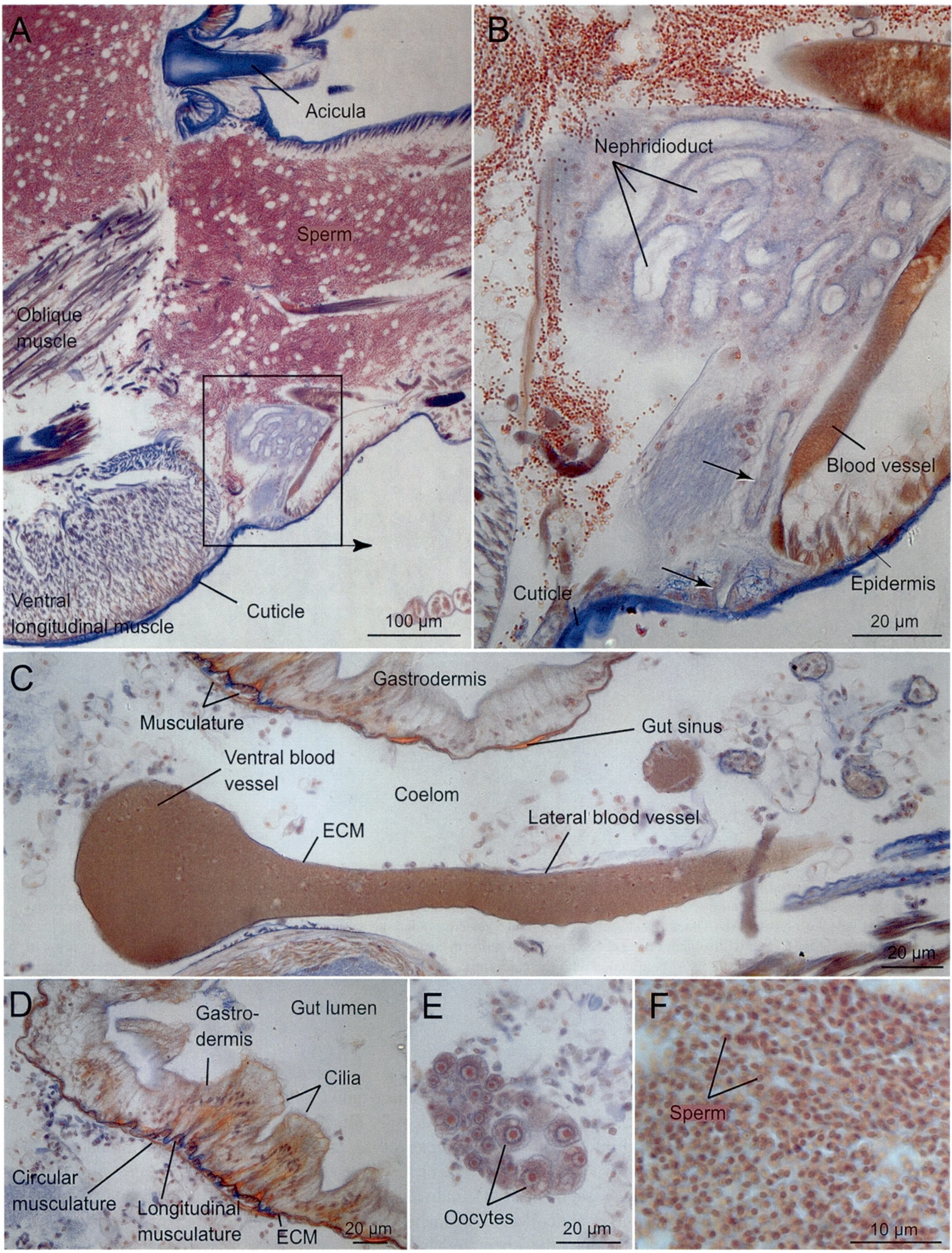

Fig. 5.8 *Alitta* sp., body cavity (coelom), nephridia, intestine, and gametes. Azan staining. **a, b,** and **f** are sexually mature males, and **c, d,** and **e** are sexually mature females. (**a**) Metanephridium is located ventrolateral at the base of the parapodia. Coelom, densely filled with sperm. (**b**) Magnification from a; ciliated nephridial canal cut multiple times; Arrows: outward leading passage; ciliary funnel not visible. (**c**) Coelom, ventral blood vessel with branching lateral vessel. Blood contains few cells (hemocytes). (**d**) Ciliated gastrodermis with isolated longitudinal and circular muscle fibers. (**e**) Ovarian structure with early oocytes. (**f**) Mature sperm with spherical head and midpiece at high magnification. Specimens from the Biological Collection of the University of Osnabrück

5.2   The Common Earthworm (*Lumbricus terrestris*)

External Organization

Lumbricus terrestris, which we use here for dissection, is one of the largest oligochaete species in Europe, with a length of up to 35 cm. Generally summarized as earthworms, they are crucial to soil ecosystems, serving as a prominent part of the decomposer communities in the cycle of organic matter and modifying the soil structure. *Lumbricus terrestris* is known under numerous trivial names: in Germany, it is known as the "rain worm," in Britain, as the "common earthworm," or "dew worm," in North America, as the "nightcrawler," sometimes also called the "Granddaddy earthworm." It is the only species that regularly forages at night for plant material around the entrance of its burrows. It also emerges from these burrows for mating and copulation on the soil surface. On these occasions, the flattened posterior end is swollen, serving as a holdfast in the tube, ready for a rapid retreat in case of disturbance and danger. The burrows are large and may extend to a depth of 1–3 m. *Lumbricus terrestris* is often chosen in textbooks for detailed study and dissection as a generalized representative of annelids. However, it should be remembered that the morphology and physiology of *Lumbricus terrestris* are not typical for either annelids or clitellates in general. Each chosen example represents a mosaic of plesiomorphies and apomorphies, mirroring that evolution has occurred.

When observing an earthworm, the uniform homonomous segmentation and the absence of any body appendages are immediately noticeable (Fig. 5.9a–d). Distinct eyes or other sensory organs, such as palps, are not recognizable. Instead, the animals have many individual photoreceptor cells distributed all over the body, allowing light-dark sensing. Since these photoreceptor cells lack dark-shading pigments, they remain unnoticed in a beginner's practical course. In addition, there are numerous small and inconspicuous sensory papillae, each comprising a few sensory cells and supportive cells. Additionally, isolated receptor cells are also present. While the front end of the earthworm is round and tapering, the rear end appears distinctly flattened (Fig. 5.9a, b). The dorsal side is easily identifiable by its darker color.

In the first third of the body, several segments are thickened. They form the saddle-shaped clitellum or girdle (Clitellata, girdle worms; Fig. 5.9a–c). The position varies among the different groups of Clitellata, but it either encompasses the female genital pores or is located behind them. Since the genital organs are restricted to the anterior body region, the clitellum is always located there. The clitellum comprises a few segments characterized by a thickened glandular epidermis. It plays a crucial role in forming the cocoon in which the eggs are laid. Like all clitellates, earthworms are simultaneous hermaphrodites, meaning that oocytes and sperm are produced simultaneously in the same individual. But how to achieve a secure transfer of sperm between partners without copulatory organs, which the

earthworm lacks? The complex anatomy of the clitellum ensures this, as it is crucial for the transmission of sperm and the cohesion of sexual partners. Upon mating, two individuals attach so that the anterior end with the spermathecae of each individual faces the clitellum of its partner. Then, the animals adhere to each other, mediated by the secretion of gland cells in the clitellum and additional accessory glands. Thereupon, the sperm is transmitted to the partner via the seminal grooves until it reaches the pores of the spermathecae. Upon closer inspection under the stereomicroscope, additional structures that play a role in mating become apparent on the ventral side (Fig. 5.9c, d). These include the relatively hard-to-see female sexual opening in segment 14, the more apparent male sexual opening in segment 15, and the sperm grooves that run from the male opening to the clitellum. Furthermore, additional gland cell areas for attachment can be more or less easily identified around certain groups of chaetae.

Locomotion and burrowing are achieved through alternating contraction waves of the longitudinal and circular muscles. If you hold the animals in your hand, you can feel the chaetae, which are invisible to the naked eye and are primarily used to anchor the body to the ground. Our movie clip collection, which includes movies showing living earthworms, can be found at figshare: sn.pub/xr9qhi.

Lumbricus terrestris: **Preparation Steps**
The characteristic features of Clitellata can be best observed by studying complete and intact animals, which are dissected afterward. Additionally, histological sections help in analyzing the internal anatomy in depth (Figs. 5.10, 5.11, 5.12 and 5.13). If available, sections through the region of the reproductive organs are preferable (Figs. 5.14, 5.15, 5.16, 5.17, 5.18, 5.19, 5.20, and 5.21). Cross-sections generally allow for easier orientation (Fig. 5.14a–d); well-guided longitudinal sections, on the other hand, will enable the examination of many structures of the reproductive organs on just a single specimen (Fig. 5.15). The vegetative organs can, of course, also be studied in sections through any region.

The animals are pinned in a dissecting dish with the ventral side down and aligned straight for preparation. Then in the middle, the body wall is cut in the median with a scalpel. Care should be taken when making the incision so that, at this stage, the intestine remains undamaged (e.g., Fig. 5.11c). The incision then continues anteriorly, including cutting through the relatively rigid dissepiments. The body wall is then folded continuously to the left and right sides and pinned down with several needles, but without "flattening" the animal. The opened animal should be covered with tap water for a closer inspection of the internal anatomy. Now, almost all structures of the internal organization

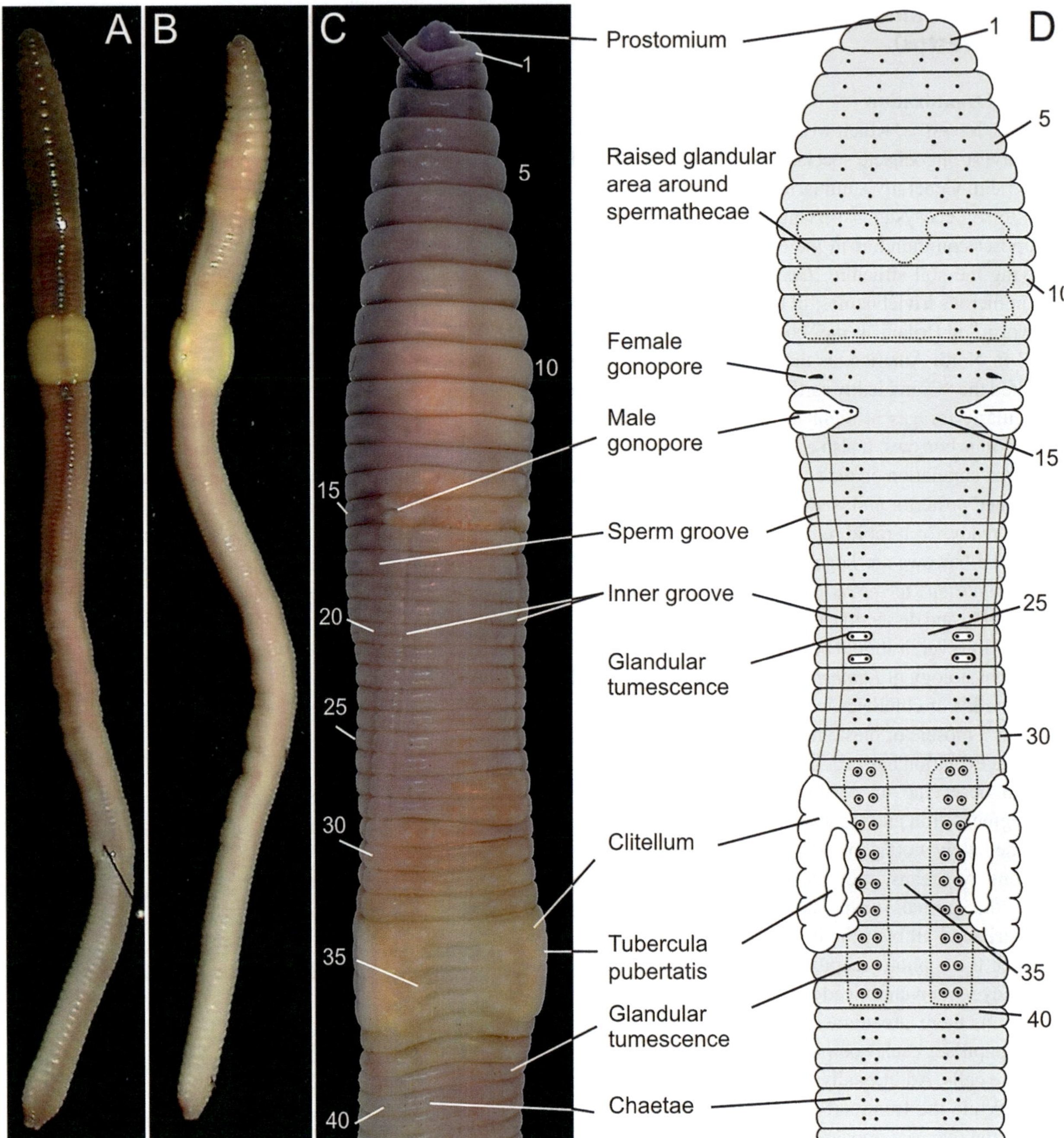

Fig. 5.9 *Lumbricus terrestris*. External morphology; length of specimen approx. 20 cm. (**a**) Dorsal view, the clitellum appears as a yellowish thickening. (**b**) Ventral view with clitellum and male genital openings. (**c, d**) Enlargements of the anterior end from the ventral. Numbers indicate the respective segment numbers. The openings of the spermathecae are located laterally in the intersegmental grooves between segments 9/10 and 10/11. The clitellum is saddle-shaped (with a ventral gap) and extends from segments 32 to 37. Glandular swellings in the area of the spermathecae, the ventral chaetae in segments 25 and 26, and the area of the clitellum, segments 31 to 39, can be more or less seen depending on sexual maturity. (**a–c**) Images of an intact animal. (**d**) Schematic drawing, from various authors

are visible (Fig. 5.10a–c). It becomes clear that the external homonomous segmentation is not reflected in the internal morphology, because the reproductive organs are restricted to a few anterior segments, and anteriorly, the intestinal canal is subdivided into several specific sections, such as the pharynx, esophagus, gizzard, and several other parts (Fig. 5.10c).

Reproductive Organs

The reproductive organs of earthworms consist of paired receptacula semines (in oligochaetes, usually called spermathecae), paired testes, seminal vesicles, spermioducts, ovaries, and oviducts (Figs. 5.10c, 5.11a, 5.12a, e, and 5.13). There are two pairs of spermathecae in segments 9 and 10, two pairs of testes in segments 10 and 11, two pairs of semi-

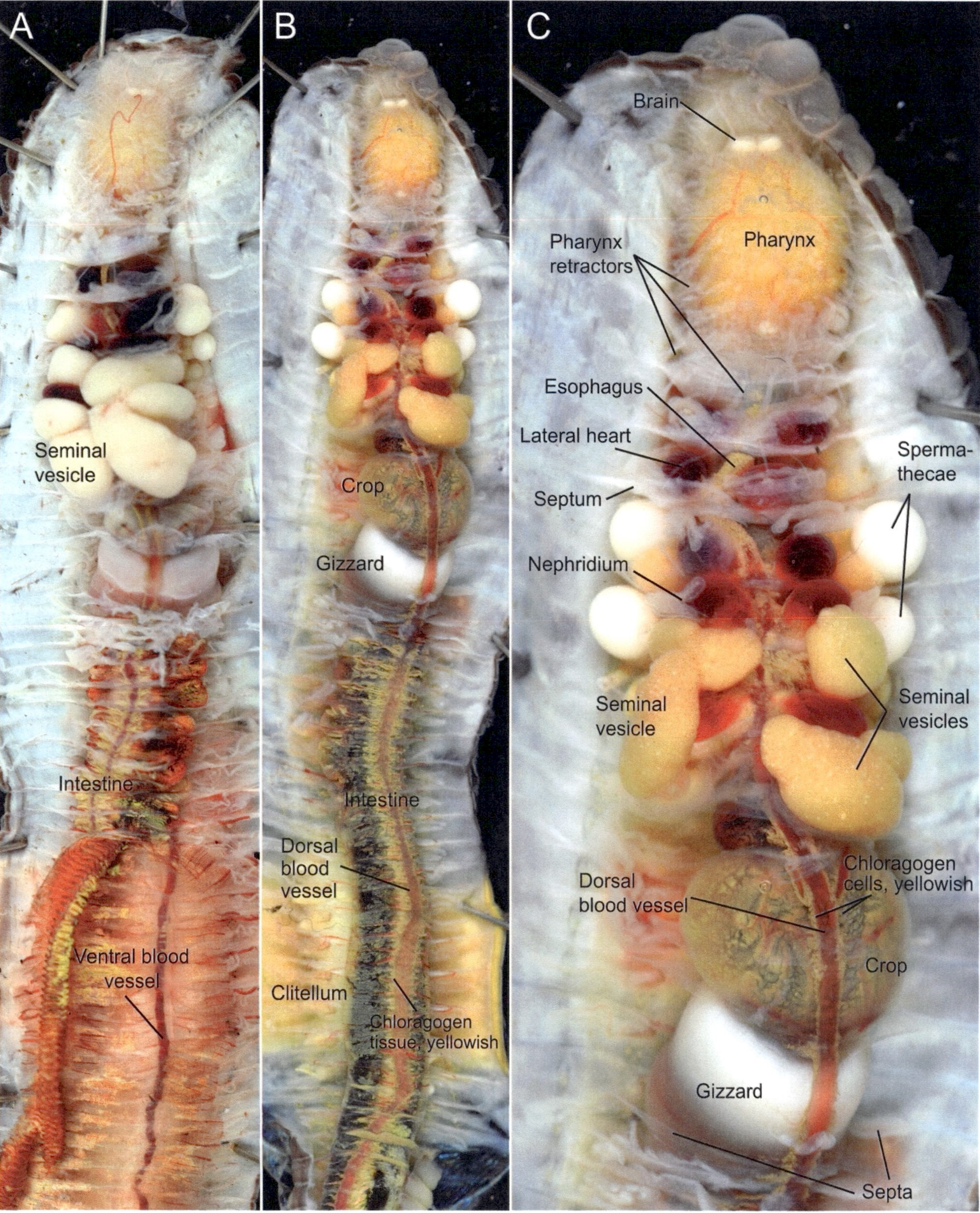

Fig. 5.10 *Lumbricus terrestris*. Adults, prepared from dorsal. (**a, b**) Overview of two complete animals to show different appearance of internal structures in size and color. In **a**, the intestine has been moved to the left body side, making the ventral vessel visible. (**c**) Close-up of the approximately 20 anteriormost segments of the specimen shown in **b** with reproductive organs and anterior differentiations of the digestive system

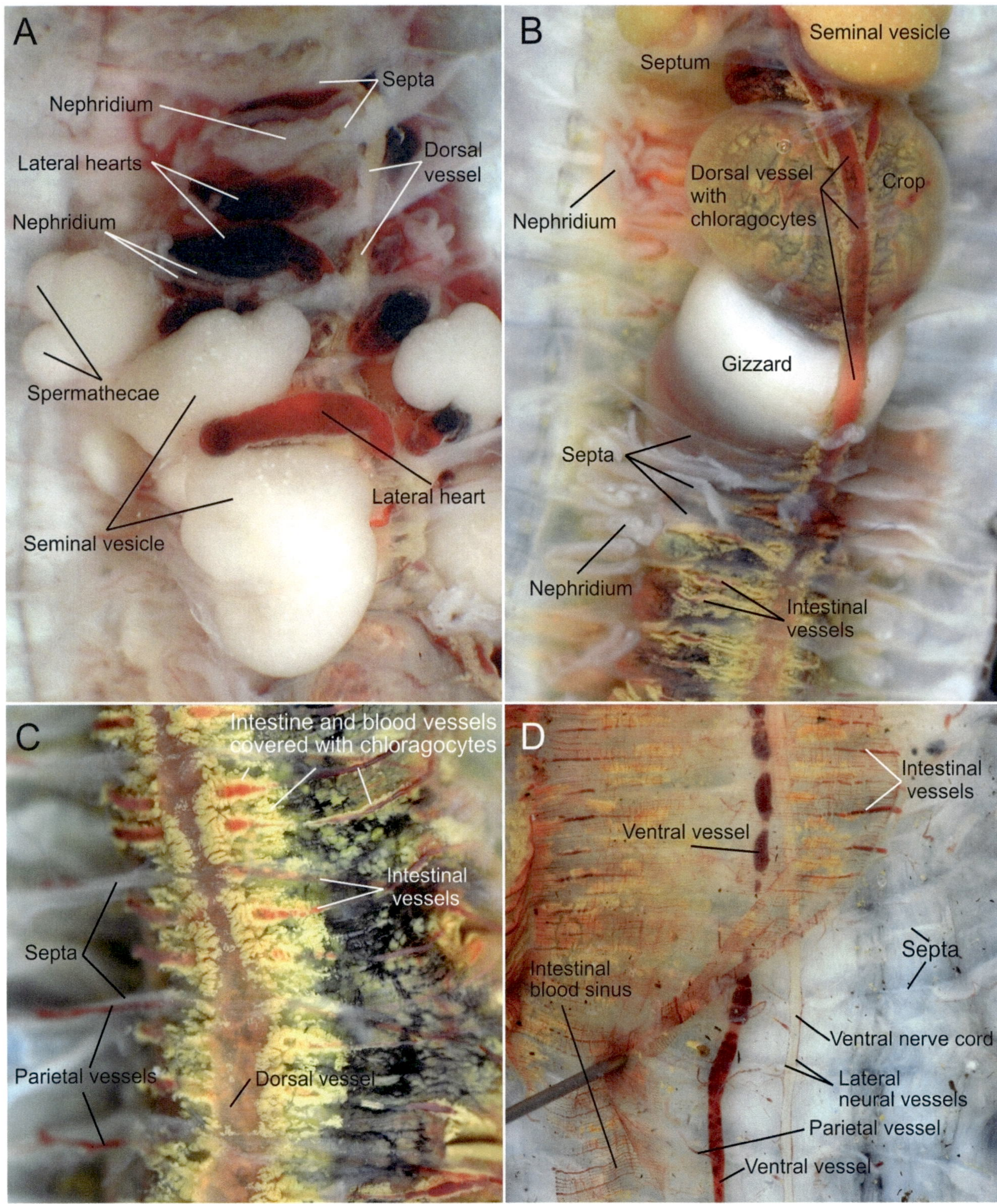

Fig. 5.11 *Lumbricus terrestris*. Adult, prepared from dorsal. (**a**) Segments 7 to 15, left side. Dorsal vessel and lateral hearts, septa, and nephridia. Spermathecae and seminal vesicles. (**b**) Segment 15 and subsequent. Crop, gizzard, and midgut. (**c**) Sterile region; intestine regularly surrounded by yellowish chloragocytes, dorsal vessel with two dorso-intestinal vessels per segment that supply the intestine, and parietal (lateral) vessels in the septa supplying the periphery. (**d**) Sterile region; intestine posteriorly opened and moved to the left, so that the ventral vessel, the intestinal vessels, the intestinal blood sinus, and the ventral nerve cord are visible

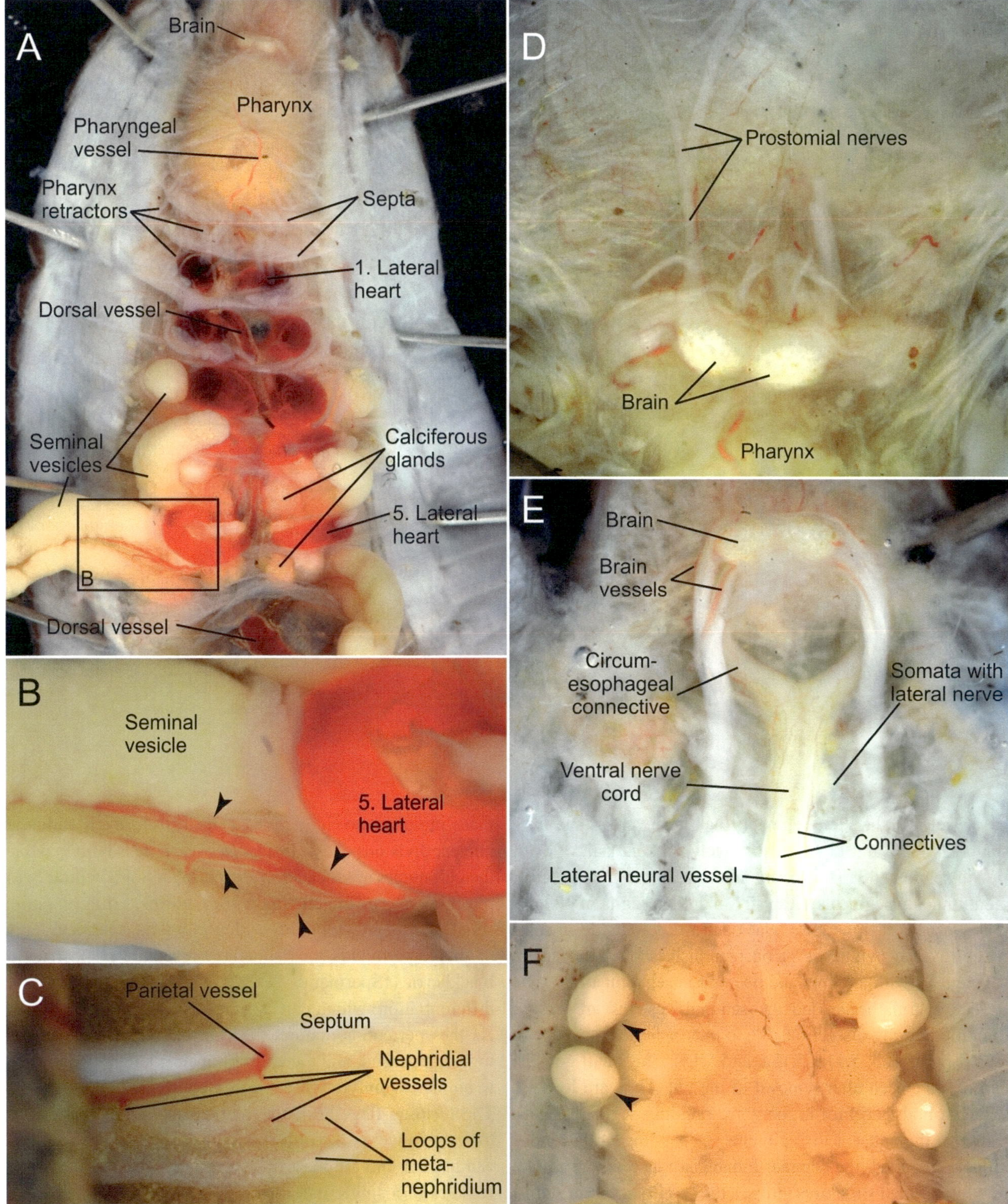

Fig. 5.12 *Lumbricus terrestris.* Adult, prepared from dorsal. Brain, pharynx, reproductive organs, and anterior blood vessels. (**a**) Segments 1–15. Dorsal vessel with five pairs of branching lateral hearts; blood vessels extending forward on the pharynx. (**b**) Enlargement of the posterior part of **a** showing blood supply of seminal vesicle. (**c**) Higher magnification of metanephridium with supplying blood vessels (arrowheads). (**d**) Brain and prostomial nerve in front of the pharynx. (**e**) The appearance of the anterior nervous system after removal of the pharynx. (**f**) The two pairs of spermathecae appear as invaginations of the ventral body wall; gut and seminal vesicles removed

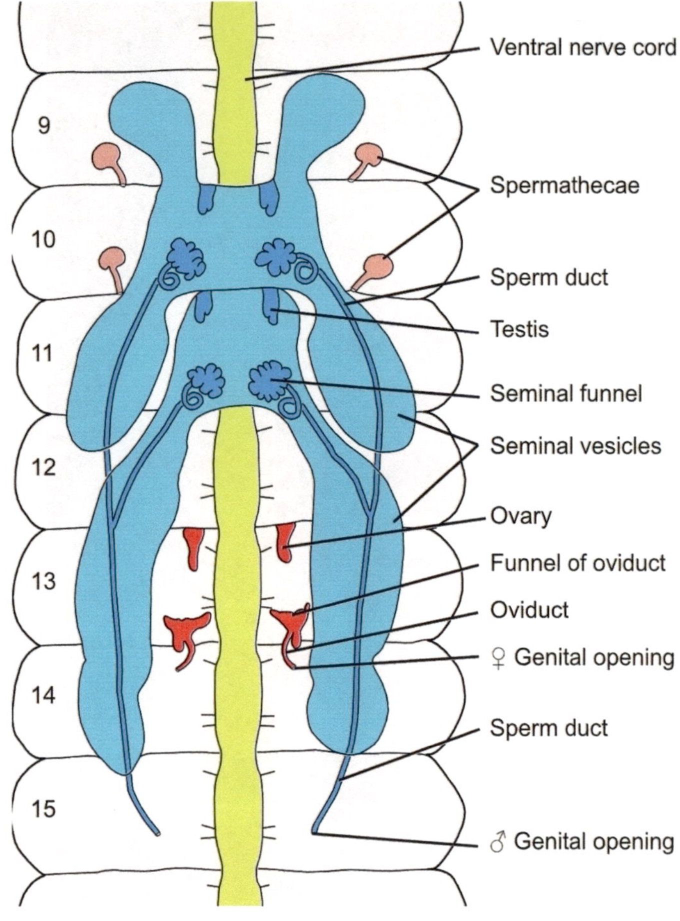

Fig. 5.13 *Lumbricus terrestris*. Schematic representation of the reproductive organs and the ventral nerve cord after removal of the intestine. Dorsal view, segments numbered, metanephridia not shown

nal vesicles in the same segments that, however, extend into segments 9–11 and 11–14, respectively, and one pair of ovaries in segment 13. The spermioducts start with a large ciliated funnel in segments 10 and 11, merge in segment 12 into a single pair of gonoducts, and open to the outside in segment 15. The oviduct is inconspicuously short, starting in segment 13 with a funnel and opening in segment 14.

The spermathecae and the seminal vesicles are easily identified during dissection, due to their spherical shape and size (Figs. 5.10b, 5.11a, 5.12a, and 5.13). The small gonads can only be seen after removal of the gut. They are located near the body's center in the corresponding segments ventrally on the anterior dissepiments. Depending on the individual examined, the seminal funnels are also recognizable as white, folded structures through the semitransparent seminal vesicle.

The testes give rise to primordial germ cells, referred to as protogonia, due to their further development (Figs. 5.15 and 5.20a). The protogonia are released into the seminal vesicles, where the entire spermatogenesis occurs. This begins with additional mitoses and is completed with the meiotic divisions to produce the haploid gametes. A distinctive feature is

that, initially, no individual cells are formed, although nuclear divisions occur. Instead, a cytophore is built in which the cell nuclei are located in the periphery of a central cytoplasmic mass, which undergoes further differentiation (Fig. 5.20c, d). Cytophores with protogonia, spermatogonia, or early spermatids thus somehow resemble raspberries or blackberries. Spermatids begin to stretch at a later stage of differentiation, which can especially be seen in their nuclei. As the spermatids mature, the chromatin becomes increasingly condensed, and the color of nuclei changes from violet to an increasingly intense red in sections stained with Azan. The forming flagella of the sperm become visible as gray threads. Since earthworms continuously form gametes, one can always find all phases of spermiogenesis in the seminal vesicles of a single individual having reached sexual maturity. The mature sperm detach from the cytophores and are collected on the ciliated funnel epithelium. There, they are attached as dense aggregates between the cilia.

Therefore, the red staining of the sperm nuclei makes the seminal funnels prominent, strongly folded structures that are easily recognizable (Figs. 5.14, 5.15 and 5.20a, e). The stages of spermiogenesis, and often parasites living in the

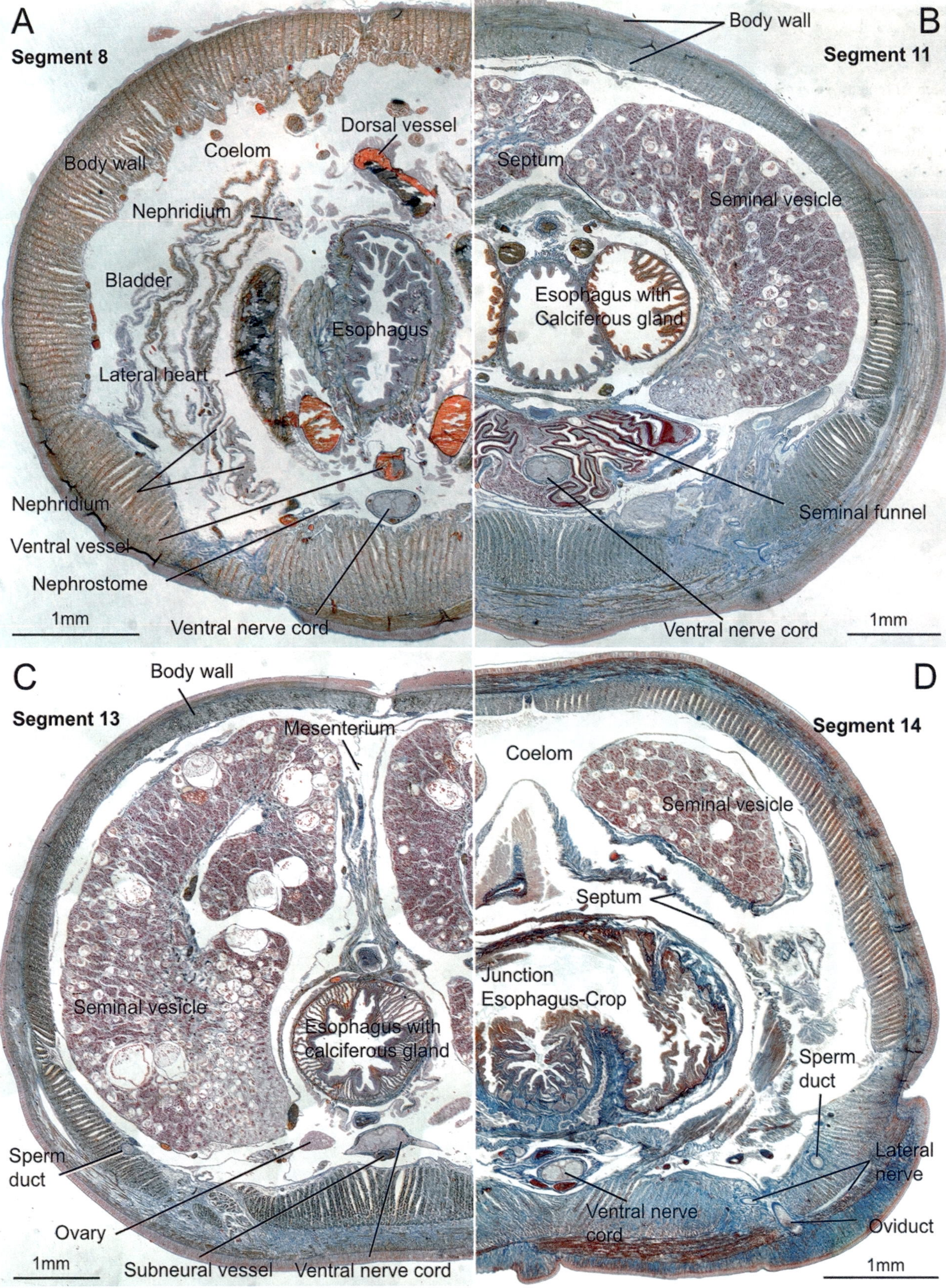

Fig. 5.14 *Lumbricus terrestris*. Cross-sections of different body regions. Azan staining. (**a**) Segment 8: sterile region in front of the spermathecae and seminal vesicles, the ciliated funnel of the nephridium of segment 9, and the nephridium of segment 8. (**b**) Segment 11 with seminal vesicle, sperm funnel; esophagus with calciferous gland. (**c**) Segment 13; ovaries, the sperm ducts have merged into a single duct, located within the longitudinal muscles. (**d**) Segment 14. Posterior region of the seminal vesicle; ventral: oviduct shortly before female pore and sperm duct. Specimens from the Biological Collection of the University of Osnabrück

Fig. 5.15 *Lumbricus terrestris*. Adult. Azan staining. Parasagittal longitudinal section through the genital region (segments 8–15). Anterior is up and ventral is left. Almost all structures of the genital apparatus, except spermathecae and ovary, are visible. The testes consist of small inconspicuous structures located on the dissepiments of segments 10 and 11, situated in front of the respective sperm funnels. Specimen from the Biological Collection of the University of Osnabrück

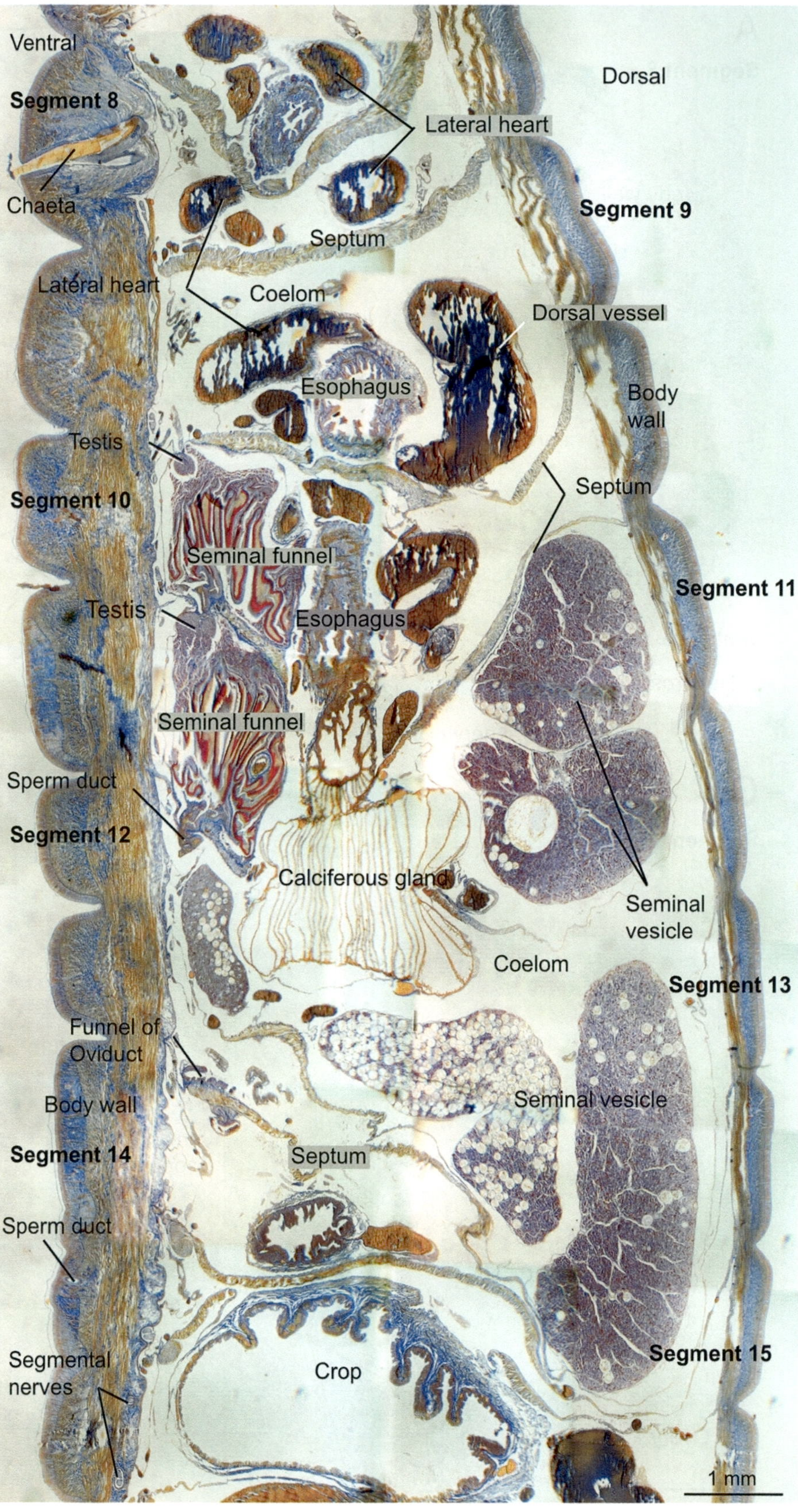

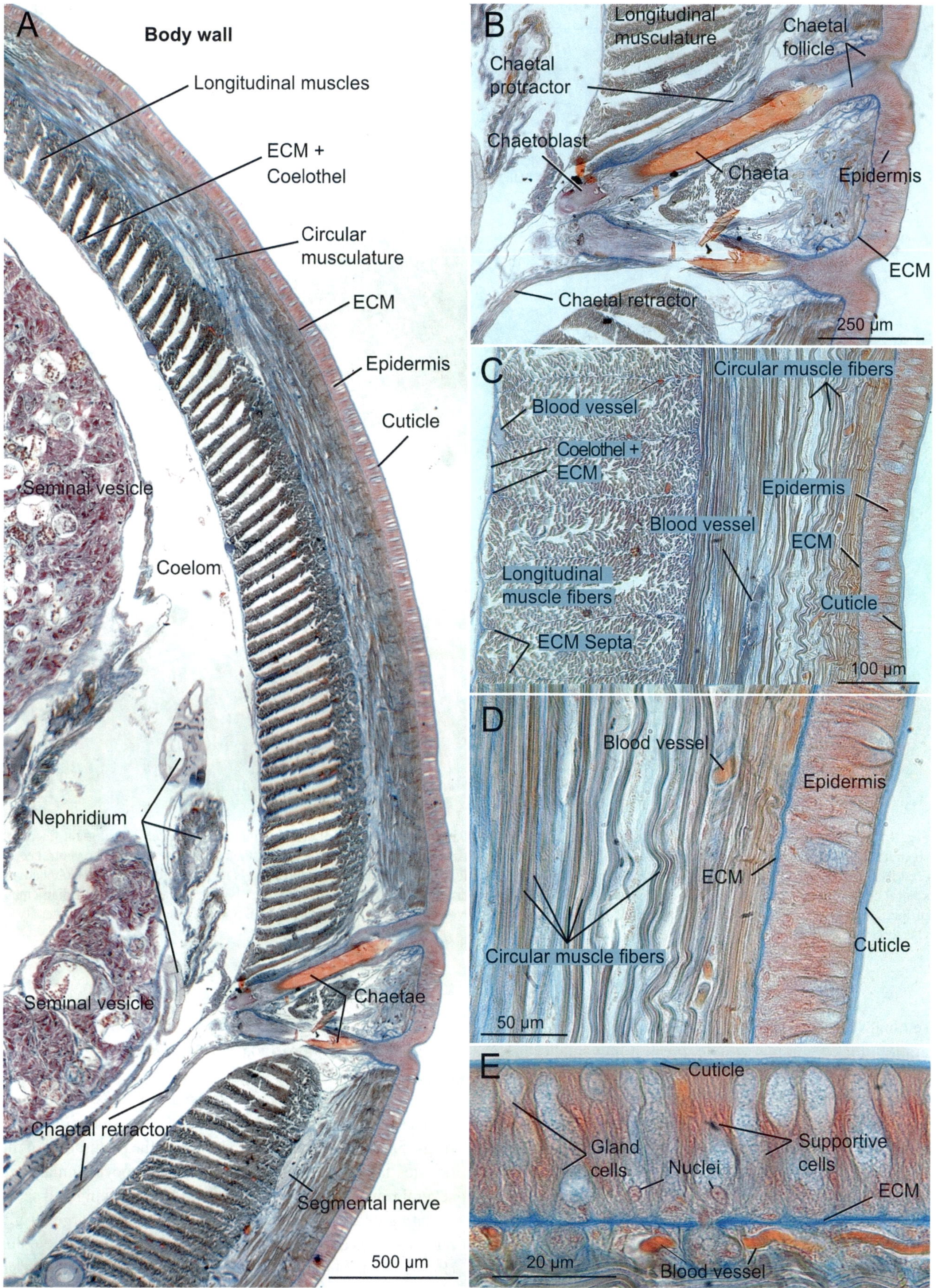

Fig. 5.16 *Lumbricus terrestris*. Body wall. Azan staining. (**a**) Lateral body wall with cuticle, epidermis, circular, and longitudinal muscles. (**b**) Chaetae—each chaeta arises in an epidermal invagination (follicle), whose base is formed by the chaetoblast (chaeta forming cell). Chaetal muscle fibers are inserted at the chaetal follicle. (**c**) Body wall musculature, ECM, epidermis, and cuticle; blood vessels within the ECM between the muscle compartments. (**d**) Epidermis and circular muscles. (**e**) Epidermis with cuticle, supporting and gland cells, and blood vessels extend up to the base of the epidermis for gas exchange. Specimens from the Biological Collection of the University of Osnabrück

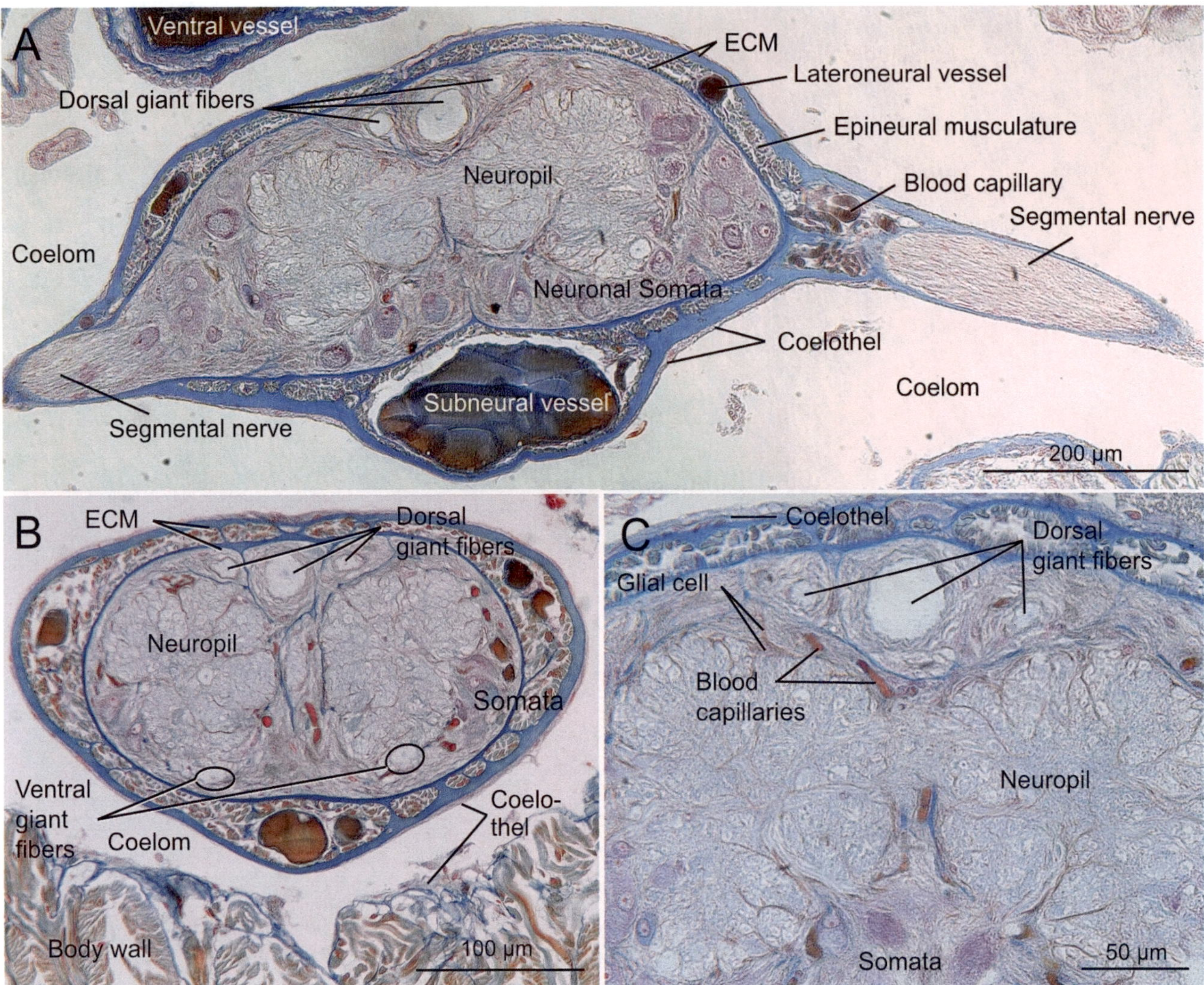

Fig. 5.17 *Lumbricus terrestris*. Ventral nerve cord. Azan staining. (**a**) Ganglion in the area of a branching lateral nerve. Dorsal giant fibers are separated by glial cells from the rest of the nervous system. Somata are located only ventrally and laterally, the neuropil centrally. The ventral nerve cord has a subepithelial position. It is surrounded by a double layer of ECM, which includes blood vessels and muscle fibers. (**b**) Connectives, ventral nerve cord with five giant fibers, ECM, blood vessels, and perineural muscles. Connectives lie directly next to each other. (**c**) Enlargement of the dorsal area with giant fibers, ECM (blue), glial cells, and blood vessels. Specimens from the Biological Collection of the University of Osnabrück

seminal vesicles, can easily be observed in a self-made preparation: Pick a seminal vesicle, break it into smaller pieces, and put it on a microscopic slide. Add a drop of 0.4% NaCl solution, place a coverslip on top, and analyze your preparation under a compound microscope, ideally with phase contrast illumination. Almost all individuals of *Lumbricus terrestris* are infested with a parasite that attacks the cytophores in the seminal vesicles and feeds on them (*Monocystis lumbrici*, Apicomplexa, Gregarinea). The cysts with gamonts, zygotes, sporocysts, or sporozoites especially stand out as round, bright structures in histological sections (Figs. 5.14b–d, 5.20c, 5.21a). Most of these contain spindle-shaped spores (also called sporocysts), each of which will release eight sporozoids, infecting new cytophores (Fig. 5.21b).

The ovaries are much smaller and less conspicuous than the testes, thus quite difficult to identify in a student's course (Figs. 5.14 and 5.22a, b). In the ovaries, the maturation of the oocytes takes place. In the process, the oocytes enlarge to several times the size of the oogonia by producing and accumulating cytoplasm and yolk (Fig. 5.22b). The large, bright nuclei always contain a prominent, orange-colored nucleolus (Fig. 5.22b). Mature oocytes detach and are retained in an expanded front part of the oviduct until egg-laying (Fig. 5.13).

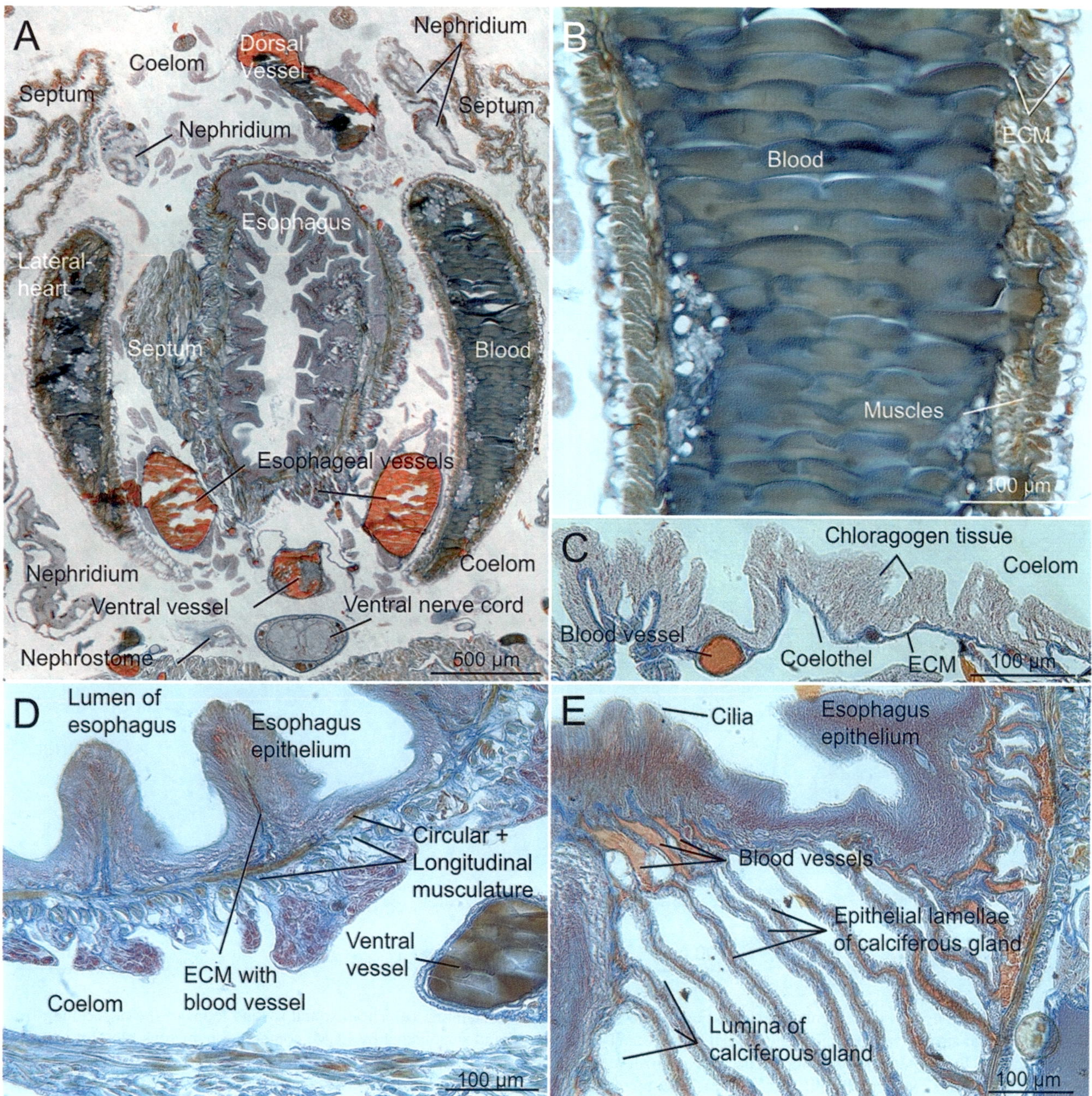

Fig. 5.18 *Lumbricus terrestris*. Digestive tract, vascular system. Azan staining. (**a**) Esophagus with the dorsal blood vessel, ventral and esophageal blood vessels, and lateral hearts. (**b**) Detail of a lateral heart with muscular wall; blood fluid broken flake-like during sectioning (artifact). (**c**) Dissepiment with blood vessels, muscle fibers, and the coelothel, differentiated as chloragogen tissue. (**d**) Esophagus epithelium, ECM, and musculature. (**e**) Calciferous glands at the esophagus. Calciferous glands are composed of double-layered lamellae. Extensive blood spaces in the lamellae. Specimens from the Biological Collection of the University of Osnabrück

During oviposition, the animals retract so far backward from the cocoon formed by the clitellum that the cocoon reaches the female sexual openings. Then the oocytes are deposited in the cocoon. The number of eggs laid at one time is relatively small, ranging from one (*Lumbricus*) to about ten in other oligochaetes. After that, the animal continues to withdraw backward from the cocoon so that it reaches the sper-mathecae, from which sperm are released. After that, the cocoon is completely stripped off and deposited in the soil. After fertilization and zygote formation, embryonic development occurs in these cocoons. Larval stages do not exist in *Lumbricus*, as is the case in all Clitellata. The cocoon provides a liquid reservoir that allows for terrestrial reproduction and a lifestyle independent of water. The cocoon thus

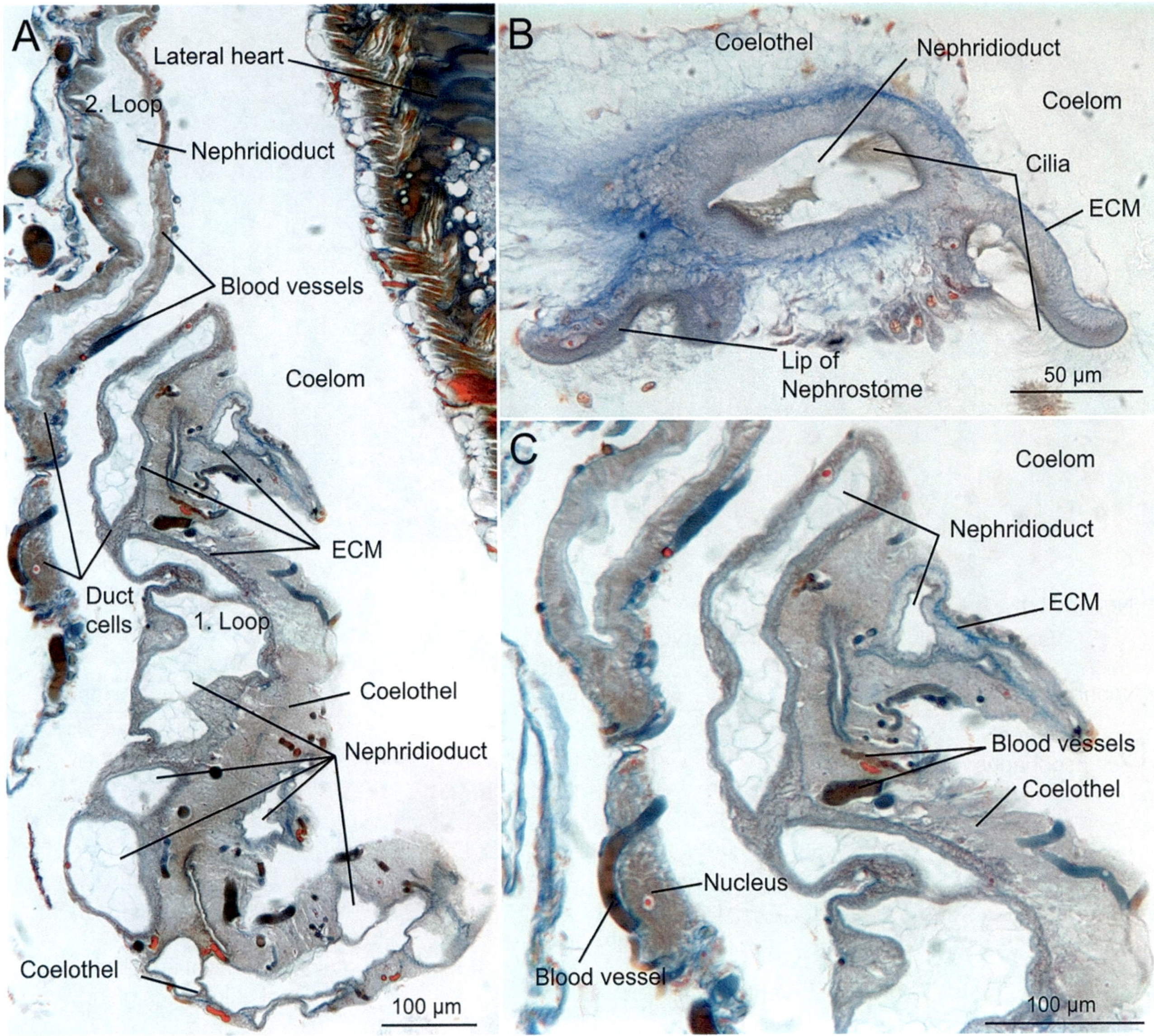

Fig. 5.19 *Lumbricus terrestris*. Nephridia (from Figs. 5.14a and 5.18a). Azan staining. (**a**) The first and second loops of the nephridium from segment 7/8 and several sections of the nephridial canal. Canal cells ciliated and surrounded by ECM, containing numerous blood vessels, the coelothel partly forms quite thick envelope cells. (**b**) Nephrostome of the nephridium from segment 8/9, ciliation of the funnel cells, in the proximal section of the canal's two densely ciliated areas. (**c**) Close-up of the first loop from **a**. Canal cells with cell nuclei, ECM, blood vessels, and envelope cells. In broad areas, the thickness of the coelothel is below the resolution limit of the light microscope. A secure assignment of the individual canal sections based on single histological sections is usually impossible. Specimens from the Biological Collection of the University of Osnabrück

fulfills the same function as our internal fertilization in combination with the so-called amniotic egg. The spermathecae are sac-like, single-layer invaginations of the epidermis, often densely filled with thread-like mature sperm (Fig. 5.22d).

Digestive System

The whitish-yellow colored pharynx forms the first part of the intestinal canal. It is a dorsally located gland-rich thickening of the foregut attached to the strong anterior dissepi-

ments and the body wall by numerous muscles (Fig. 5.10a). The pharynx leads into the esophagus. Its posterior part is provided with the so-called calciferous glands located approximately at the level of the posterior seminal vesicle (Fig. 5.10b). The calciferous glands are chambered, richly vascularized diverticula of the esophagus. They secrete calcite when there is an excess of calcium in the diet (Figs. 5.14b, c, 5.15, and 5.18e). Recent studies show that this regulates the pH and CO_2 content in the blood and the coelomic fluid. Behind the reproductive organs in segment 16, the esopha-

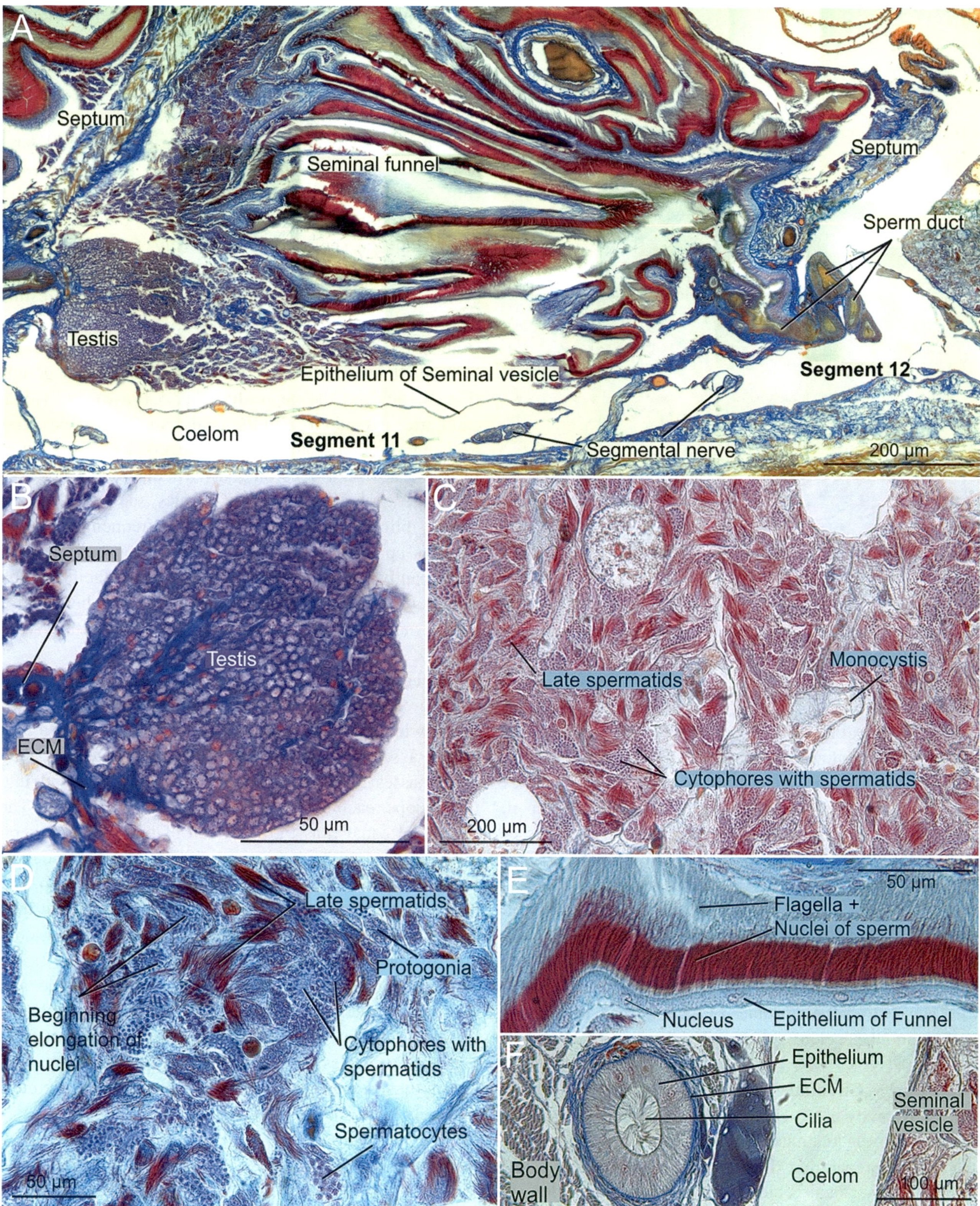

Fig. 5.20 *Lumbricus terrestris*. Male reproductive organs. Azan staining. (**a**) Ventral side, segment 11, testis attached to the dissepiment 10/11, folded seminal funnel; proximal section of the spermioduct runs through dissepiment 11/12, longitudinal section. (**b**) Testes with isodiametric spermatogonia. (**c**, **d**) Seminal vesicles with various stages of spermiogenesis and developmental stages of the parasite *Monocystis lumbrici* (Apicomplexa). At higher magnification, all phases of spermiogenesis are visible. (**e**) Seminal funnel. Single-layered, flat, and ciliated epithelium, on which the mature sperm are oriented in parallel arrangement with the nuclei toward the epithelium (red "band"). (**f**) Cross-section, spermioduct, single-layered ciliated epithelium. Specimens from the Biological Collection of the University Osnabrück

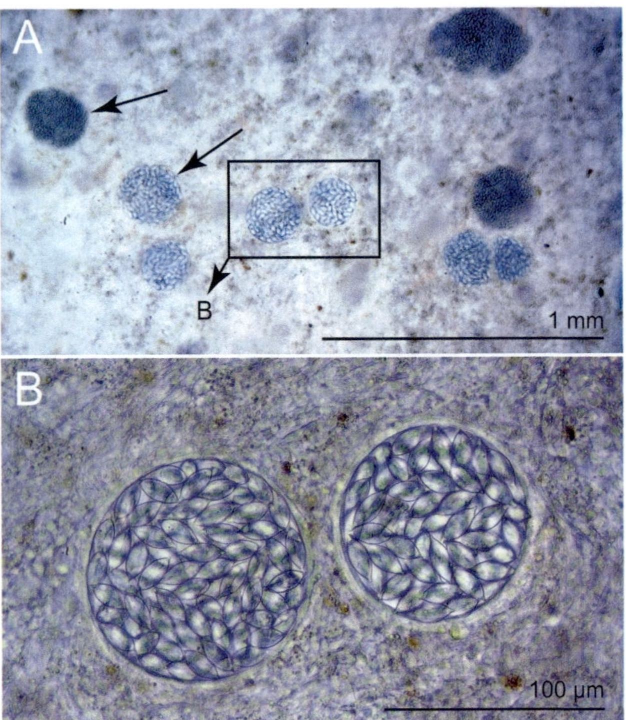

Fig. 5.21 *Lumbricus terrestris*. Squeeze preparation of a seminal vesicle. (**a**) Low-power micrograph showing cysts of *Monocystis lumbrici* with spores, sporozoites, and other stages (arrows). (**b**) Higher magnification of the boxed area in **a** with cysts containing numerous mature spores (sporocysts), each giving rise to eight sporozoites

gus opens into the crop, where ingested material can be stored. This is followed by the muscular stomach (gizzard), in which material is ground up or mixed, and the midgut where final digestion takes place (Figs. 5.10, 5.11b–d, 5.14d, and 5.15). The yellowish chloragogen tissue is especially seen in more posterior regions (Fig. 5.11b, c). This metabolically active tissue consists of club-shaped peritoneal cells (Fig. 5.18c). Its function relies on storing and releasing nutrients when needed. Thus, it reminds us of the function of the vertebrate liver. Chloragogen tissue is also associated with many blood vessels (Figs. 5.10c and 5.11b, c). They rest on the coelomic ECM, and their cell bodies project into the coelom.

Body Cavity, Circulatory System, and Respiration
The body cavity is a coelom. The dissepiments and mesenteries are strong and muscular and, therefore, easy to identify during preparation (Figs. 5.10, 5.11a–c, 5.12a, 5.15). The vascular system is closed, as is typical for coelomate metazoans. It consists of clefts in the extracellular matrix between adjacent epithelia (coelothelia or coelothelia and gastrodermis or epidermis). The term coelothelium is used for the cell layer that covers a coelom. An inner epithelium of the blood vessels, called the endothelium, does not exist in this case; that is restricted to vertebrates. The dorsal and ventral longitudinal vessels form the main blood vessels in

the earthworm. They are connected by segmental circular vessels (Figs. 5.10a–c, 5.11a, and 5.12a, b). The circular vessels in segments 7–11 are massive and muscularized and, therefore, are also called lateral hearts (Figs. 5.11a, 5.12a, 5.14a, 5.15, and 5.18a, b). The vessel musculature is formed by the coelothelia, and in the case of the lateral hearts, it is multilayered and supplied by its blood vessels (Fig. 5.18a, b). The subneural blood vessel, which runs in the longitudinal direction through the body, is located underneath the ventral nerve cord (Fig. 5.14a, c). It supplies the ventral body wall (Fig. 5.11d) and the nervous system (Fig. 5.17a, b). The vessels supplying the intestine mainly originate from the dorsal and ventral vessels (Fig. 5.12e). Other branches supply the pharynx (Fig. 5.12a, b) and all other organs, such as the calciferous glands (Fig. 5.18e) and the nephridia (Fig. 5.19a–c). Individual capillaries extend close to the epidermis (Fig. 5.16e). Gas exchange in almost all clitellates occurs across the entire body surface, and accordingly, particular organs for gas exchange are lacking. In larger species, such as the earthworm, blind-ending blood vessels frequently extend between epidermal cells, and thus diffusion distances are significantly reduced. Gas exchange is also facilitated by secretions of the numerous epidermal glands, which keep a film of moisture around the body.

Osmoregulation and Excretion
The paired metanephridia occur in all body segments. They begin with an open ventral ciliated funnel, located just before a segment's posterior dissepiment. The following canal passes through the dissepiment and is wound into three loops, each extending far dorsally (Fig. 5.14a). The nephridial canal runs through each of these loops several times. The nephridia appear as whitish, irregularly delimited structures in our dissection (Figs. 5.10b, 5.11a, b, and 5.12a, c). In histological sections, it can be seen that they consist of the ciliated canal cells resting on an ECM, covered by the peritoneum and blood vessels (Fig. 5.19). An expansion in the posterior loop acts as a urinary bladder and is visible on sections as a large cavity. These large nephridia are likely correlated with the limnic or terrestrial lifestyle of many oligochaetes, living in a hypotonic medium. This causes passive water influx, and the animals must excrete a relatively large amount of water.

Body Wall
The body wall of *Lumbricus* is relatively robust and consists of a cuticle, an epidermis, and circular and longitudinal muscles (Figs. 5.14a–d, 5.15, and 5.16a–e). The circular and longitudinal muscles form a closed cylinder. This is only interrupted by the chaetal follicles extending far into the body cavity (Fig. 5.16a, b). The collageneous cuticle appears as a uniform, blue-colored layer above the epidermis in Azan-stained histological sections. The epidermis is single-

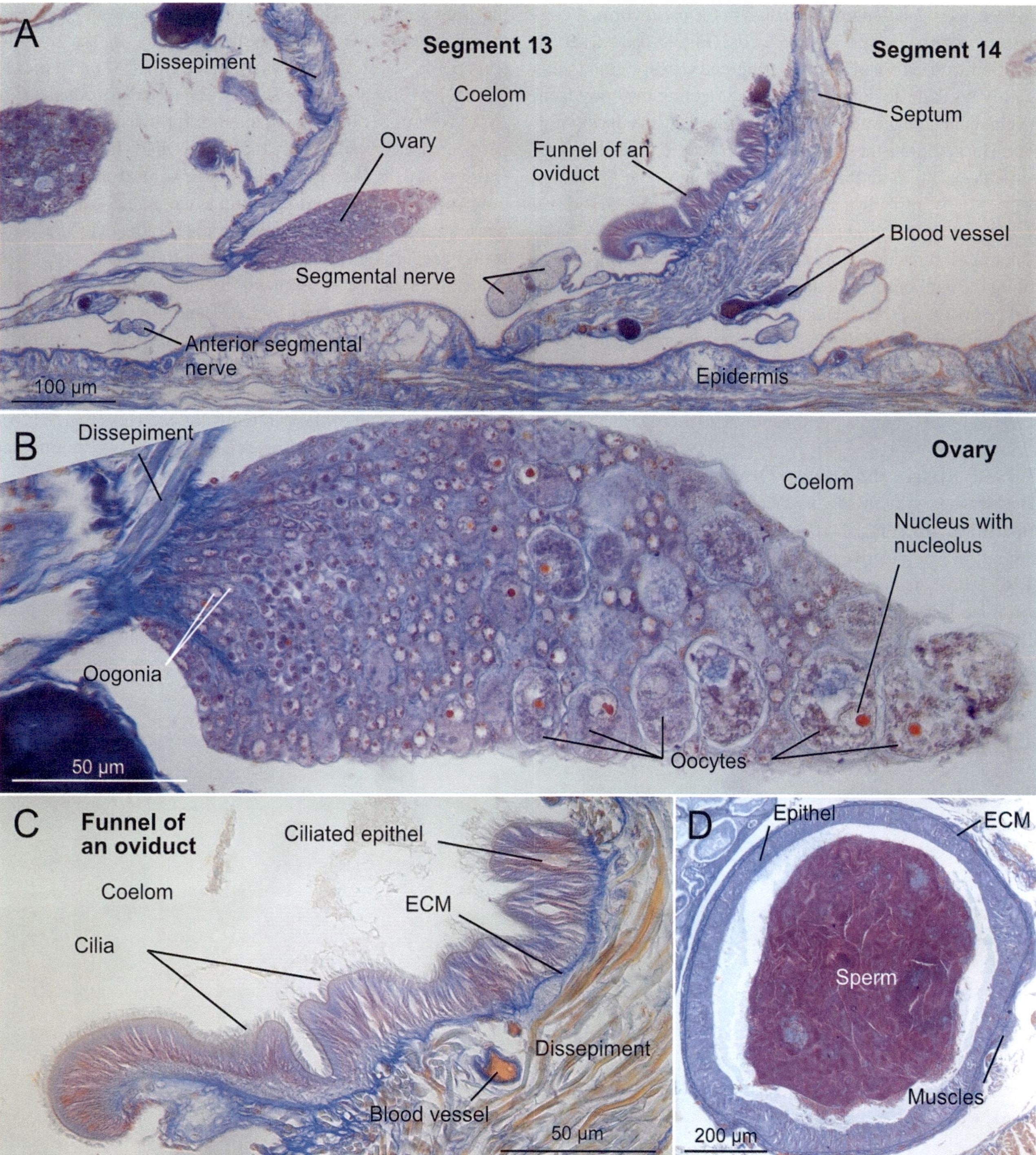

Fig. 5.22 *Lumbricus terrestris*. Female reproductive organs and spermathecae. Azan staining. (**a**) Longitudinal section, ovary, situated at the dissepiment, ciliated funnel of the oviduct opposite. (**b**) Ovary, with ECM between the basal cells. Differentiation to oocytes progressing from anterior to posterior: At the dissepiment, oogonia, followed by previtellogenic oocytes and vitellogenic oocytes. The latter are the largest cells released into the coelom from the terminal end of the ovary. (**c**) Large ciliated funnel of the oviduct with single-layered, high epithelium. (**d**) Spermatheca, its cavity with sperm (red), is enclosed by single-layered epithelium. Specimens from the Biological Collection of the University of Osnabrück

layered, as is usual in invertebrates. In the earthworm, it consists of relatively tall, slender cells (Fig. 5.16a, d, e). In addition to the so-called supporting cells, many gland cells mainly produce acidic mucus, thus protecting the body from mechanical damage and entry of microorganisms during digging. In addition, the mucus is hygroscopic, and thus prevents desiccation and aids in gas exchange. The epidermal cells usually rest on a recognizable ECM (Fig. 5.16d, e). Occasionally, the chaetal follicles are cut, and the huge basal chaetoblast (chaeta-forming cell) is visible. The cells constituting the wall of the follicle are relatively flat where the pro- and retractors of the chaetae insert (Fig. 5.16b). The circular and longitudinal muscles lie below the epidermis. Each muscle consists of numerous fibers, with the longitudinal muscles being much more robust. The longitudinal muscle fibers are covered by ECM and a peritoneum (Fig. 5.16c).

Nervous System and Sensory Organs

The nervous system of *Lumbricus terrestris* consists of a bipartite brain (Figs. 5.10c, 5.12a, d, e), paired circumesophageal connectives, and the ventral nerve cord (Fig. 5.11e), from which three pairs of segmental nerves originate in each segment (one anterior and two posterior; Figs. 5.20a and 5.22a). Unlike in *Alitta (Nereis)* and many polychaetes, the brain of *Lumbricus terrestris* is not located in the prostomium; it is shifted backward instead and lies in the third segment. The brain, ventral nerve cord, and segmental nerves are located subepithelially in the coelom (Fig. 5.14a–d), so they are surrounded by ECM and a mesodermal envelope (Fig. 5.17a–c). This sheath consists of ECM with intervening musculature, followed by ECM and peritoneum. In addition, there are also blood vessels (Figs. 5.11d, 5.14c, d, 5.17a–c). The ventral nerve cord is only vaguely divided into ganglia and connectives. The two main strands lie directly next to each other and are only weakly separated. Dorsally in the ventral nerve cord, three giant fibers can be recognized, located apart from the rest of the ventral nerve cord by ECM and glial cells, and thus indicating the existence of an additional median longitudinal nerve (Fig. 5.17a–c). In addition to the three dorsal giant fibers, there is a slightly thinner, ventrally located giant fiber in each of the two main connectives (Fig. 5.17b). In the posterior area of the segments, the two segmental lateral nerves also originate within the ganglia (Fig. 5.17a). Most of the somata form a lateral and ventral cortex around the centrally located neuropil (a dense network of neurites or nerve cell processes). In the ganglia, certain nerve fibers are recognizable that connect the two longitudinal connectives as commissures with each other. In addition to glial cells, blood vessels also penetrate the ventral nerve cord (Fig. 5.17b, c). Complex sensory organs are absent in clitellates.

Mollusca (Mollusks)

► **Summary** **The mollusks (Mollusca) comprise over 120,000 recent species. This very diverse group of animals includes bivalves (Bivalvia), gastropods (Gastropoda), and cephalopods (Cephalopoda), but also largely unknown representatives such as tusk shells (Scaphopoda), chitons (Polyplacophora), and aplacophorans (Aplacophora, worm-like mollusks). They are all characterized by a soft body. Most mollusks produce a shell of calcium carbonate to protect their soft bodies, which, in some taxa, has been reduced or internalized as a supporting element. In some cultures, the shells of mollusks served as raw materials for manufacturing tools and were considered a substitute for money or jewelry. Mollusks are also an essential part of the nutrition in many cultures.**

Possibly the largest living invertebrate species belong to the mollusks. The giant squid, *Architeuthis dux*, has a body length of 6 m and a total length of about 18 m, including arms. But there are also microscopic mollusks known that are less than 1 millimeter in size. Also, the lifespan of mollusks varies enormously. While the river mussels discussed in this chapter often live to be 10 years old, the Roman or Garden snails from breeding farms can reach an age of well over 15 years. Some clams are exceptionally long-lived species. Scientists recently discovered that the ocean quahog (*Arctica islandica*) can reach an age of up to 400 years. Groundbreaking insights into learning and memory ability have been gained from the up to 70 cm long sea slug *Aplysia californica* (California sea hare).

Although most taxa are marine, certain mollusks have colonized freshwater and terrestrial biotopes successfully. Among the latter, only gastropods, which represent about 80% of all recent mollusk species, have conquered land. In the food webs, mollusks are essential prey for numerous other animals, such as birds, starfish, sea urchins, fish, and many marine mammals. The tiny sea angels of the genus *Gymnosomata*, also known as sea butterflies, are an essential component of the Whalaat. Whalaat refers to the masses of tiny marine organisms that serve as food for baleen whales. Mollusks are, of course, also consumers, and in addition to herbivorous and detritivorous forms, there are also numerous carnivores and scavengers. The marine, brightly colored *Conus* species (Gastropoda) even possess a venomous tooth in their radula, a tooth-bearing cuticular plate, with which they kill their prey. However, many mollusks are microphagous browsers or filter feeders, which may represent the primary form of nutrition in mollusks.

The Body Structure of Mollusca

The body of the Mollusca is bilaterally symmetrical and divided into four morphologically different regions: head, foot, visceral mass, and mantle. The head is distinctly separated from the rest of the body in gastropods and cephalopods. It always carries the mouth, the buccal apparatus with the radula, the brain, and the most important sensory organs. The term "buccal apparatus" collectively describes the structures necessary for food uptake, extending from the mouth opening to the esophagus. The muscular foot, located on the ventral side of the body, is multifunctional. It is used for locomotion, as a holding organ, or to dig into the substrate. The head and foot form a single functional unit, which is therefore referred to as cephalopodium. The visceral mass contains all internal organs. In snails, it extends into the spirally coiled shell. The mantle (pallium) forms a multisectional fold found among the head, foot, and visceral mass, and thus forms the mantle cavity. The delicate organs for gas exchange, the branchiae, are located under and protected by this fold. Dorsally, the mantle encloses the visceral mass. The mantle secretes the protective dorsal body covering—primarily a chitinous cuticle, into which calcareous spines are usually incorporated. In contrast to arthropods, the cuticle is not molted.

However, in most mollusks, such as gastropods and bivalves, the mantle epithelium secretes a shell made of calcareous material. All mollusks primarily forming such a shell are called Conchifera, indicating their common ancestry. This includes gastropods, cephalopods, bivalves, and scaphopods, but not the chitons (Polyplacophora) or the worm-like mollusks (Aplacophora). The shell is exceptionally diverse in Conchifera. In gastropods, for example, we find asymmetrically coiled shells of varying size, shape, and color, and in bivalves, multifaceted bipartite shells. As with the abovementioned sea hare, many marine snails and cephalopods no longer have an external shell. Our native red slug (*Arion rufus*) has lost its shell due to adaptation to a largely underground lifestyle. In cephalopods, the shell is transformed into a chambered hydrostatic organ. For example, cuttlefish (*Sepia*) and squids, as well as octopuses (Octobrachia), either have an internalized shell (cuttlebone of the cuttlefish, gladius of the squids = hydrostatic organs) or they have entirely lost their shell during the course of evolution. This is the case with benthic octopuses. The only recent representatives of cephalopods with an external shell are the nautiluses, which belong to the Nautilidae.

The mantle (pallium) forms an epithelial duplication above the visceral mass, called the mantle cavity. The mantle cavity is water-filled in all aquatic mollusks and protects the sensitive gills. The ciliation of the gills drives the respiratory water flow. As originally paired double comb-like gills, they consist of a single shaft and numerous gill leaflets arranged in two opposing rows. Therefore, they are called ctenidia, which means combs. In many gastropods, only one ctenidium is still formed; in pulmonate gastropods, it has been replaced by a lung. The water, after having passed through the gills, flows to the anus and leaves the mantle cavity, thereby eliminating the feces. As the excretory pores and the gonads open into the mantle cavity as well, the respiratory water carries excrements, metabolic end products, and gametes to the outside. The mantle cavity is air-filled and forms a lung in terrestrial gastropods, such as the Roman or Garden snails. Gas exchange occurs through the tissues of the mantle roof, which has an excellent blood supply. The visceral mass and the mantle envelope form a single unit, called the visceropallium, which protects all internal organs. Most shell-bearing gastropods can completely retract their head and foot into the mantle cavity and thus effectively protect themselves from predators.

Very different forms of food acquisition have become prevalent among mollusks. Most gastropods feed on food particles, which they scrape off of the substrate with their radula, a chitinous tooth plate. The radula has proven to be a versatile organ in evolution. It is primarily found in all mollusks and shows specific adaptations in its structure depending on the type of diet. For example, there are predatory gastropod species with radula perfectly adapted to hold prey. Some gastropods are scavengers who can crush their food with their radula. The exclusively marine cephalopods are active hunters, grasping prey with their tentacles. The captured prey is then bitten with the help of the radula and the additionally present parrot beak-like jaws. In contrast, using their gills, bivalves filter their microscopic food particles from the surrounding water. This filter-feeding coincides with a more or less sessile lifestyle, and consequently, a dedicated head region and the radula have been lost in the course of evolution.

Sensory organs are found in large numbers and diverse forms in mollusks, for example, on the tentacle-like body appendages. Mechano- and chemosensory receptor cells can occur individually or are grouped into sensory fields or organs. Statocysts are only present in conchiferans. They are located in the foot and innervated by the cerebral ganglia. They arise from ectodermal invaginations that eventually form vesicles, and are lined on the inside by an epithelium with ciliated receptor cells. In gastropods, the vesicles may be closed, filled with fluid, and contain a single statolith. In other mollusk groups, such as certain bivalves, the statocysts are connected to the outside via a thin duct and contain several statoconia. These are sediment particles taken up from the outside, instead of a single statolith. Statoliths are tiny spheres of calcareous material secreted by the statocyst cells. Depending on the animal's orientation, all types of inclusions follow gravity and stimulate different mechanoreceptor cells distributed all around the statocyst wall. Thus, the animal perceives its orientation in a 3D space. Eyes are missing in some mollusk groups, but in many others, they are differentiated into more or less complex organs. Light-sensing organs are often distributed regularly along the mantle edge in clams and mussels. In terrestrial snails and slugs, called Stylommatophora, the eyes are located on long movable eye stalks. In particular, the eyes of cephalopods are capable of considerable visual performance, efficiently achieving the image quality of a vertebrate eye. As an adaptation to life in the low-light deep sea, some giant squids have developed huge eyes with an extraordinarily high number of photoreceptor cells. For example, the eyes of the giant squid (*Architeuthis dux*) have a diameter of about 19 cm, and the eyes of the colossal squid *Mesonychoteuthis hamiltoni* have a diameter of up to 27 cm.

A sensory organ unique to aquatic mollusks is the originally paired osphradium. The osphradia are located in the mantle cavity near the gills. In some species, the osphradia sit directly on the gills. They are equipped with numerous chemo- and mechanoreceptors for sensing food. The osphradia may also play a role in finding mates and contribute to releasing eggs and sperm within the same time window.

The nervous system of the Mollusca is derived from the typical protostomian nervous system with an anterior brain and main ventral nerve cords. In mollusks, the brain may form a ring around the esophagus, which gives rise to two pairs of longitudinal nerve cords (tetra-neuralian nervous system). The ventromedial pair is called the pedal (or ventral) cords. They supply the foot. The lateral pair of nerves, the pleurovisceral (or side) cords, project toward the mantle and the viscera. Transverse commissures connect the longitudinal pairs of nerve cords and form a ladder-like nervous system, not divided into ganglia and connecting nerve fiber bundles (commissures and connectives). However, this primary type of nervous system is only found in the worm-like mollusks (Aplacophora), the chitons (Polyplacophora), and the deep-sea monoplacophorans, which we will not discuss further here. The nervous system of all other mollusks is divided into a few paired ganglia: the cerebral, buccal, pleural, pedal, parietal, and visceral ganglia, and their connecting nerves. A ganglion consists of a cortex of nerve cell bodies (somata) and a medulla of nerve cell processes (axons and dendrites) interconnected via synapses. Therefore, these groups are united as Ganglioneura, comprising gastropods, cephalopods, bivalves, and scaphopods. Cephalopods have developed an extraordinarily complex nervous system, enabling astonishing sensory performances. As experiments with octopuses (*Octopus octopus*, *Octopus vulgaris*) have shown, these animals can display memory and learning performances that are typical of those of many vertebrates. The total number of neurons in an octopus, including the brain and the whole nervous system, is as high as 500 million, two-thirds of which are distributed among its arms. This means that there are about 40 million neurons in each tentacle. That's more than two times the number of neurons that the average frog, a vertebrate, has in its entire body!

Mollusks are primarily gonochoristic. This means, sexes are separated, and we have females and males. This is also true for most marine gastropods, bivalves, and cephalopods. In contrast, almost all pulmonate snails (terrestrial and limnic snails and slugs) and numerous freshwater mussels and clams are consecutive or simultaneous hermaphrodites. The gonads, ovaries, and testes can be present as individual organs or merge into a unique hermaphroditic organ, which produces both oocytes and sperm. Such a mixed gonad is called an ovotestis and is otherwise rare in metazoans. The gonads are always located in a paired, separate coelomic space, the gonocoel.

Mollusks may have an indirect or direct development. In the former case, a free-swimming trochophora larva develops, showing the characteristic features of this type of larvae: a prototroch (a girdle of ciliated cells anterior to the mouth), an apical tuft and an apically located sensory field. In most molluscan groups, the trochophore is followed by a unique molluscan larval stage, the veliger. The swimming organ, the velum (usually two ciliated lobes originating from the prototroch), is the most characteristic feature of veliger larvae. While in most marine mollusks, such as those that are free-swimming, ciliated larvae appear in the life cycle, freshwater and land-dwelling snails secondarily show a development

without a free larval form. Therefore, they are independent from an aquatic habitat at any stage of life. The young animals leave the egg capsules in the so-called crawling stage.

> **Recommended Animals for Dissection**
> **Bivalvia**: All European pond and river mussels (*Anodonta* sp., *Unio* sp.) are suitable for dissection. Fish farms and pond fish suppliers regularly offer them in sizes of 5–20 cm. A good alternative is the marine blue or common mussel (*Mytilus edulis*), available all year, either from specialized food suppliers or several marine biology stations, such as the BAH-Helgoland. Living mussels are anesthetized and killed by incubating them for at least 10 h in a 1% chloral hydrate solution. The shells should remain open. The animals are now ready for dissection or can be fixed for later use in a 4% formaldehyde solution.
>
> **Gastropoda**: Roman or Garden snails are increasingly difficult to obtain in European countries, as the shipping of live animals may only be possible with considerable restrictions. In some countries, including Germany, snail farms reject live shipping and only deliver frozen snails, which are unsuitable for dissection. Here, picking up the animals directly from the farm should be considered. To dissect the animals in an extended state, they are first anesthetized and afterward killed in a 0.5 to 1% hydroxylamine solution. After 10–20 h, the animals are ready for further processing. We continue with incubating the animals for several hours or overnight in 40% ethanol to wash off mucus.

6.1   The Duck Mussel (*Anodonta anatina*)

With about 15,000 species, bivalves or clams are the second largest group of mollusks and exclusively gill-breathing aquatic inhabitants. They have secondarily adopted a sessile or hemisessile lifestyle. Many features of their body organization are directly related to this lifestyle. They are surface-dwelling or infaunal, colonizing all substrates from soft to hard bottoms. Their transition from being grazers and hunters to detritivores and filter feeders is associated with an extensive reduction of the head, including the buccal apparatus and radula, and all species rely on their gills for food collection. The mantle forms two large lobes that completely enclose the body. The mantle margins are initially free, but can be almost wholly fused ventrally, so only a narrow passage remains for the foot. At the posterior end, the fused mantle margins leave additional openings for the dorsally exhalant and the ventrally located inhalant openings for respiratory water. Exhalant and inhalant openings can be extended into tubular structures (siphons) in species living buried in the sediment. Some clams burrow rather deep into the ground, such as the soft-shell clam or sand gaper (*Mya arenaria*), which burrows up to 30 cm deep into the sand and forms correspondingly long in- and exhalant siphons that, in this species, are fused into a shared organ. Some boring clams even colonize wood or rocky substrates, into which they bore their tunnels using their shells and with the help of secreted chemical substances.

Anodonta anatina, discussed in more detail in this book, belongs to the family of the river and pond mussels (Unionidae) and was formerly widespread throughout Europe, like its closest relative, the Swan Mussel, *Anodonta cygnea* (Fig. 6.1). However, in recent decades, all *Anodonta* species have become rare due to the loss of suitable habitats and temporarily high water pollution. Like all our native freshwater clams, *Anodonta* species are protected. Therefore, only animals from certified breeding farms are available for zoology courses. The large pond and river mussels live well over 10 years in cold and nutrient-poor waters. The animals usually colonize clean freshwater lakes or slowly flowing rivers. They burrow into the muddy-sandy ground so that only their posterior ends protrude above the surface. This can often be seen in the discoloration of the shell. With their ciliated epithelium, the gills generate a respiratory and feeding water current, which the mussel uses to collect suspended food particles (detritus) from the water. While the water

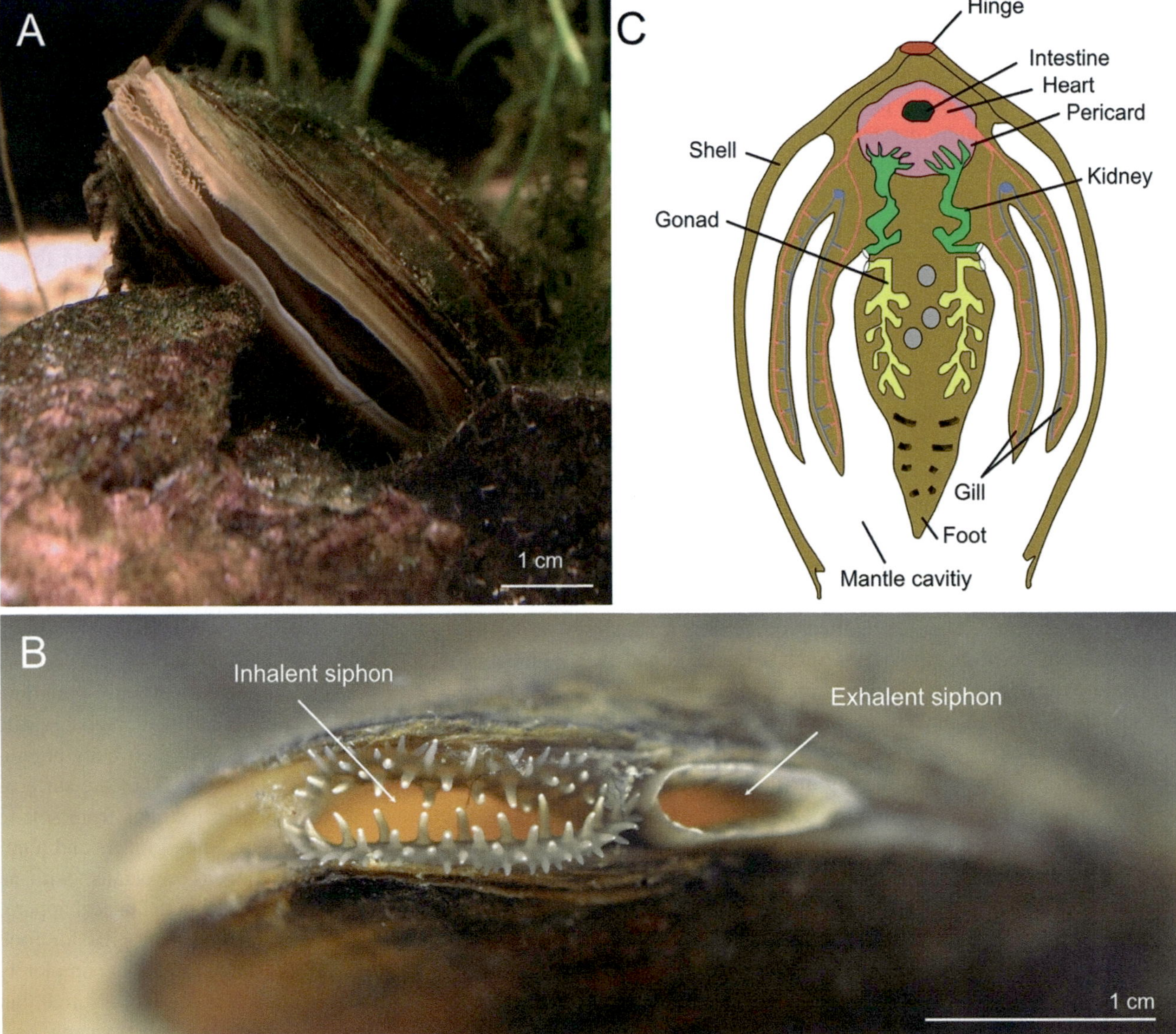

Fig. 6.1 *Anodonta cygnea.* (**a**) In their habitat, animals grow up to 20 cm large and live for more than 10 years. The mussel buries itself in the ground with the short inhalant and exhalant siphons always sticking out. Aquazoo Löbbecke Museum Düsseldorf. (**b**) Inhalant and exhalant siphons at high magnification. (**c**) Cross-section through a pond mussel, internal structure, schematic, according to various authors

passes through the often leaf-like gills, the particles are retained and transported to the labial palps, where they are pre-sorted. Clams must filter large quantities of water to obtain sufficient amounts of food. The filtration rates per individual, for example, are up to 1.9 liters per hour for the blue mussel (*Mytilus edulis*), up to 15 liters per hour for oysters, and even up to 40 liters per hour for large river mussels.

External Features

The soft parts of the mussel are protected by the shell, consisting of two similar valves hinged together. The shell of mussels, like that of all other mollusks, is made of calcium carbonate (Fig. 6.2). A toothless lock dorsally connects the two shell halves (*Anodonta* means "the toothless"), consisting of a locking hinge ligament (Fig. 6.2b, c). The epithelial cells of the mantle rim ensure continuous growth of the shell. Since annual and, in marine dwellers, even tidal periodic growth activity can be observed, the growth lines of the shells can be used for age determination. The oldest part of each shell is the umbo, also called the beak. It usually represents the most prominent, highest part of each valve of the shell (Fig. 6.2b). The shell of mussels, and conchiferans (the shell-bearing mollusks) in general, consists of three layers. The outer colored layer is formed by the periostracum, which corresponds to an organic cuticle and mechanically protects the underlying layers. If the mussel dries out, the periostracum peels off very quickly and, for this reason, is often missing from the mussels we find on the beach. The periostracum is followed by the massive and very break-resistant prismatic layer, the ostracum. It consists of calcium carbonate, predominantly in the tough crystallization form aragonite or calcite, and various organic components, significantly increasing the shell's strength (Fig. 6.2c). The organic component is called conchin or conchiolin and contains fibroin-like proteins and various carbohydrates, such as chitin. Conchiolin also makes up the main part of the periostracum.

The inside of the shells is formed by the hypostracum, often consisting of very thin layered aragonite lamellae composed of conchiolin and a low calcite content. The hypostracum, with its unique chemical composition, is also called mother-of-pearl or nacre, and gives off the typical iridescent shine (Fig. 6.2c). This shine comes from interference phenomena of light on thin layers (structural color!). However, if the hypostracum consists only of prismatic calcite crystals, this shine is missing, and the hypostracum appears porcelaneous, not nacreous. The formation of pearls represents a defense reaction of the mussel after the penetration of small foreign particles into the mantle epithelium. In this process, the shell-forming epithelial cells of the mantle edge enclose and overgrow the debris, forming a so-called pearl sac. The epithelial cells deposit layers of nacre around the foreign body, which ultimately leads to the formation of a pearl. In breeding farms, pearl formation is induced by deliberately introducing foreign particles, a procedure that not all individuals in these farms survive. Pearls can occur in all mussels; they are most attractive in species that form an iridescent and thick layer of mother-of-pearl.

Anodonta: Opening the Shell

(a) Start by removing the *left* valve of the shell. This allows for an initial orientation concerning the internal anatomy and the details of the shell structure. To open the clam completely, work along the valve edges of the shell with a sharp knife or scalpel.

(b) The two shell valves are now slightly pulled apart to about 1 cm with a blunt object. Identify the short but potent retractor muscles and their insertion surfaces inside the shell valves (Fig. 6.3a).

(c) The mussel is now placed on its right side in a dissecting dish. With the help of a probe and scalpel, the mantle, connected to the shell, is separated from the inside without damage. Now, with the scalpel, the large adductor muscle on the anterior side is cut, as well as other smaller retractor muscles, which are nearby. The best way to do this is to place the knife tip between the shell valves and the mantle. With the blade angled, the muscles are cut at the base of the shell. Repeat this procedure at the posterior end. When all muscles are cut, and no structures are still attached to the shell, the mussel can be carefully opened (Fig. 6.3). For a careful mussel dissection, it is advisable to fix the animal in the dissection dish. The easiest way is to use a hot glue gun from craft stores. The mussel is generously glued horizontally to the dry wax base of the dissection dish. The organs are then covered with water. For better orientation, a schematic cross-section is shown in Fig. 6.1c.

Mantle, Foot, and Visceral Mass

For a more detailed examination of the opened mussel, the shell valve is filled with tap water to cover all organs (Fig. 6.3b). With the help of tweezers and probes, an initial examination can now be carried out. First, the cut adductor muscles should be identified. Together with the rigid shell, they provide the most essential protection against predators, as they tightly press the two shell valves together. One of the most likely predators in marine habitats is starfish, which can develop up to 10 kg of pulling force with their arms but often have to capitulate against the force of a mussel's retractor

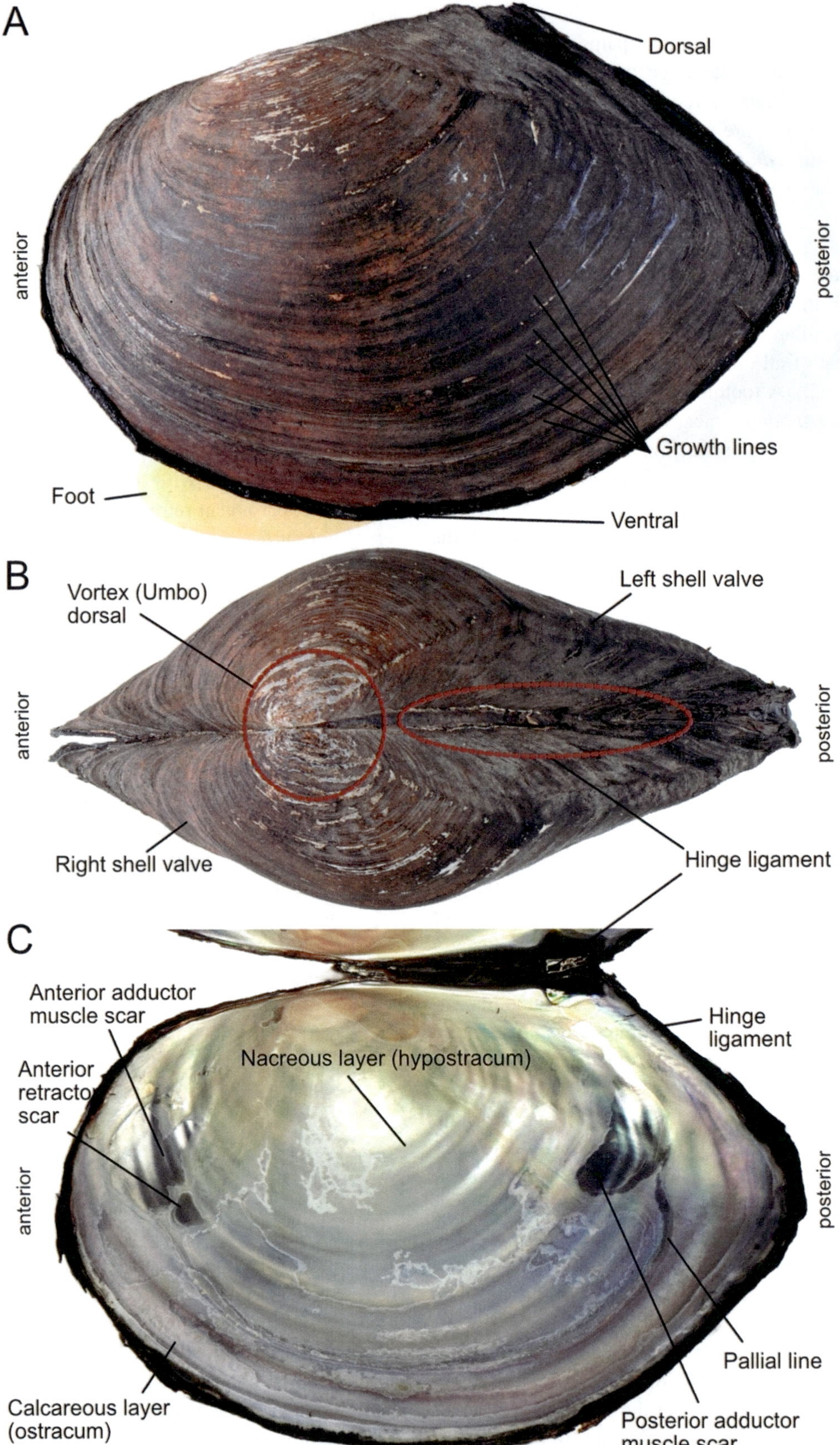

Fig. 6.2 *Anodonta anatina*. (**a**) Lateral view. (**b**) Dorsal view. (**c**) Opened shell viewed from the inside

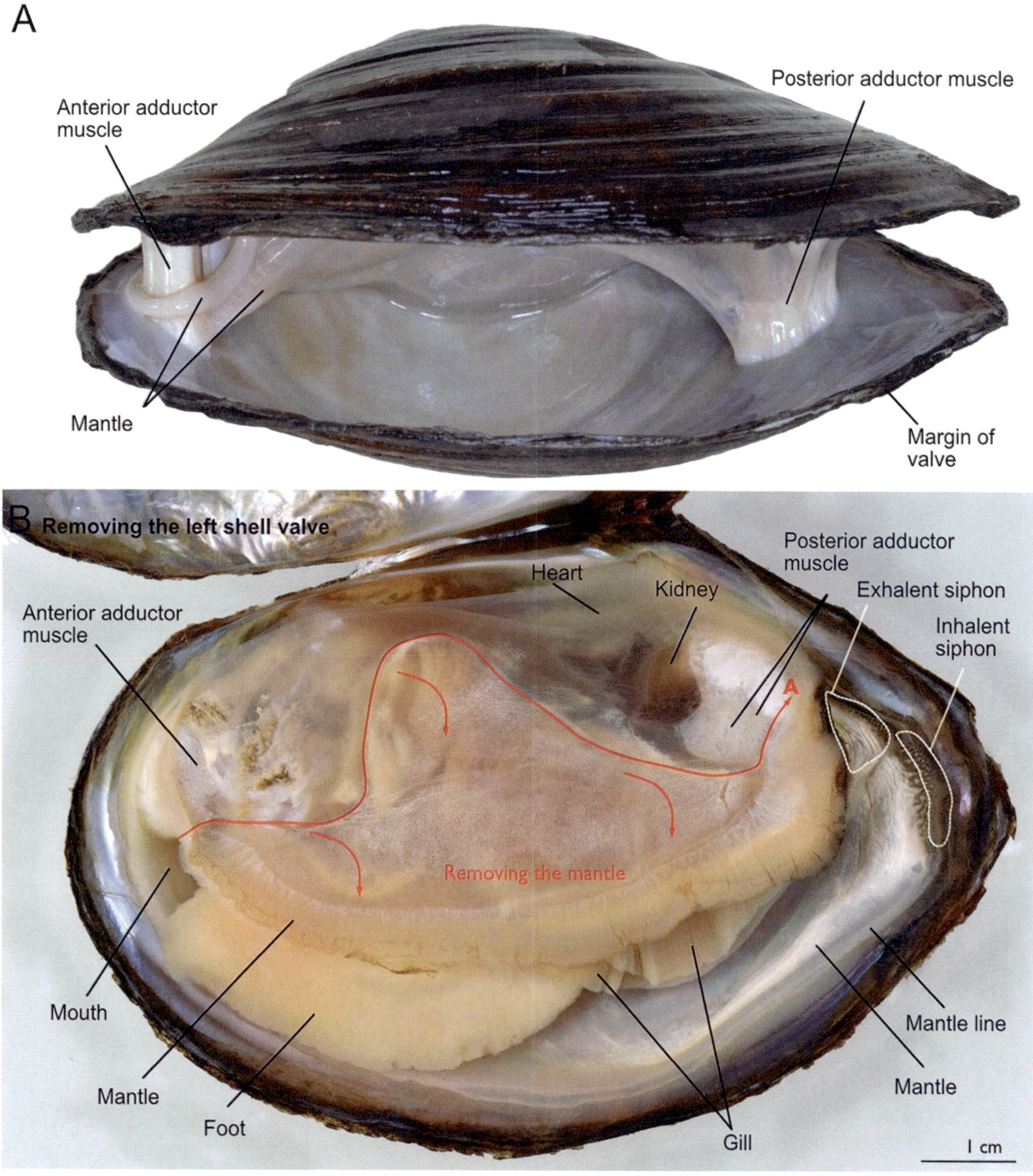

Fig. 6.3 *Anodonta anatina*. (**a**) View inside with slightly opened shell valves. (**b**) Mussel after removal of the left shell valve. The red line shows a possible cutting path to open the mantle

muscles. However, the starfish waits. After a while, the mussel must open its shell to avoid suffocation, and that is the moment the starfish anticipates. Then, the starfish pushes its stomach into the open mussel and begins digestion there. Other predators of our native clams and mussels are muskrats and otters.

We now look at the mantle separated from the shell, the former of which has slightly contracted and retracted (Figs. 6.3 and 6.4). The central part of the foot and the anteriorly located mouth opening are usually easily recognizable. With the help of the muscular, shape-variable foot, the mussel can burrow into the sediment or hold onto a suitable sub-

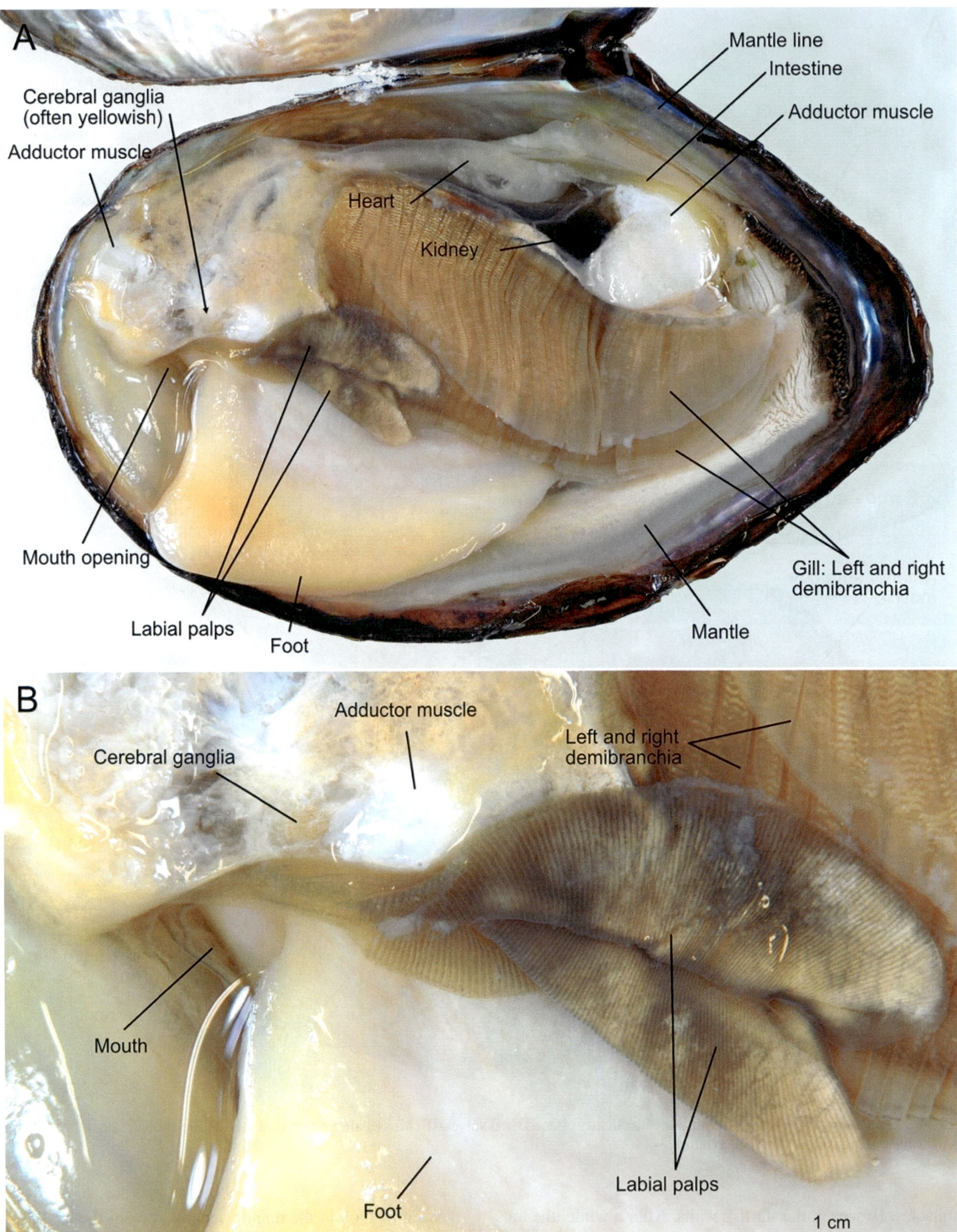

Fig. 6.4 *Anodonta anatina*. (**a**) After the mantle is loosened and folded back, internal organs can be studied. The heart, kidney, gills, and foot are well visible. (**b**) Labial palps and the mouth opening at higher magnification. The cerebral ganglion is recognizable as a yellowish organ

strate. A movie showing a clam digging into the ground can be found at figshare: sn.pub/xr9qhi. A bipartite opening is visible on the posterior side of the opened mussel, the ventral inhalant and the dorsal exhalant openings. The inhalant opening is also referred to as the respiratory or ingestion opening, as it allows for the entry of fresh feeding and respiratory water into the mantle cavity, where the large leaf-like gills are located. The inhalant opening is surrounded by fine, fringe-like papillae, which are equipped with numerous receptor cells and prevent the entry of larger particles. After having passed the inhalant chamber, the gill filaments, and the exhalant chamber, the respiratory current leaves the mussel via the exhalant opening, also called the egestion opening.

> **Anodonta: Opening of the Mantle**
> The mantle is slightly lifted with tweezers and opened along the cutting line (Fig. 6.3b, red line). Fold back the mantle and pin it down.

Gills, Labial Palps

When the mantle has been lifted and set aside, the underlying organs become visible (Fig. 6.4). Here, the large gills, which lie freely in the mantle cavity, are noticeable. *Anodonta*'s right and left gills each consist of two ciliated lamellae. They are formed by numerous, closely adjacent threads firmly connected via tissue bridges. Thus, the gills form an extensive grid of regularly perforated leaves, through which the water current flows from the inhalant chamber to the exhalant chamber. The gills have three essential functions.

First, they serve for the filtration of food particles. Numerous cilia are located on the gills, which generate the water flow and particle transport. In addition, the epithelium of the gills produces mucus for trapping food particles between 1 μm and 40 μm in size. The particles are then either moved up or down on the filaments, and, upon reaching the edges of the filaments, they are transformed into a string of mucus and transported by ciliary activity toward the mouth opening. Then, the two labial palps take over for the final separation of the captured particles and transport of the food into the mouth (Fig. 6.4b). Particles larger than 40 μm in diameter are directly sorted out by specific long cilia on the frontal side of the gills, grouped into cirri. Usually, these particles are moved downward to the edge of the gill, where an expulsion channel ensures that these particles are released to the outside. Thus, these particles do not even reach the mouth opening. The gills and palps produce the so-called pseudofeces, which mainly consist of sediment particles enclosed in mucus. The feces is moved posteriorly by ciliary tracts in the mantle skirt toward the inhalant opening. From time to time, the adductor muscles contract and produce an opposite water current, which transports the rejected particles to the outside.

The second important function of the gills is gas exchange and, thus, respiration. Through the inhalant opening, water flows into the mantle cavity and passes the gills before entering the exhalant chamber. The rectum (anus), the gonoducts, and the paired nephridia also open into this cavity. The excrements and urine leave the mussel with the respiratory water current through the exhalant opening.

In *Anodonta*, the gills have an additional important function for reproduction. The gametes enter the mantle cavity via the paired egg or sperm ducts. While the sperm leave the mussel directly with the respiratory water, the eggs become trapped in the gill network of the mussel and remain there for some time. The eggs of the mussels are fertilized by sperm from another mussel that enter the mantle cavity of the females with the inhalant current. The first steps of embryonic development occur within the mantle cavity. This kind of parental care represents an adaptation to the unique life cycle of our river mussels, which will be discussed later in a separate section. In most clam and mussel species, sperm and egg cells leave the mantle cavity and fertilization occurs in the open water.

> **Anodonta: Exposing the Heart and Kidney**
> The remnants of the mantle are removed to reveal the heart and kidney. In Fig. 6.5a, a possible cutting line is marked in red.

The Circulatory System and Heart

Clams and mussels possess a dorsally located, transparent, three-chambered heart (Fig. 6.5b), which surrounds the intestine. It consists of a pair of atria and a single ventricle into which oxygen-rich blood flows from the left and right gills. Contractions, which occur at a frequency of 1–3 s, force the blood from the ventricle into the anterior and posterior arteries and to the various tissues. A movie showing the heart of a duck mussel can be found at figshare: sn.pub/xr9qhi. A lacunar system later returns the deoxygenated blood into a ventral sinus below the heart, from where it passes via the kidneys to the gills. The term blood sinus refers here to a nontubular cavity formed by tissue gaps and bounded by ECM. In the gills, oxygenation of the blood occurs. After that, the blood is transported back to the pair of atria and, finally, to the ventricle. The foot of the mussel is very well vascularized. The amount of blood, together with

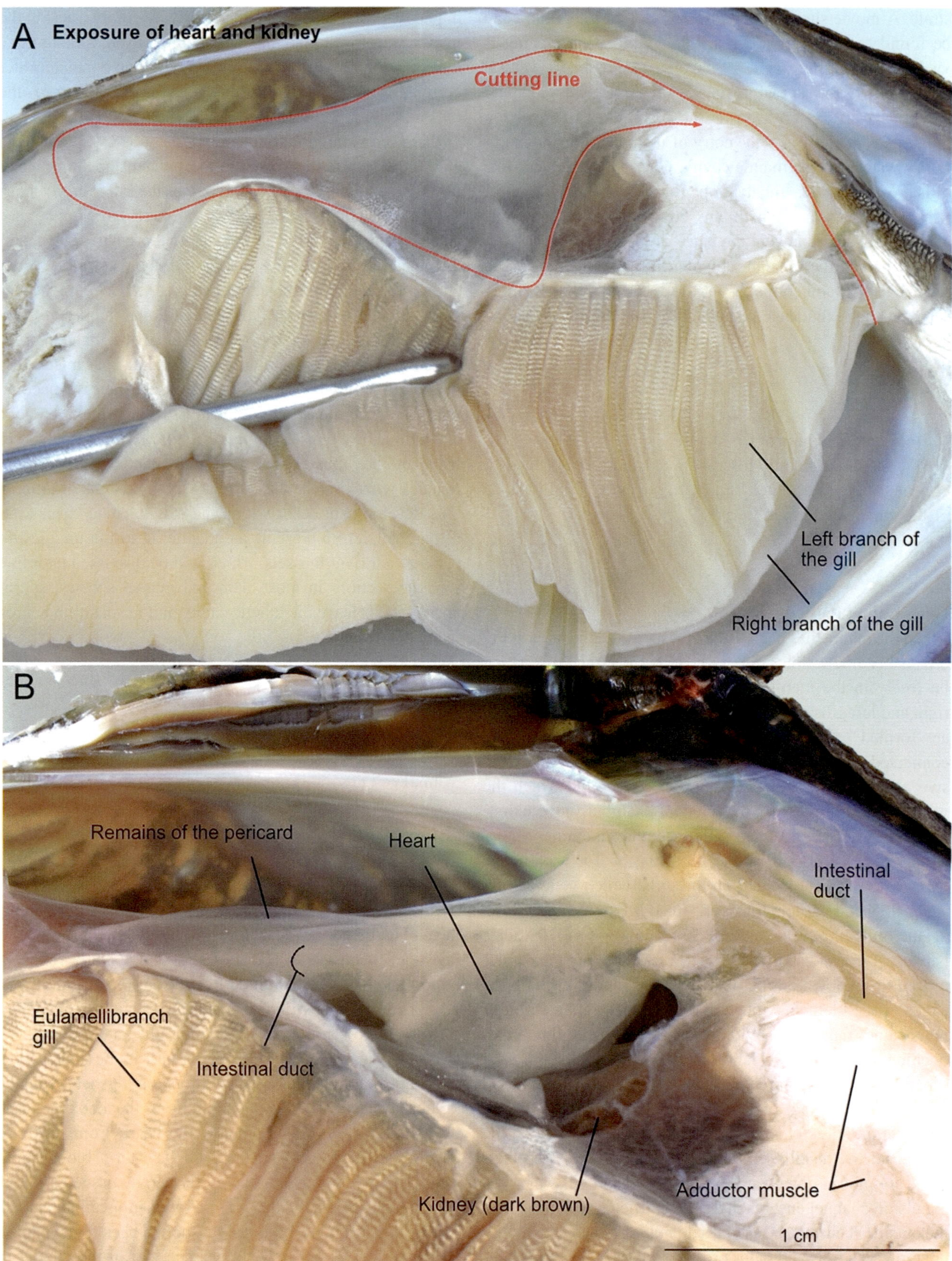

Fig. 6.5 *Anodonta anatina*. (**a**) Mantle remnants impede the view of the heart and kidney. (**b**) View of the heart and kidney after removal of the mantle remnants

the musculature, controls the variable shape and mobility of the foot in its lacunar system. All mollusks use a copper-containing respiratory pigment, the blue hemocyanin.

Excretion

The paired metanephridia (kidneys) lie near the heart, just below the pericardial cavity (Figs. 6.4a and 6.5a, b). They are directly connected to the pericardium (heart sac, a coelomic space) via a ciliated funnel. The coelomic fluid or primary urine is formed through ultrafiltration directly at the heart via podocytes. This fluid enters the nephridia via the ciliated funnels. The ciliated renal ducts (pericardioducts, renopericardial ducts) conduct the urine via the nephridial sac and an excretory canal (ureter) to the mantle cavity. Thereby, the definite urine is formed by processes of reabsorption and secretion. There, the definite urine is discharged with the water flow. In addition to excretion, the nephridia play a vital role in osmoregulation in *Anodonta* and other freshwater mussels. Mussels, like all freshwater animals, have a higher osmotic value than their environment. They, therefore, absorb the electrolytes dissolved in water through their gills against a concentration gradient and excrete the excess water through their kidneys. Due to the osmotic gradient, there is also a considerable passive influx of water, balanced, again, via the nephridia. As a consequence, freshwater mussels produce highly diluted urine.

> **_Anodonta_: Removal of the Left Gills and Exposure of the Gonads**
>
> (a) The foot and visceral sac are between the two gill cavities and the gill branches. The left gills are removed first to get a clear view of the internal organs (Fig. 6.6a).
>
> (b) Using fine scissors, the visceral sac is opened. A possible cutting line is shown in Figs. 6.6a and 6.7a. The dissection should be performed under the stereomicroscope. The blade of the scissors is inserted into the mouth. Now, cut centrally in a dorsal direction across the stomach. Insert the scissors slightly deeper. The cut follows the course of the stomach (tilted somewhat to the left). Before cutting, the scissors should be lifted slightly to avoid accidental injury to other tissues. Carefully expose the midgut step by step. Use forceps and a water-filled squirt bottle to free the midgut glands from floating tissue debris. Several blind sacs are located between the midgut glands. Following the recommended cutting line shown in Fig. 6.6a, gonads and other intestinal loops are exposed. The more tissue of the visceral mass is removed, the better the gonads and intestinal loops can be studied.

Digestive System

The anteriorly located mouth opening leads into a narrow intestinal canal, extending through the entire body. Radula or jaws are absent. As already mentioned, food is captured with the help of the gills. The mouth forms two ciliated labial palps. They protrude into the mantle cavity and help to sort the food and to guide it to the mouth opening (Figs. 6.6a and 6.7a). The mouth is connected to a spacious stomach via a short esophagus. The paired digestive glands are directly connected to the stomach (Fig. 6.7b). The stomach of the mussels contains ciliated sorting fields and septa as guiding structures. They sort the ingested food and partly transport it into the diverticula of the digestive gland. The sac-like posterior end contains the crystalline style. The crystalline style is a rod-like mass of proteins and enzymes, mostly amylases, constantly rotated and pushed forward by cilia in the surrounding sac. The enzymes are slowly released and mixed with the ingested food. The digestive glands consist of numerous blind-ending epithelial tubes. The tube wall comprises basophilic, secretory, resorptive, and endocytic active cells. The basophilic cells possess a highly developed rough endoplasmic reticulum and numerous secretory granules. They play an essential role in enzyme production and secretion. The second dominant cell type in the digestive glands is responsible for nutrient uptake and intracellular digestion. Thus, the digestive gland of the mussels, like in other mollusks, has multiple functions in the production of digestive enzymes, extracellular digestion, endocytosis, and the storage of nutrients. The term "gland" is only partially correct; this is the digestive part of the intestine. Mussels first digest extracellularly in the diverticula of the digestive gland and then intracellularly in its epithelium. There are also calcite cells, which absorb and store calcium carbonate. Calcium carbonate is needed for the construction of the mussel shells.

Nervous System and Sensory Organs

The nervous system of the mussel is relatively simple compared to those of gastropods or cephalopods. It consists of three large, paired ganglia connected by vigorous nerve cords. The simple construction of the nervous system and the absence of complex sensory organs, except for the statocysts, are due to the sessile lifestyle. The yellow-orange cerebral ganglion, approximately 1–2 mm in size, is most likely to be detected in a student's course, because it is located near the anterior adductor muscle and the esophagus (Figs. 6.4 and 6.7b). The second pair of ganglia (pedal ganglion) is fused and is located anteriorly in the foot. The third pair of ganglia, the visceral ganglia, is located posteriorly and ventrally beneath the posterior adductor muscle. Like the pedal ganglion, it is fused into a single ganglion. The nervous system of a river or pond mussel is challenging to prepare in the context of an introductory zoological practical course. In addition to the osphradia and the statocysts located in the foot, other macroscopically visible sensory organs are

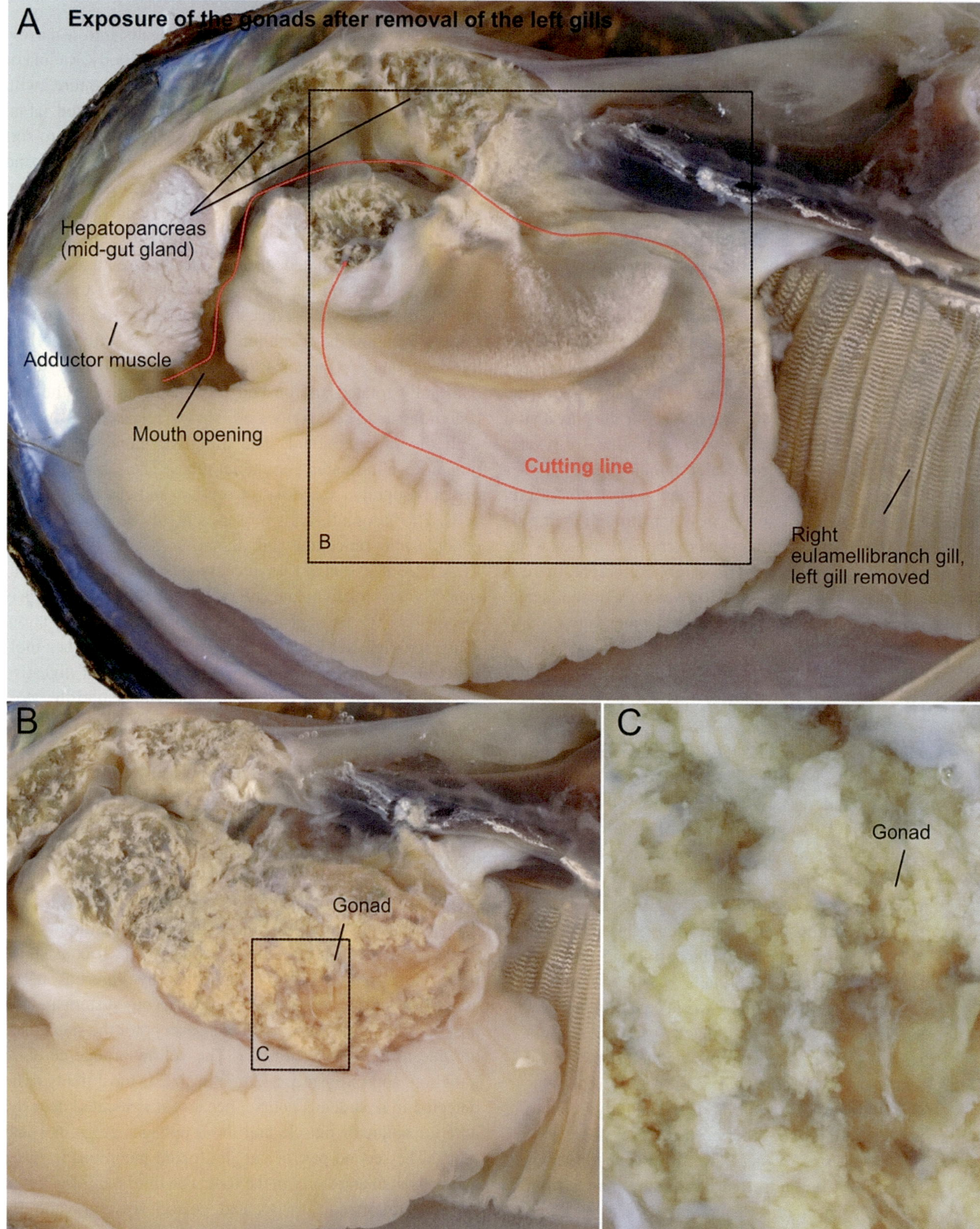

Fig. 6.6 *Anodonta anatina*. (**a**) View of the anterior region with mouth opening and digestive glands. The gills have already been removed. Remains of the visceral mass and foot still cover the gonad. (**b**) Gonad after removal of the overlying tissue. (**c**) Gonad at high magnification

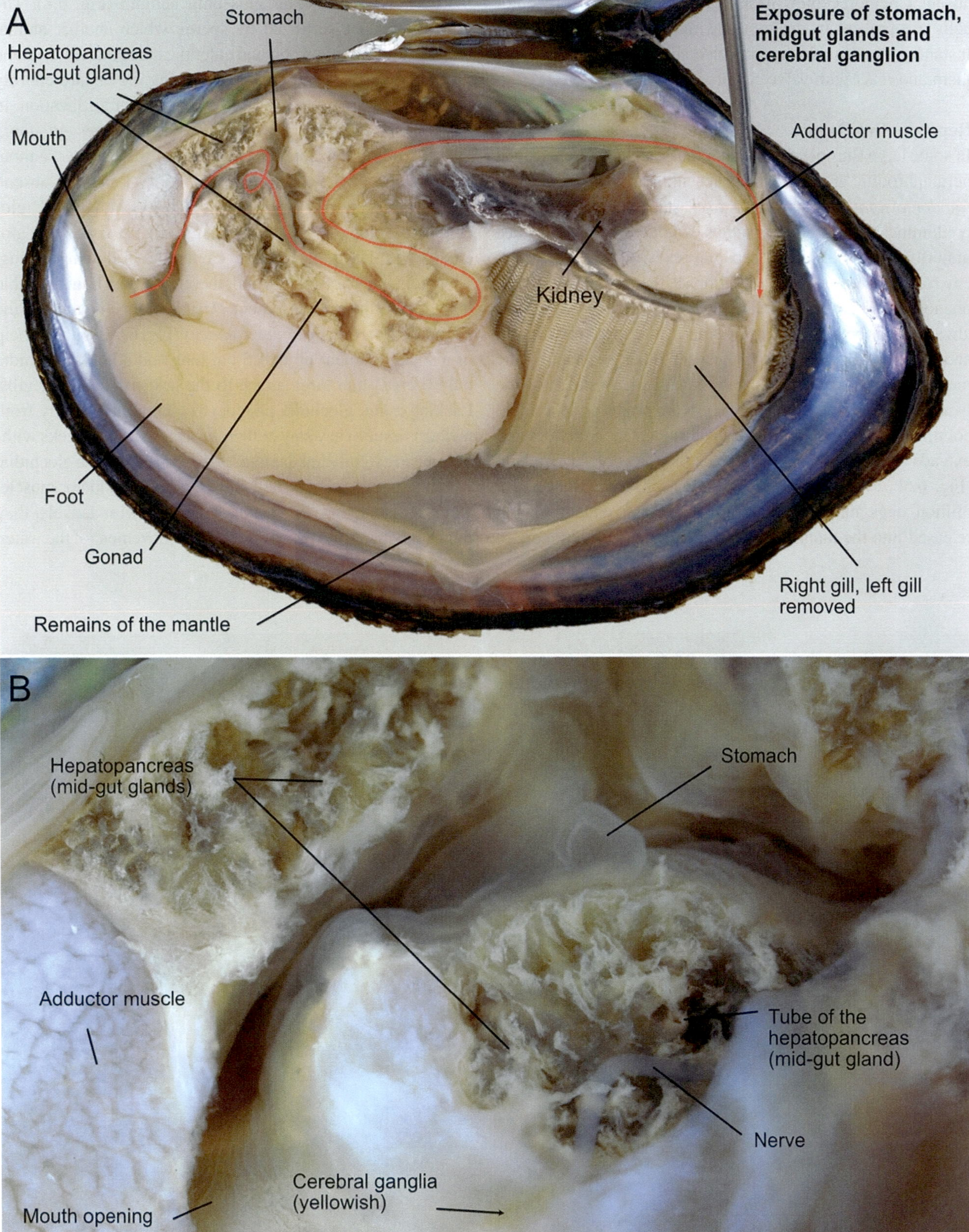

Fig. 6.7 *Anodonta anatina.* (**a**) After removing the mantle and the right gills, the visceral sac is opened. The red line shows the path of the intestinal tract. (**b**) Stomach, Cerebral ganglion, and nerve at high magnification

absent. Chemoreceptor, mechanoreceptor, and photoreceptor cells are present in the foot and mantle edge, so that, for instance, the animals can perceive shadows that glide over them and react with valve closure.

Reproductive Organs

In some large European freshwater mussels, female, male, and hermaphroditic animals can occur within a given population. *Anodonta anatina* populations from stagnant waters consist predominantly of hermaphrodites, while in rivers, the vast majority of animals are of separate sexes. It is assumed that environmental conditions may impact the sexual geno- and phenotype. The closely related *Anodonta cygnea* is almost always described as hermaphroditic in the literature. Sperm and eggs are not, as with almost all other hermaphroditic metazoans, produced in separate gonads, the testes, and the ovaries. Instead, both sperm and eggs are produced in a single common gonad, the ovotestis, which is located in the foot and extends between the diverticula of the digestive gland (Fig. 6.6b, c). In one reproductive season, more than half a million eggs mature in a single animal, eventually being released into the mantle cavity. They initially remain there in the network of the outer gills until autumn (Fig. 6.8). The oocytes are then fertilized by sperm, which another animal releases into the free water and enter the mantle cavity with the inhalant water current. Particular mating organs are entirely missing; they are not necessary for this type of fertilization. In *Anodonta anatina* and the closely related species *Anodonta cygnea*, early development occurs in the mother animal between the gills. A free, planktonic larval stage is absent. However, secondary larvae called the glochidium (pl. glochidia) develop within the mother's gill network. The glochidia will leave the mantle cavity the following spring. Glochidia live parasitically and, therefore, must find a suitable host for further development. Usually, this is a fish, the European bitterling, *Rhodeus amarus* (Fig. 6.9). With the help of a sticky filament up to 1.5 cm long and a hook, glochidia only 0.2–0.3 mm large attach to the host fish's skin or gills. Lacking a gut, glochidia phagocytose nutrients directly from the host tissue. They stay at their hosts for several weeks without causing significant damage to the host. Since the glochidia, which also possess two shell valves, a single adductor muscle, and sensory organs (Fig. 6.9b, c), cannot swim actively, they must find their hosts passively with the movement of the water.

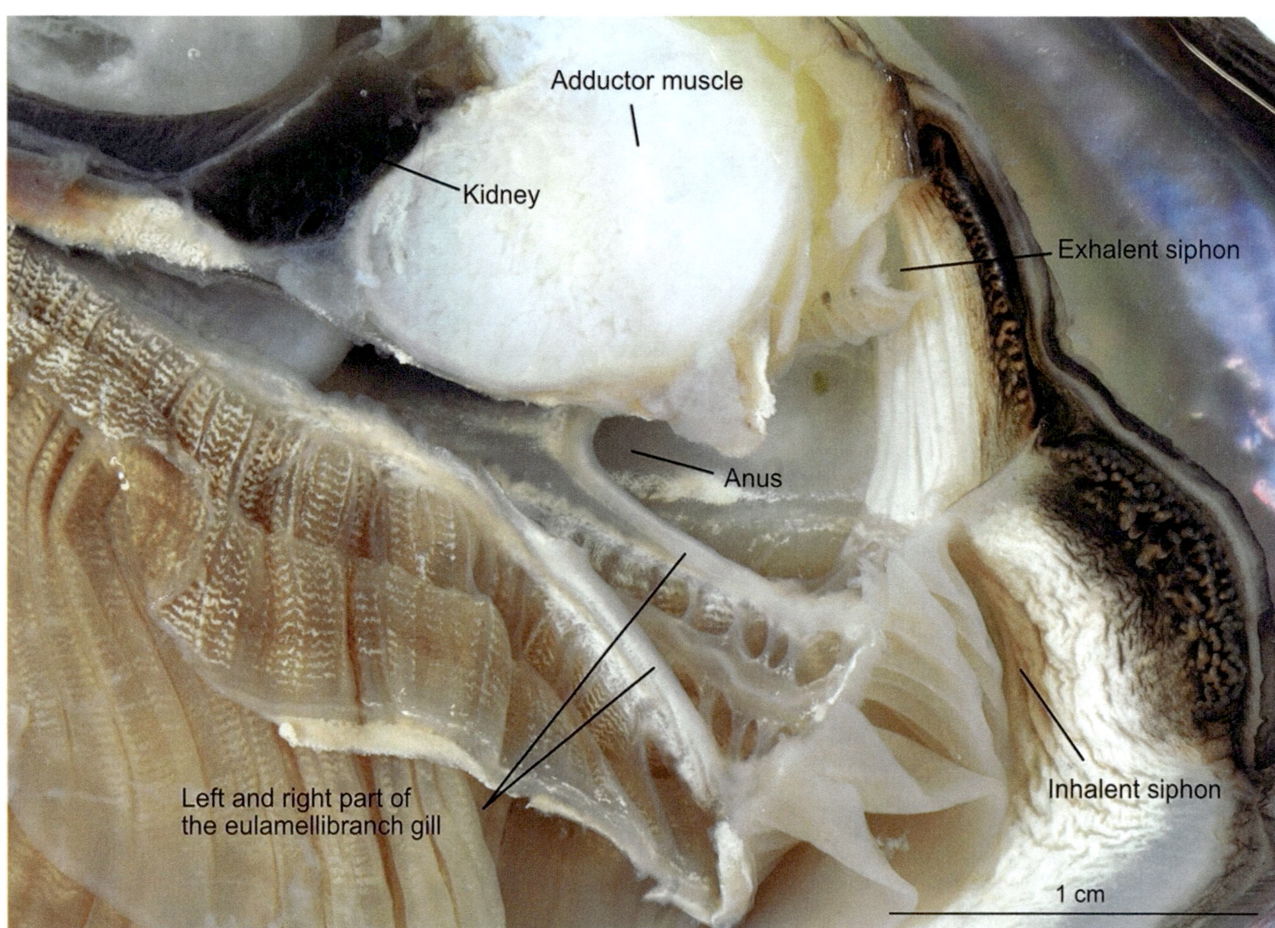

Fig. 6.8 *Anodonta anatina*. View of the inhalant and exhalant siphon, anus, and the entrances to the right and left gills

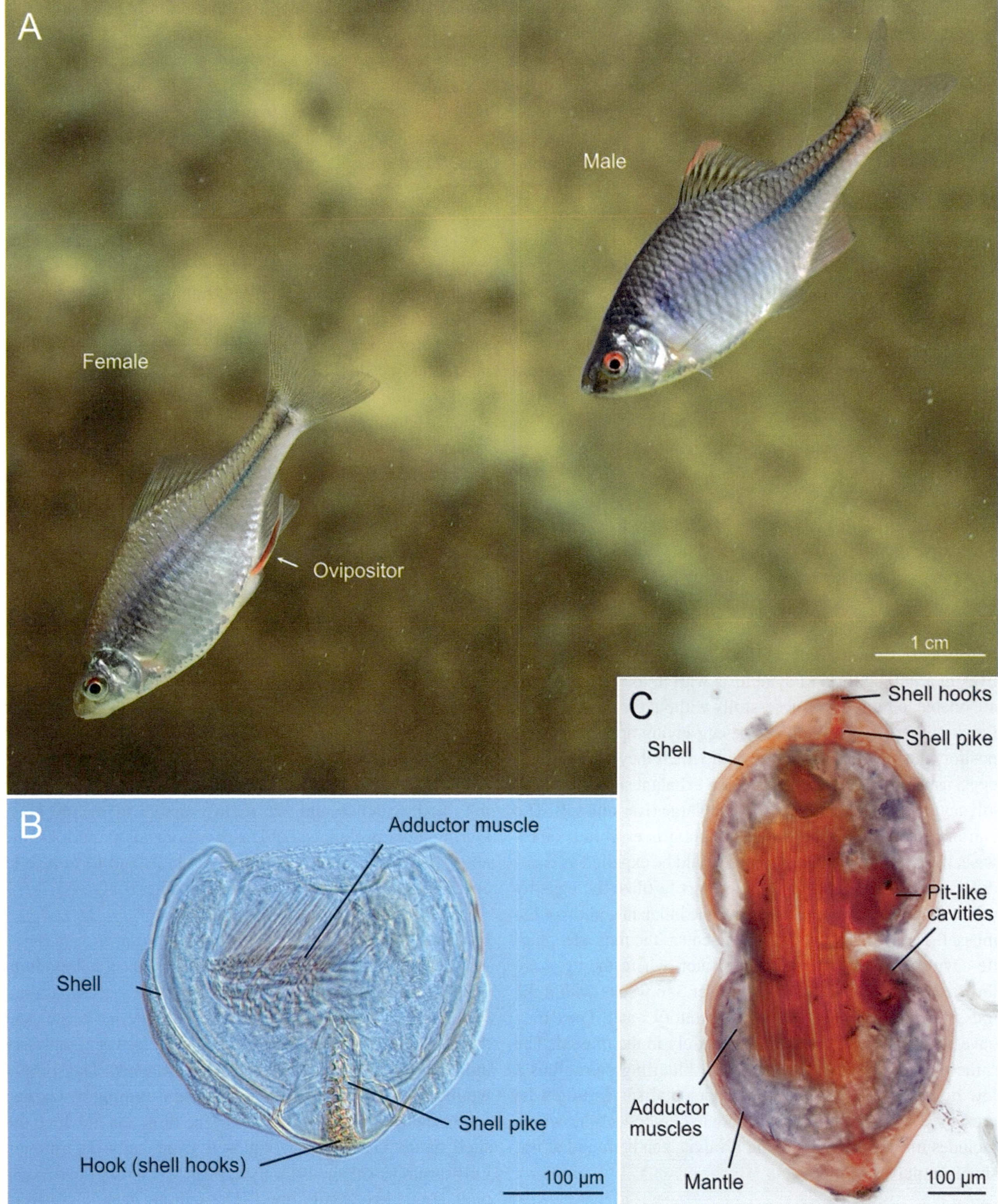

Fig. 6.9 *Anodonta* embryogenesis. (**a**) *Rhodeus amarus*, European Bitterling. Female with ovipositor (left fish) and a male. (**b**) Glochidium, total, unstained, Nomarski interference contrast. Specimen of the Biological Collection of the University of Osnabrück. (**c**) Glochidium of *Anodonata*. Nuclear Fast Red staining, specimen from Johannes Lieder, Ludwigsburg

After 2–3 months, they undergo metamorphosis and detach from the host. If the glochidia do not find a suitable host, they die. The complex mode of development bypasses the planktonic stage and, in this way, prevents the offspring from drifting away into unfavorable waters. At the same time, new habitats can be colonized, since the hosts move around but stay in environments suitable for mussels. This developmental mode also explains the high number of eggs, as only the smallest number of glochidia will successfully develop into a mature mussel. The developmental mode is very similar to that of our native river mussels (*Unio* spp.), which usually are gonochoristic. In the mantle cavity of female painter's mussels (*Unio pictorium*) about 300,000–400,000 develop simultaneously for about 4–6 weeks. Finally, 0.25 mm small larvae are released and sink to the ground. When fish, for instance, the afore mentioned bitterling, try to ingest these larvae, they attach to the fish' branchiae, become encapsulated and live there as parasites. Young mussels become free-living after about 6 (3–11) weeks.

Bitterling and Duck Mussel: A Mutualistic Coexistence

Our native duck mussels depend on fish for successful reproduction, as the development of the glochidia occurs in the hosts' skin or gills. Among the potential hosts are various species of the carp family, but the preferred host fish is the European bitterling (*Rhodeus amarus*; Fig. 6.9a). The mussel lives in a unique form of symbiosis with it, as the bitterlings are unable to reproduce successfully without the mussels. The female bitterlings develop a long egg-laying apparatus (ovipositor) during mating season, with which they release their eggs, usually only a very few, via the exhalant siphon into the gill cavity of the duck mussel or other large river mussels. The entrance of eggs via the exhalant siphon is essential; otherwise, the comparatively large eggs would be expelled as pseudofeces by the mussel. Male bitterlings fertilize the eggs by releasing their sperm directly above the inhalant siphon of the mussel. Thus, the bitterling's sperm enters the mussel's mantle cavity through the respiratory water, where the eggs are fertilized. The fish larvae develop over 3–6 weeks within the mussel before leaving it with the respiratory water. Thus, they have spent their early development safely in the mussel. The same mussel will be used by several bitterling pairs, thus a few dozen eggs at different developmental stages might be present in a single mussel. Our movie clip collection, which includes movies showing living bivalvia, can be found at figshare: sn.pub/xr9qhi.

6.2 The Roman Snail (*Helix pomatia*) or the Garden Snail (*Helix aspersa*)

Gastropods, slugs, and snails are the most species-rich group of recent mollusks, with over 100,000 known species. The asymmetry of their bodies is particularly striking, a phenomenon that can be explained by the torsion of the visceropallium, the visceral mass, and its overlying mantle and shell by as much as 180 degrees. Torsion occurs during all gastropods' development, starting at the end of the veliger larval period. An external shell is primarily present in all gastropods, but it was lost in several recent groups during evolution, for example, in many slugs.

Among mollusks, only gastropods have conquered all aquatic and various terrestrial habitats. The gills are regressed in the terrestrial gastropod groups, and the epithelium of the mantle cavity now serves as a lung for gas exchange. Land slugs and snails carrying a lung are summarized as Pulmonata. This group of mollusks includes, for example, the well-known Roman and Garden snails (*Helix* spp.), which carry an impressive shell to protect the body, and the red or black slugs (*Arion* spp.), in which the shell is almost entirely reduced except for a few calcareous grains overgrown by the mantle, the so-called concretions (Fig. 6.10). *Helix pomatia*, the common Roman snail, and *Helix aspersa*, the Garden snail, with a body length of 8–10 cm and a shell diameter of 4–5 cm, are among the most prominent European land snails. The distribution of *Helix pomatia* covers all of Central Europe and extends to Eastern Europe. *Helix aspersa* is smaller than *Helix pomatia* and prefers a milder climate, such as the entire Mediterranean area. Both species are bred for culinary purposes and can live up to 20 years as farm snails, but in the wild, they usually only reach an age of 6–7 years. Both species are pure herbivores that crush their food with their radula. The radula is equipped with 20–30,000 small chitin teeth, which are constantly renewed. The animals avoid direct sunlight to protect against dehydration during food searches and are mainly active in the evening. During more prolonged cold or very hot and dry periods, the animals withdraw entirely into their shells and form a solid cover made of lime, called epiphragm (Fig. 6.11a).

External Features

The Garden snail has a long, spindle-shaped cephalopodium (complex of foot and head) (Fig. 6.10). The visceropallium, which is the visceral hump containing all internal organs and the mantle, is not visible from the outside, as it is completely hidden in the shell, and therefore well protected. The spirally wound shell (Fig. 6.11b) is always right-turning in Garden snails and most other shell snails. The direction in which the shell of the Garden snail winds is genetically determined. Left-turning snail shells rarely appear; when they do, the animals are called "snail kings." The famous geneticist Alfred Henry Sturtevant, a student of Thomas Hunt Morgan, had already discovered by the 1930s that it is a maternal (matrocline) inheritance. This means the mother's genotype is decisive in hermaphroditic snails, thus of the animal from which the egg comes.

Numerous epidermal grooves and furrows are responsible for the rough body surface of the snail. There are four mov-

Fig. 6.10 *Helix aspersa*, Garden snail. (**a**) Crawling on the substrate, note the open pneumostome. (**b**) Crawling on a glass plate, note the contraction waves of the muscular foot

Fig. 6.11 *Helix aspersa.* (**a**) Two snails producing an epiphragm. (**b**) To show the inner windings of the *Helix aspersa* housing, part of the shell was grinded away. (**c**) Hypothesis on the evolutionary origin of gastropod asymmetry through torsion. According to Hickmann and other authors, modified

able tentacles on the head of the animals. The posterior dorsal and longer pair of tentacles carries the eyes at the tip, and the anterior ventral, shorter pair of tentacles is exclusively tactile. Right at the front is the mouth opening, which leads into the radula-bearing oral or buccal cavity. At the transition between head and foot on the right side of the body lies the genital opening. Here, the sperm are transferred to the sexual partner, and fertilized eggs are released and laid in a nest

(Fig. 6.10a). The sole is ventrally flattened, and many glands continuously produce mucus, upon which the snails move by sliding (Fig. 6.10b). If one observes a snail crawling over a glass plate, the contraction waves in the central area of the foot can easily be seen. They reflect the activity of the enormous foot musculature (Fig. 6.10b). The foot is characterized by excellent mobility and variability in shape. A snail's foot consists of spongy connective tissue interspersed with blood lacunae and musculature. The foot tissue thus serves as a highly flexible swelling body and can dynamically change its shape. Together with the activity of the muscles, the foot adapts perfectly to the ground in this way. On the right side of the body, at the transition between mantle and shell, is the pneumostome, the breathing opening, which, in the living animal, rhythmically opens and closes. It leads into the mantle cavity, a space formed by a duplication of the body wall, thus forming the mantle skirt between the visceral mass and the shell. In pulmonates, the roof of the mantle cavity is differentiated into a lung. The opening of the mantle cavity is relatively small, and the mantle edges are widely fused so as to minimize water loss, which is unavoidable during respiration.

The snails, with their visceral mass wound into the shell and the mantle cavity (= lung cavity) located on the right side, have partly lost the bilateral body symmetry of their ancestors. The asymmetric shape of most gastropods is caused by a process called torsion, the rotation of the visceral mass during embryonic development. The evolutionary origin of the gastropod asymmetry is illustrated in Fig. 6.11c. Through torsion, the mantle cavity with the anus shifts to the anterior end and opens just above the head. After torsion, the initially left kidney and the left gills are located on the right side (and vice versa) and the gills are situated in front of the heart. The same applies to all other organs inside the visceral mass. According to various authors, the mantle cavity gains more space through torsion so that the snails can retract their entire cephalopodium into the mantle cavity. However, this is only possible after the second step of gastropod body evolution, the elongation of the visceral mass and the shell. The animals are now very well protected from predators and harsh environmental conditions. Likewise, the osphradia located pre-torsional at the posterior end come into a more favorable position near the front end of the animals and are situated in the direction of primary movement. However, this torsion has disadvantages, because the nerves that innervate the visceropallium (the pleurovisceral nerve cords) cross. This leads to stimuli from the post-torsional right half of the body being projected to the left, and vice versa. Another disadvantage is that the so-called pallial complex, the anus, and the excretion pores, as well as the gills and osphradia, are located directly above the head. Since the paired gills generate respiratory currents that enter the mantle cavity laterally and release them medially, the respiratory current would also expel the feces and urine above the head. However, this situation does not exist in recent snails. Those, which still possess two gills, have a mantle slit that opens outward far behind the head so that the feces and urine leave the body at some distance from the head. In other gill-bearing snails, only the left gill is retained, while the right gill, including the corresponding atrium and the nephridium, is lost. At the same time, the pallial complex has shifted to the right. In these species, the respiratory water enters the mantle cavity laterally on the left and leaves it on the right side behind the head. In other gastropods, which are summarized as Euthyneura, the crossing of the pleurovisceral nerve strands is eliminated, either by untwisting the visceral mass or, in the case of pulmonates like our Roman snail, by locating the ganglia in the head area.

> ***Helix*: Opening of the Shell**
> The animals should be in an outstretched position for a successful dissection. Animals that have withdrawn their bodies into the shell are less suitable. See the "Recommended animals for dissection" section for details.
> (a) Remove the shell. Hold the animal by the foot and turn it very slowly and entirely out of its shell. If the animal does not detach, the shell must be removed with a coarse tweezer bit by bit until the last whorl, without damaging the animal.
> (b) The roof of the mantle cavity, with its blood vessels, the heart, the kidney, and several other organs, can now be directly examined.

The Mantle Cavity, Circulatory System, and Heart

After we have carefully removed the shell, the roof of the mantle cavity, with its blood vessels (pulmonary plexus, lung), can be examined. Several additional organs inside the visceropallium are already visible (Fig. 6.12). A metal probe from the dissection toolkit is inserted through the pneumostome into the lung to examine its size and extent. If the lung has collapsed, air is blown in carefully using a pipette to record the size and shape of the lung. Already with the naked eye or under slight magnification, the numerous blood vessels in the tissue of the lung can be easily examined (Fig. 6.12a, b). For breathing, the snail lowers the bottom of the mantle cavity when the pneumostome is open, creating a negative pressure—oxygen-rich air flows in. Gas exchange takes place at the roof of the mantle cavity. Oxygen can diffuse there through the respiratory cells of the lung epithelium into the hemolymph of the pulmonary blood plexus. In the opposite direction, carbon dioxide diffuses from the hemolymph into the air of the mantle cavity. As a respiratory protein, snails possess hemocyanin dissolved in the blood, with copper as the central atom of the prosthetic heme group. Hemocyanin is the second most common oxygen transporter in Metazoa after hemoglobin and is widely spread in arthropods and mollusks. In addition, the blood contains various cells, referred to as hemocytes and blood cells, responsible for immune defense.

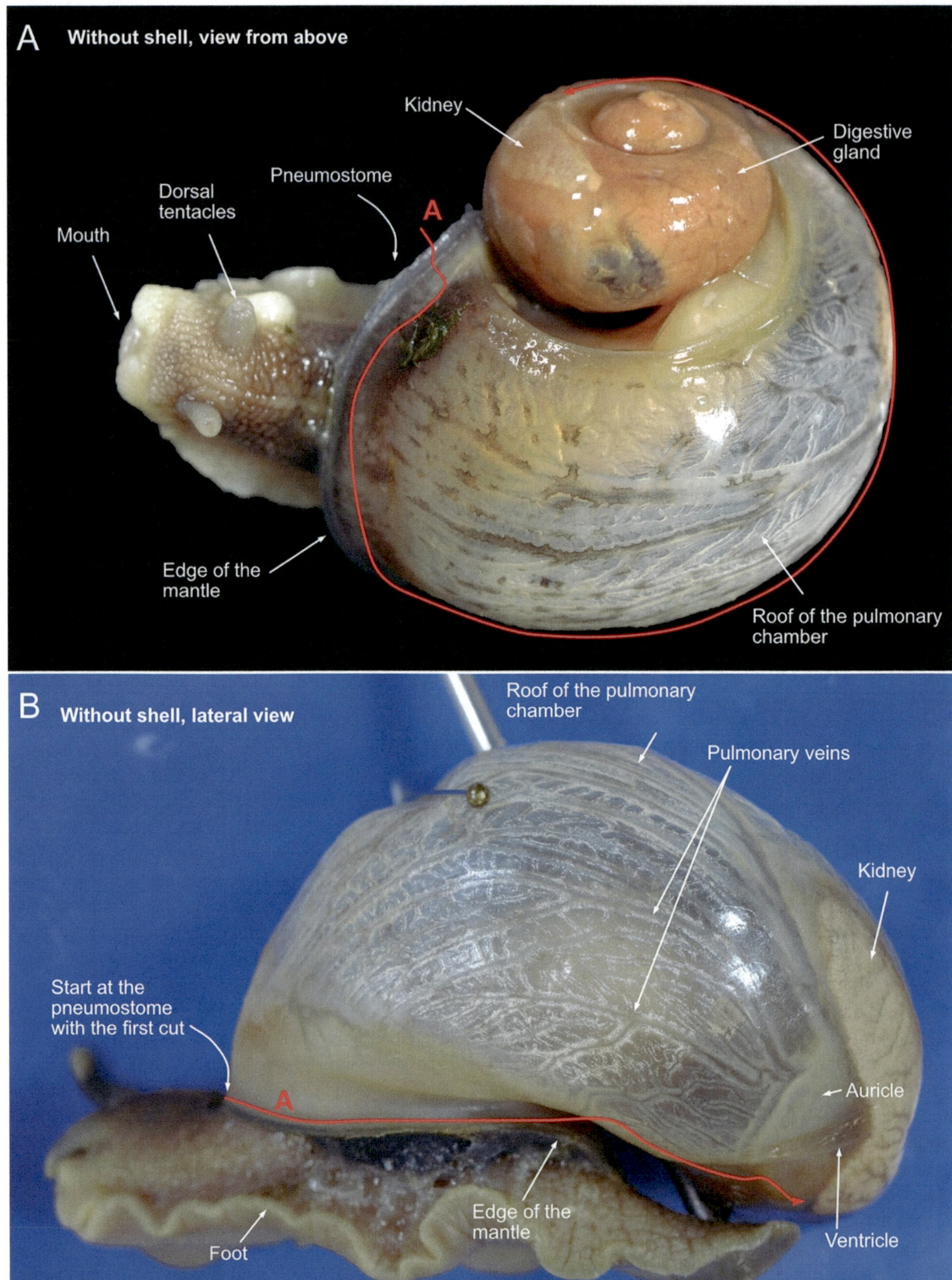

Fig. 6.12 *Helix aspersa*. (**a**) After removal of the shell, dorsal view. (**b**) After the removal of the shell, lateral view, looking at the roof of the mantle cavity

The heart is located at the posterior end of the mantle cavity roof on the left side (Fig. 6.12b). It consists of an atrium (also called auricle) and a ventricle with a valve in between. The heart is located in a fluid-filled coelomic chamber, the pericardium, also called the pericardial sac. The pericardium of the snail is bipartite and forms the posterior pair of the two coelomic cavities. It directly connects with the renal lumen via the renopericardial duct (Fig. 6.12a).

The atrial chamber of the heart receives the oxygen-rich blood from the pulmonary veins of the mantle cavity and passes it on to the ventricle. From there, the blood first enters the unpaired anterior and posterior aortae, branching into smaller arteries and finally leading to the so-called blood sinuses. The sinuses are irregularly shaped cavities delineated by ECM. They are mainly located in the body's connective tissue, but also reach the epidermis and gastrodermis. The back-flowing blood passes through the nephridia and the lung. All blood spaces are lined by ECM, as is the lumen of the heart and the aorta. Such a vascular system with extensive sinuses is often incorrectly called "open" in several textbooks, although coelom and blood spaces are separated. The decisive criterion for a true open vascular system is the lack of separation of the secondary body cavity (coelom) from the blood vascular system (the remnants of the primary body cavity). Such an open circulatory system is found, for example, in insects and all arthropods, and is there also called hemocoel. Thus, like all mollusks, the Roman and the Garden snails have a closed circulatory system. Furthermore, the vascular system of land snails is even extended into distinct vessels. Invertebrate blood vessels are usually lined by ECM only; a distinct cellular endothelium is restricted to vertebrates.

Renal System, Kidney

The elongated, faintly yellowish kidney can be seen directly next to the heart (Fig. 6.12b). Excretion is like that in all mollusks. The primary urine is formed through ultrafiltration at podocytes, creating part of the walls of the atrium and ventricle and entering the pericardial cavity. The renopericardial duct starts with a nephrostome at the pericardial cavity and leads to the nephridium (kidney sac or simply kidney). The primary urine is modified through selective absorption and secretion by the nephridial epithelium (Fig. 6.18), where it is converted into the final urine. The urine is released through the ureter, which opens near the anus and pneumostome. This long secondary ureter is another adaptation of our snail to the terrestrial environment, because a release of urine into the mantle cavity near the kidney (as in non-terrestrial gastropods) is impossible due to the lack of water inside the mantle cavity. As in many animals whose living conditions are characterized by water shortage, the kidney produces uric acid, which is excreted as a whitish mass or stored in the kidney in case of severe water shortage. Such a situation occurs for the snails during drought periods or hibernation, which they spend buried in their sealed shell in the ground.

The uric acid is stored in large quantities in the kidney during this time. In histological sections, this is visible due to numerous granules (Fig. 6.18).

> ***Helix*: The Final Coils**
> Using a probe and a pair of tweezers, the last coils of the visceral mass are unrolled. The bipotential gonad, the ovotestis, the combined vas deferens/oviduct, and the albumin gland become visible from the outside.

Ovotestis

Before the visceral mass is opened and the internal organs are examined in more detail, a look should first be taken at the intact visceral mass (Fig. 6.13a, b). The whitish ovotestis differentiates as a bipotent gonad, in which the maturation of sperm and eggs takes place simultaneously (Fig. 6.14). Garden and Roman snails are simultaneous hermaphrodites that fertilize each other. The gametes, which mature in the ovotestis, are discharged via a combined vas deferense/oviduct (Figs. 6.15, 6.16, and 6.17).

> ***Helix*: Opening the Mantle Cavity**
> (a) The tip of a pair of dissecting scissors is inserted into the respiratory opening, and the mantle margin is cut. If the lung were still filled with air, it would collapse at this point. Now cut to the right along the mantle margin until just before the tip of the kidney (Figs. 6.12a, b and 6.13a, red cut line A). In doing so, we cut the large aorta leading away from the heart.
> (b) The tissue along the kidney is very delicate (Fig. 6.13b, yellow cut line B). Using two fine tweezers for dissection is better in this instance than a scalpel. Avoid damaging the internal organs. By this point, at the latest, the dissecting dish should be filled with water. This causes the organs to float and makes precise dissection easier.
> (c) The mantle cavity is opened along the coil up to the end using two forceps. The lung roof can be placed aside and pinned down.

After Opening the Mantle Cavity

First, the dissected snail is pinned down in the dissecting dish in the anterior and posterior parts of the foot. Then the lung epithelium can be folded aside and pinned down in place (Fig. 6.14). The position of the heart and kidney and the venous vessels leading to the heart are easily visible.

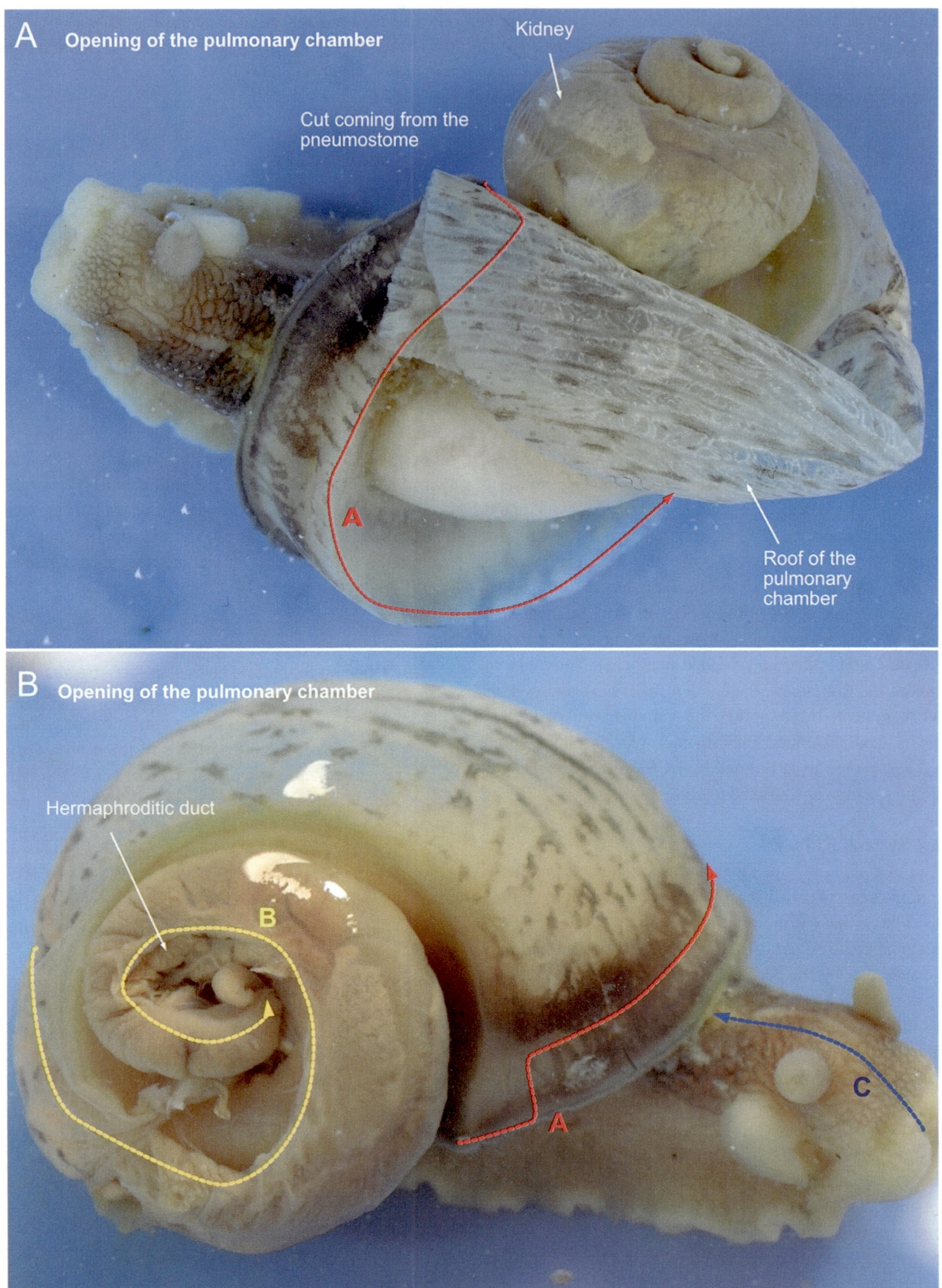

Fig. 6.13 *Helix aspersa*. (**a**) Opening of the mantle cavity, dorsal view. (**b**) Opening of the mantle cavity, lateral view from the right

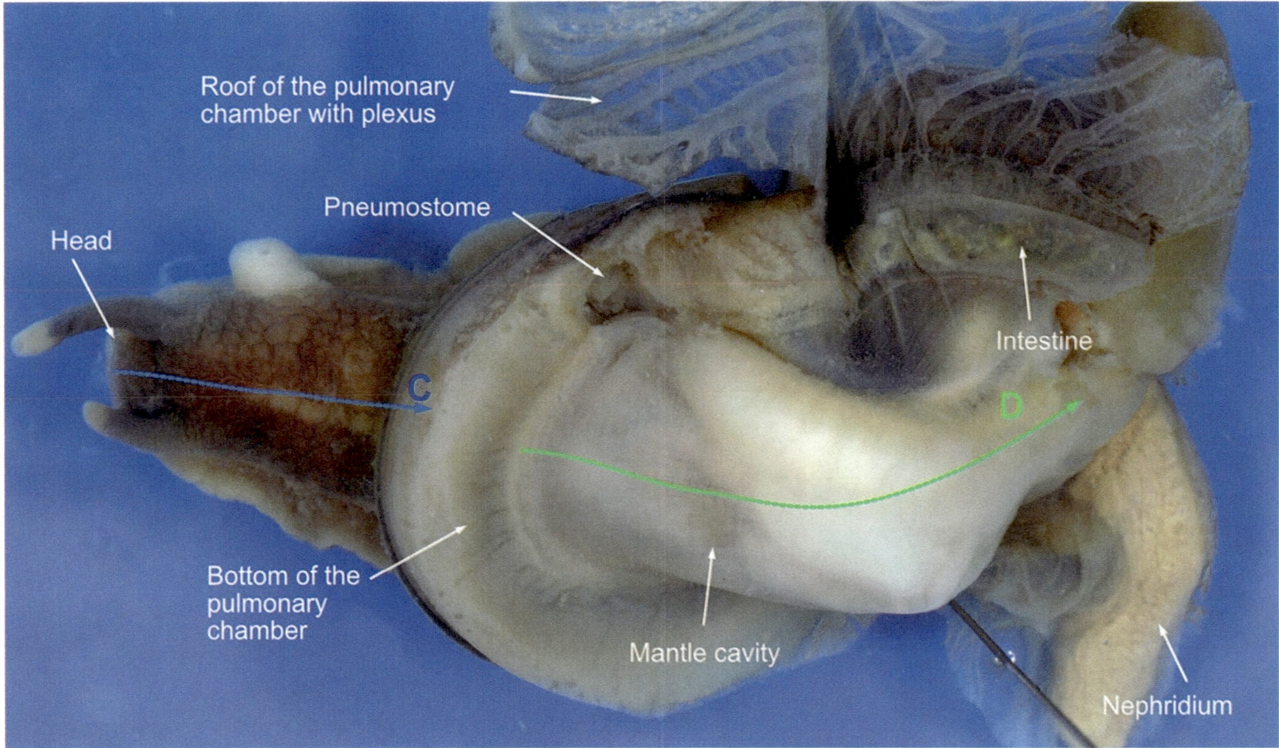

Fig. 6.14 *Helix aspersa*. Opening of the visceral mass (visceropallium)

Helix: Opening of the Visceral Mass

(a) With careful preparation, the visceral mass remains undamaged (Fig. 6.14). To open the visceropallium, first place the scissors' edge in the snail's mouth and cut dorsally median between the eye tentacles up to the mantle margin (Figs. 6.13b and 6.14, blue cut line C). The mantle margin is cut through. The scissors must always be flat to avoid damaging internal organs.

(b) The cut is continued medially over the visceral mass (Fig. 6.14, green cutline D). The sides of the body are pinned down with needles.

(c) All internal organs can be freed from the spirally twisted visceral mass by careful plucking, pulling, and pushing with forceps and probes. Work carefully, with the specimen always under water, to avoid organ damage. The organ systems are spread out in the dissecting dish (Figs. 6.15 and 6.16). Special attention must be paid to all the tubular connections among the organs.

(d) Next, expose the reproductive organs and pin them down.

(e) Then expose the digestive tract and fix it with needles.

Digestive Tract

The digestive system begins with the mouth and the adjoining chitinous dorsal jaw, with the radula on the ventral side. The radula lies on the muscular radula cushion. Like a conveyor belt, the radula moves back and forth, rasping and breaking food into small chunks. The radula represents the lower jaw, so to speak, and can be exposed and microscoped with a circular cut (Fig. 6.19). The radula wears away continuously at the tip and is constantly renewed at the back in the so-called radula sac. Like in sharks, the teeth are continually replaced according to the degree of wear. Next comes the crop and the stomach. The ducts of the paired salivary glands open into the crop. They produce the enzymes necessary for predigesting. The food passes from the stomach into the midgut, and then to the hindgut. The digestive gland, whose function has already been discussed in detail in the case of mussels, opens into the midgut and ensures digestion and absorption of nutrients. Food residues pass from the midgut to the hindgut, and finally through the rectum to the anus, which is located next to the respiratory opening.

Reproductive Organs

Roman and Garden snails are simultaneous hermaphrodites, mating mutually after several hours of copulation (Fig. 6.20). Copulation is usually preceded by a courtship involving circling, tentacular and oral contacts, and intertwining. For

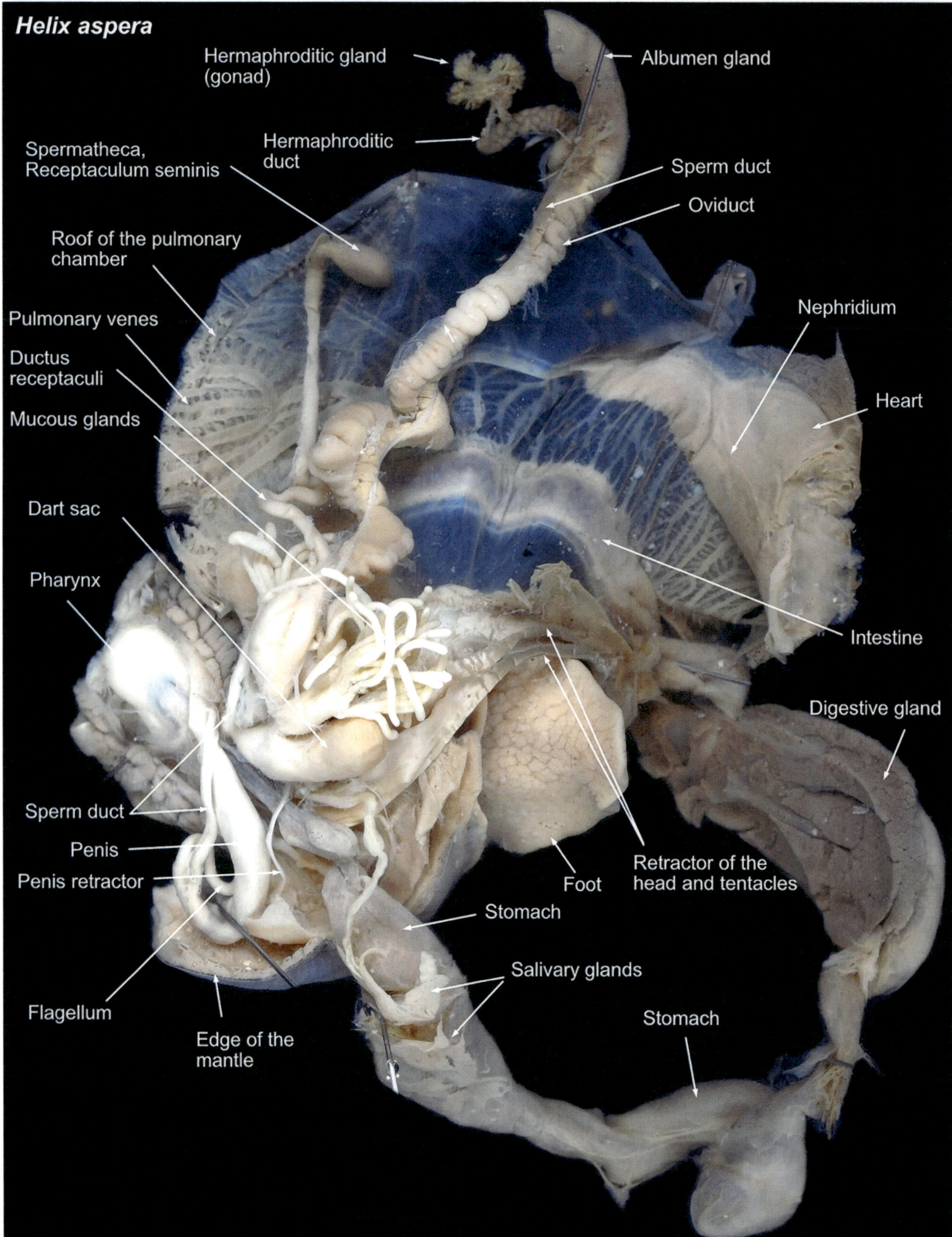

Fig. 6.15 *Helix aspersa*, Garden Snail. Prepared and organs laid out

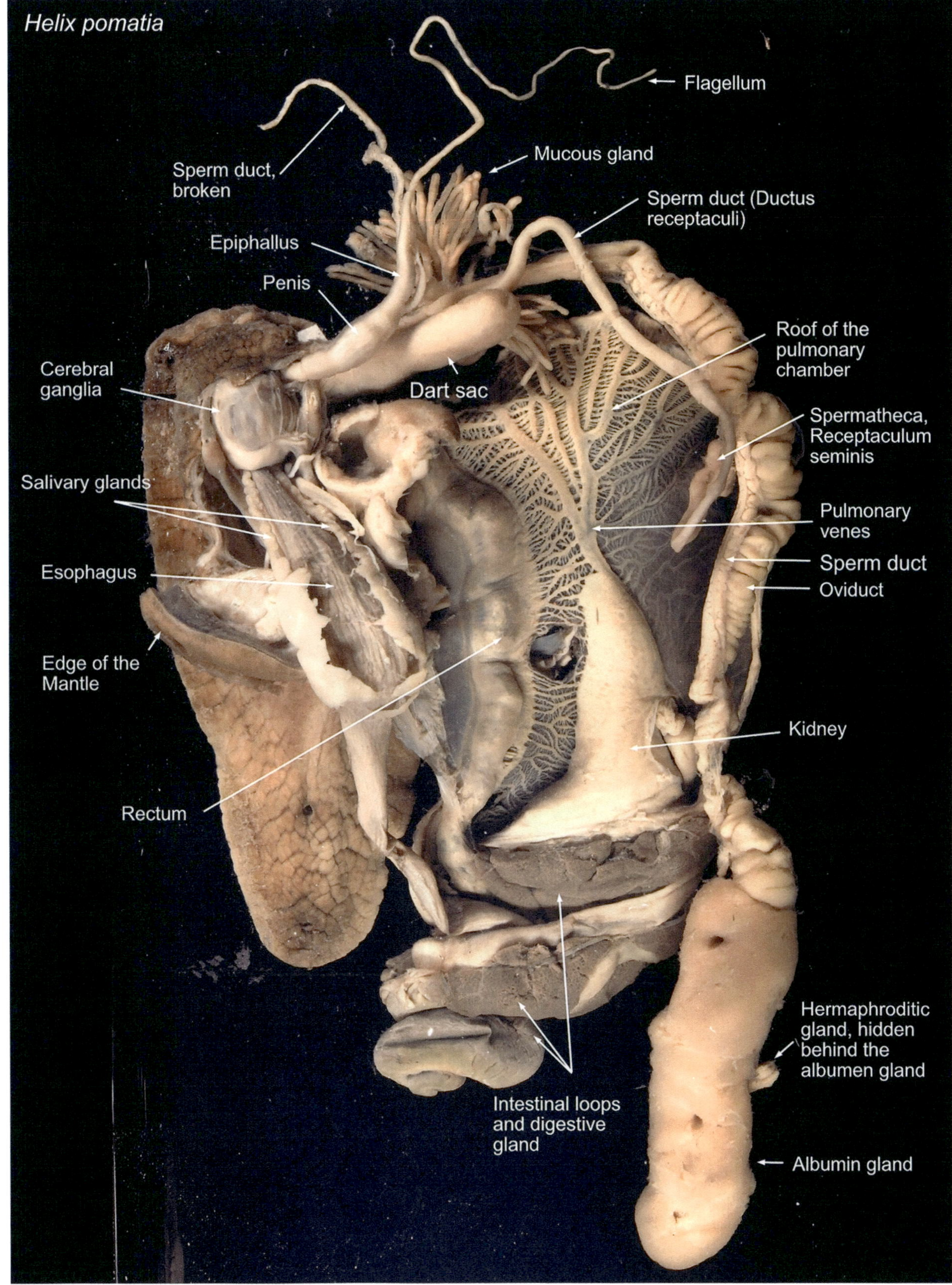

Fig. 6.16 *Helix pomatia.* Prepared and organs laid out. Wet specimen from the Biological Collection of the University of Osnabrück

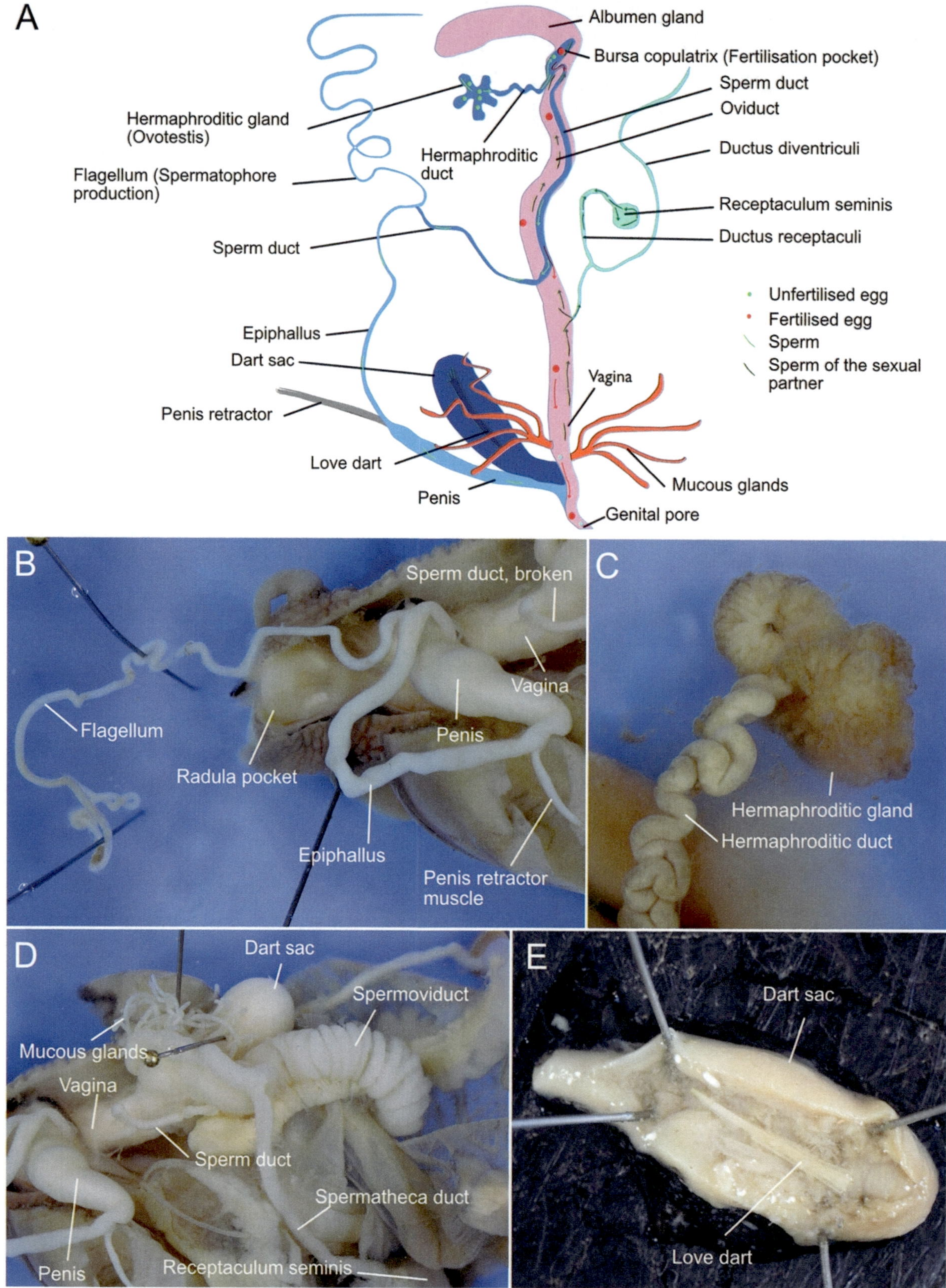

Fig. 6.17 *Helix aspersa.* (**a**) Schematic representation of the reproductive system, according to various authors. (**b**) Penis, flagellum. (**c**) Ovotestis and the combined oviduct/sperm duct. (**d**) Oviduct/sperm duct, dart sac, finger-shaped mucous gland. (**e**) Love dart

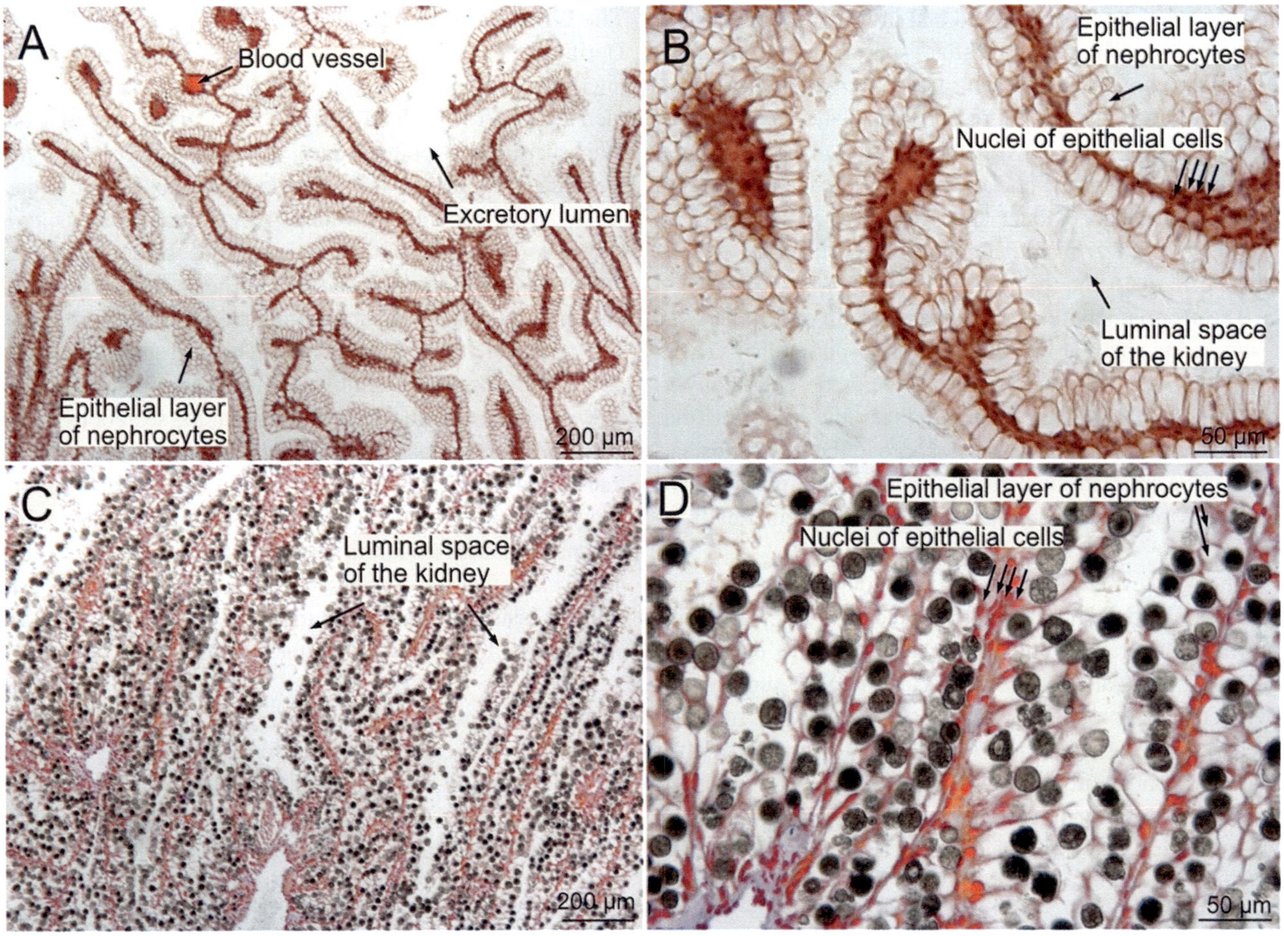

Fig. 6.18 *Helix pomatia*. Histology. (**a**) Cross-section through a summer kidney. (**b**) Detail. (**c**) Cross-section through a winter kidney. (**d**) Detail. Preparations of the Zoological Teaching Collection, Philipps University Marburg

Fig. 6.19 *Helix aspersa*. Scanning electron microscopic image of a radula. Image by Ekaterini Psathaki, University of Osnabrück

copulation, the penis is everted and inserted into the vagina of the partner. The snails transfer a spermatophore, a gelatinous mass containing numerous spermatozoa. The sperm are passed through the ductus receptaculi to the receptaculum seminis (sperm pouch), where they are initially stored. Later, they leave the seminal pocket, travel up the hermaphroditic duct, a combined oviduct and seminal duct, and fertilize oocytes in the bursa copulatrix (fertilization pouch) (Figs. 6.15, 6.16 and 6.17).

The spermatophore contains sperm originating in the ovotestis during spermatogenesis and is transported first via the hermaphrodite duct, then via the vas deferens, and finally to the flagellum. The flagellum produces the secretion that encloses the sperm in a spermatophore in the epiphallus and penis. The intricate anatomical details of the reproductive apparatus of the Garden snail are shown schematically in Fig. 6.17a.

The oocytes formed in the ovotestis during oogenesis are fertilized in the bursa copulatrix and then supplied with nutritive material by the albumin gland. The eggs later acquire a thin calcareous shell as they migrate through the oviduct. The reproductive organs can be well exposed in the preparation (Figs. 6.15 and 6.16). Four to six weeks after mating, 50–60 approximately 5 mm large eggs are placed in a hole in the ground. The development is direct, and after about 2 weeks, the young snails hatch (Figs. 6.17, 6.18, 6.19, 6.20).

The Love Dart

During the early phase of copulation, when the partners are intertwined, Roman and Garden snails usually make use of their love dart (Fig. 6.20). This is a calcareous, up to 1 cm long sharp-edged pointed dart, which also occurs in other Helicidae (Fig. 6.17e). It is secreted into a muscular love dart sac and, together with pheromone-containing secretions, produced by the finger-shaped mucous glands. Upon copulation, the love dart is forced into the foot of the partner animal. The exact role of this mechanical stimulation and the pheromone transfer is not yet fully understood. The pheromones are supposed to break down existing sperm from previous matings. However, recent studies have shown that the secretions of the finger-shaped mucous gland prevent the breakdown of the transferred sperm. Presumably, it is both, removing old sperm and protecting the new ones. Other observations suggest that the snail that uses its dart first is stimulated to copulate, while its partner often remains passive. This led to the hypothesis that this behavior has evolved to induce one individual to act as a male and the other as a female.

Nervous System and Sensory Organs

Within the framework of an introductory zoology course, only the two ring-shaped cerebral ganglia, which lie on the esophagus, can be presented. From here, nerve cords extend to the eye stalks and the anterior tactile tentacles, as well as to the trunk ganglia. Our movie clip collection, which includes movies showing living snails, can be found at figshare: sn.pub/xr9qhi.

Fig. 6.20 Reproduction, *Helix aspersa and Helix pomatia* (**a**) Two copulating Roman snails (*Helix pomatia*). (**b**) *Helix aspersa* with everted penis. (**c**) Copulating Garden snails (*Helix aspersa*). (**d**) *Helix aspersa*. Penis and genital pore

▶ **Summary** **Nematodes are bilaterally symmetrical, unsegmented animals without a coelom. Their body cavity is the primary body cavity, which is not lined by epithelia but an extracellular matrix (ECM). Nematodes have colonized all conceivable habitats, including the deep sea, the soils of aquatic and terrestrial biotopes, and, as parasites, the body cavities or tissues of animals and plants. Some researchers estimate that there may be as many as one million nematode species on Earth. However, only about 80,000 species have been taxonomically recorded so far. Although not the most species-rich group, nematodes are considered the most abundant metazoan animals on the planet.**

One square meter of mud in the Wadden Sea may contain more than 4 million nematodes. The majority are tiny and often only visible under a microscope. They feed on organic matter (saprobiontic) or microorganisms like bacteria. Only a few free-living species are predatory, while parasites are common. Due to their preferred lifestyle, almost all species can tolerate a lack of oxygen very well. Some species even occur in the oxygen-free (anoxic) reduced black layer of marine sediments, characterized by its high content of toxic hydrogen sulfide. Tolerance of oxygen deficiency and other harsh conditions explains why their transition to an intestinal parasitic lifestyle was relatively easy. It likely occurred independently several times in evolution, because the intestinal lumen of their hosts is also anoxic. Several parasitic nematode species can be clinically relevant. It is estimated that about 90% of all humans become infected by parasitic nematodes at some point in their life. Without exception, all larger nematode species are adapted to parasitic life. For example, the 30–40 cm long roundworms *Ascaris suum* (large roundworm of pig) and *Ascaris lumbricoides* (human roundworm) live under anaerobic conditions in the small intestine of large vertebrates, including pigs, cattle, sheep, and humans. They feed exclusively on the contents of the intestine. With a body length of up to 9 m and a diameter of 2.5 cm, the females of *Placentonema gigantissima* are the largest known nematode representatives. *Placentonema* is an extreme habitat specialist found exclusively in the placenta of sperm whales. Very little is known about the reproduction of these animals. It is suspected that the juvenile stages of *Placentonema* can already be found in unborn whale calves and that this is how they reach their next host.

Nematoda (roundworms) are an important subgroup of a diverse taxon called Cycloneuralia. The name refers to a common characteristic, the ring-shaped brain. Arthropoda (arthropods) and Cycloneuralia form the two main groups of a particular taxon, the Ecdysozoa. They all have a solid cuticle formed by the epidermis that is regularly molted during growth. While the cuticle is essentially made up of chitin in arthropods, it mainly comprises collagens in nematodes. In addition to these two main components, numerous proteins and lipids are incorporated into the cuticle, which is essential for determining the physical properties of the animal's outer cover. Due to the enormous variety of habitats in which nematodes colonize, the cuticle has adapted to the conditions of the respective environment. However, the advantageous solid cuticle of arthropods and nematodes means continuous growth is impossible in arthropods, and only minimal in nematodes. Therefore, molting (ecdysis) always accompanies these animals' growth during their individual development. In other words, the animals form a new cuticle for each further growth stage and then strip off their old one. This is easily observable in crustaceans, such as crayfish and lobsters, which grow throughout their lives by recurrent molting cycles. In nematodes, however, the molts only occur during the juvenile stages, and molting ceases after adulthood has been reached. Like all nematodes, *Ascaris suum* passes through four juvenile stages and molts four times. After the last molt, the animals live in the small intestine of their hosts, but continue to grow to lengths of up to 30 or 40 cm (Fig. 7.1). The absence of chitin and the corresponding proteins (arthropodin) in the cuticle of the nematodes may be the reason why

A. Paululat, G. Purschke, *Metazoa – Morphology and Evolution of Animals*, https://doi.org/10.1007/978-3-662-69904-1_7

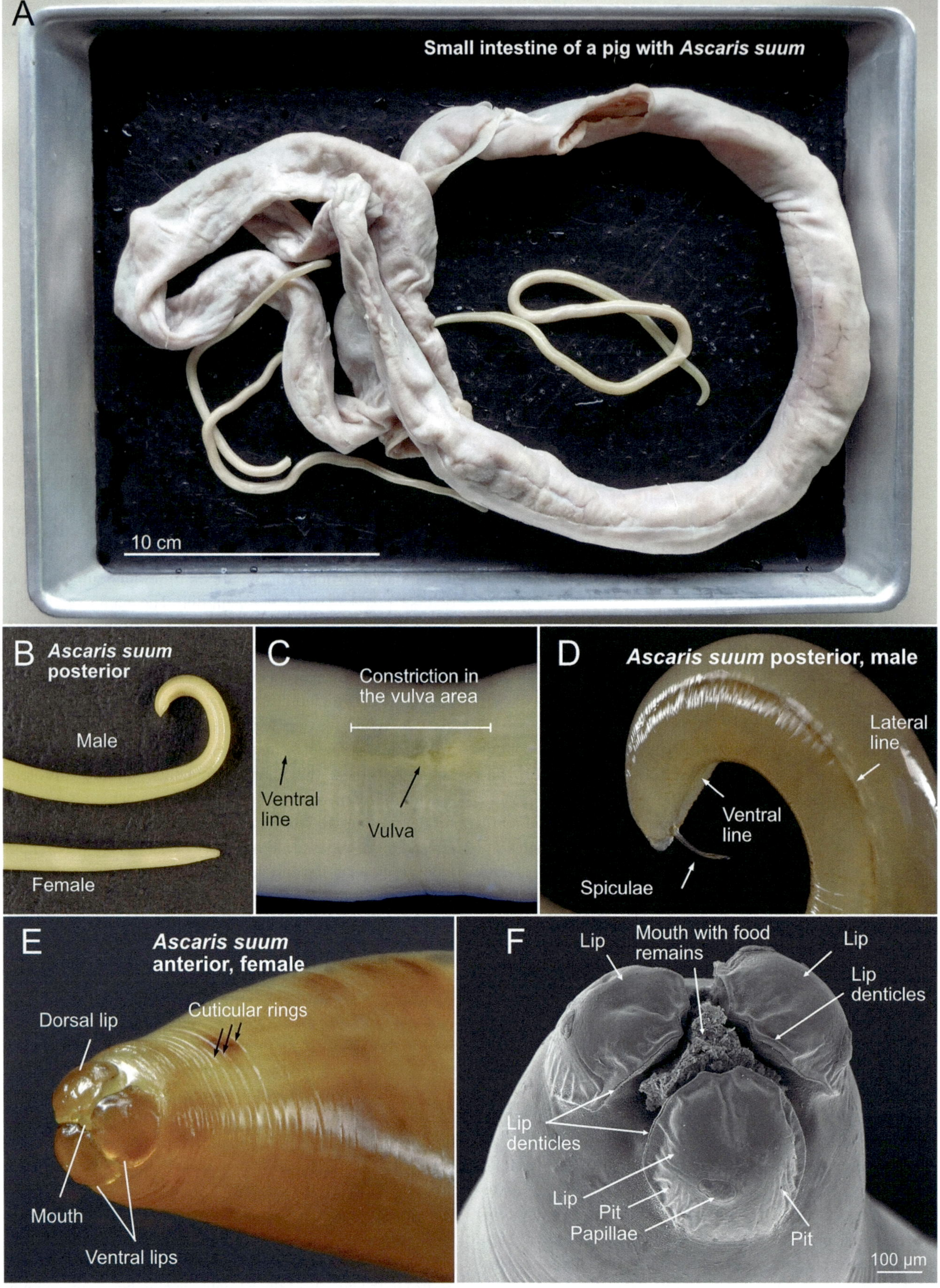

Fig. 7.1 *Ascaris suum*. (**a**) The small intestine of a pig with severe roundworm infestation. (**b**) Male *Ascaris* can be identified by the spirally curved body end. (**c**) Close-up of the female genital opening. (**d**) The posterior end of a male with spiculae. (**e**) Three lips surround the central mouth opening. (**f**) Scanning electron micrograph of the mouth of *Ascaris suum*. SEM image captured by Ekatherini O. Psathaki, University of Osnabrück

the animals can still increase in size after their last molt. The formation of the cuticle results in another peculiarity. The cells of the epidermis lack motile cilia in nematodes and arthropods. Also, the sperm of nematodes are flagellum-less and move in an amoeboid manner. The only cells with immobile cilia in nematodes are the receptor cells of the sensilla.

Recommended Material
- **How to get *Ascaris*:** The veterinarians at local pig slaughterhouses often set aside roundworms for student courses. After being "rescued" from the pig's small intestine, the animals are fixed in 4% formaldehyde and stored in ethanol. Unfortunately, the animals become very hard and brittle due to fixation and storage. This makes the preparation more difficult. The hardening can be alleviated if the water that the animals are soaked in for 1 to 2 days is frequently changed. Even better results are obtained if the animals are ultimately placed in boiling water for a few minutes. The "Kükenthal" recommended fixation in 80% isobutanol to avoid hardening of the animal (Storch V & Welsch U, 2014).
- **Histological sections:** Sections from different body regions should be used in addition in the student courses.

External Examination of Ascaris suum
Like all nematodes, *Ascaris suum* has a thread-like elongated body with a round cross-section tapering at both ends (Fig. 7.1a). The males are overall thinner and smaller than the females, and posteriorly, their body forms a hook (Figs. 7.1b, d and 7.2). In females, at the end of the first third of the body length, roughly, a slight annular constriction marks the ventral opening of the vulva (Figs. 7.1c and 7.3). The anterior end is characterized by three lips that enclose the terminal mouth opening (Fig. 7.1e). The animal's side lateral ridges can be felt when palpating the body. Viewing histological sections, it will become clear that these ridges form a muscle-free epidermal cord, each harboring an excretory canal (Figs. 7.4, 7.5, 7.6, 7.7, and 7.8). Along the anterior-posterior axis, two other epidermal cords are present, one ventrally and one dorsally. These can be felt with fingertips as slight elevations of the cuticle. A single nerve cord runs through each of these two epidermal ridges.

The body wall of the nematodes consists of a cuticle, epidermis, and longitudinal muscles, followed by the fluid-filled primary body cavity. The body cavity is under high internal pressure. Combined with the solid cuticle, it leads to a round and highly stable body shape. The body cavity thus forms a uniform, tubular hydroskeleton against which the muscles work. As already mentioned, the musculature of the body wall lacks circular muscle fibers and is only composed of longitudinal fibers, which are arranged in four groups, each separated from one another by an epidermal cord. The dorsal nerve cord innervates the muscle cells of the dorsal body side. The muscle cells of the ventral body side are innervated by the ventral nerve cord. Thus, all muscles are divided into two functional groups that contract alternately during locomotion. These contractions cause the typical wriggling movement of nematodes. These two functional muscle cell groups act antagonistically. Since there is a ventral and a dorsal muscle group, the body inevitably falls to one side and pushes itself off the substrate, causing the animal to move forward. In some textbooks, it is stated that the antagonist of the muscles is the internal body pressure or the cuticle. This is not correct.

Dissection of a roundworm after external examination
For dissection, the roundworm is pinned down at the anterior and posterior ends with needles in a dissection dish filled with water. If the animals have already burst when removed from the pig's small intestine or during fixation, one should start cutting at these points. Otherwise, carefully open the body, starting anteriorly or posteriorly, and slowly work your way longitudinally with scissors. The scissors must always be applied flat to avoid damaging any internal organ. Particular care must be taken near the vulva and anus to prevent damaging the ducts of the uteri, vas deferens, and intestine. During the dissection, the opened body sides are continuously pinned down with needles (Figs. 7.2 and 7.3).

Observing living nematodes in a petri dish without substrate, their movement looks awkward, as the animals need abutments. In contrast, nematodes move skillfully and effectively in soils, mud, sandy sediments, or the intestine. The way in which nematodes move using eel-like sinuous undulations of the body is reminiscent of how snakes or eel-like fish move around. However, the bending axes of nematodes and snakes are offset by 90 degrees. Nematodes crawl on the side of their body, thus bending in the dorsal-ventral plane and not in the sagittal plane as in vertebrates. Thus, snakes

Fig. 7.2 *Ascaris suum.* A dissected male displaying the internal organs

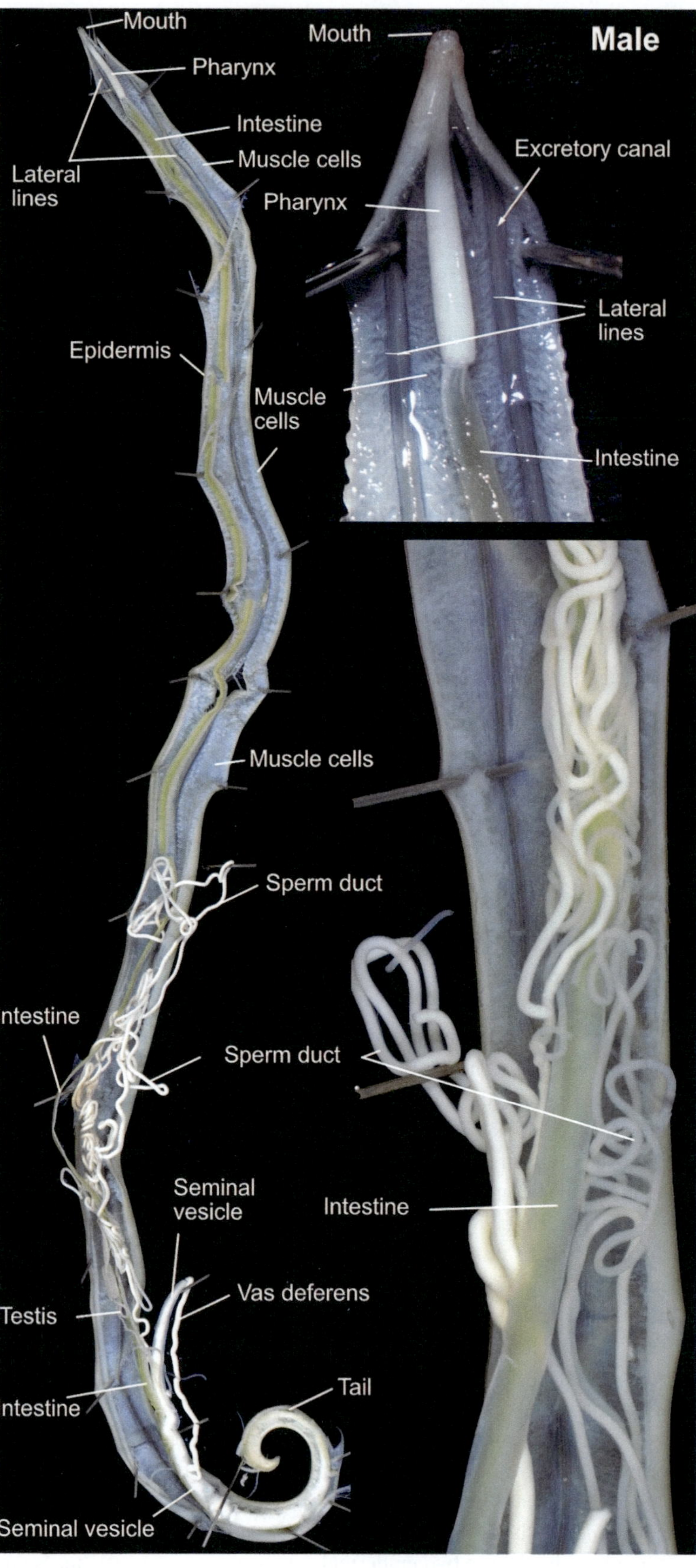

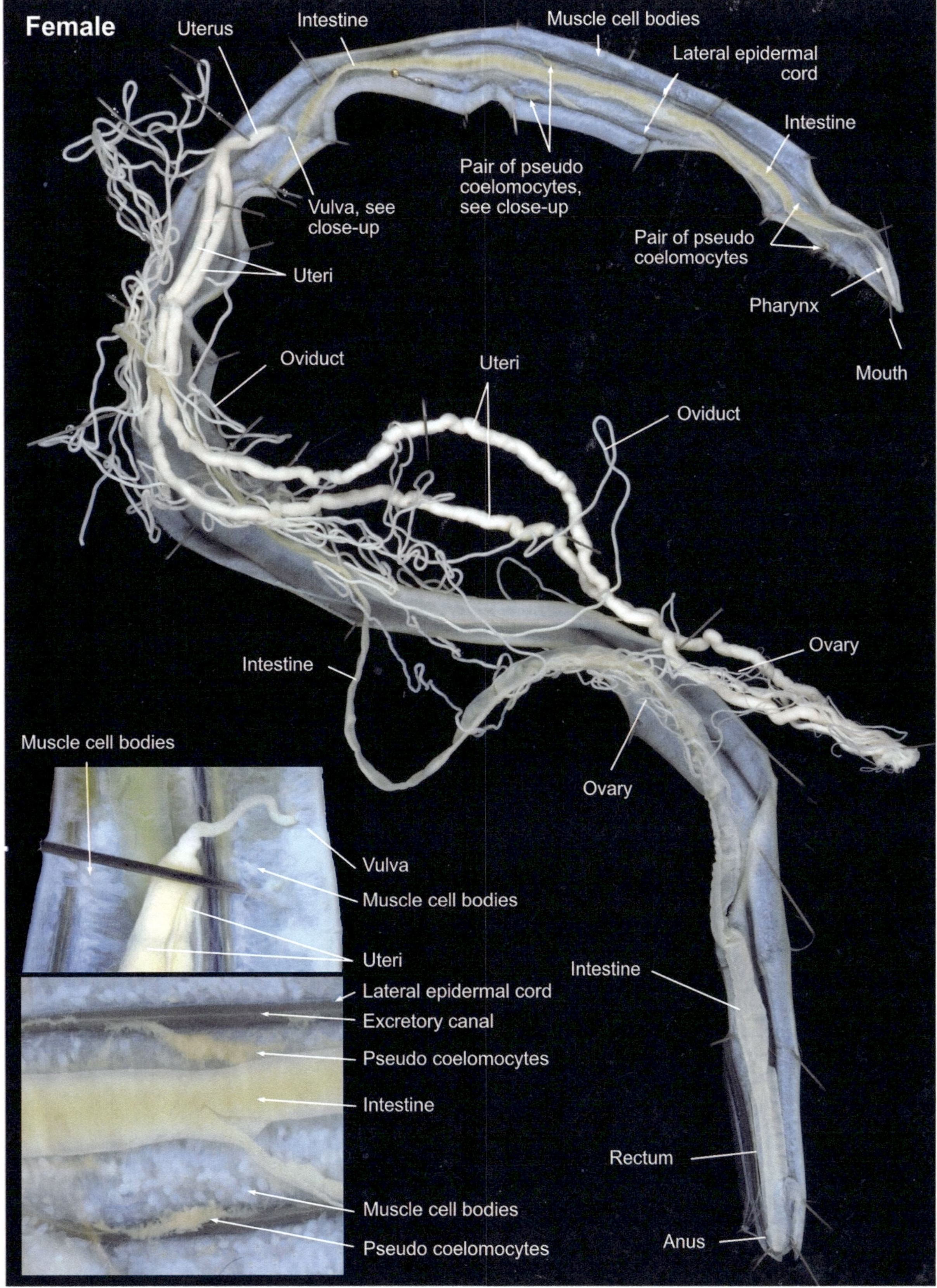

Fig. 7.3 *Ascaris suum.* Prepared female. This animal was fixed with formaldehyde and stored therein for more than a year

Ascaris suum Female

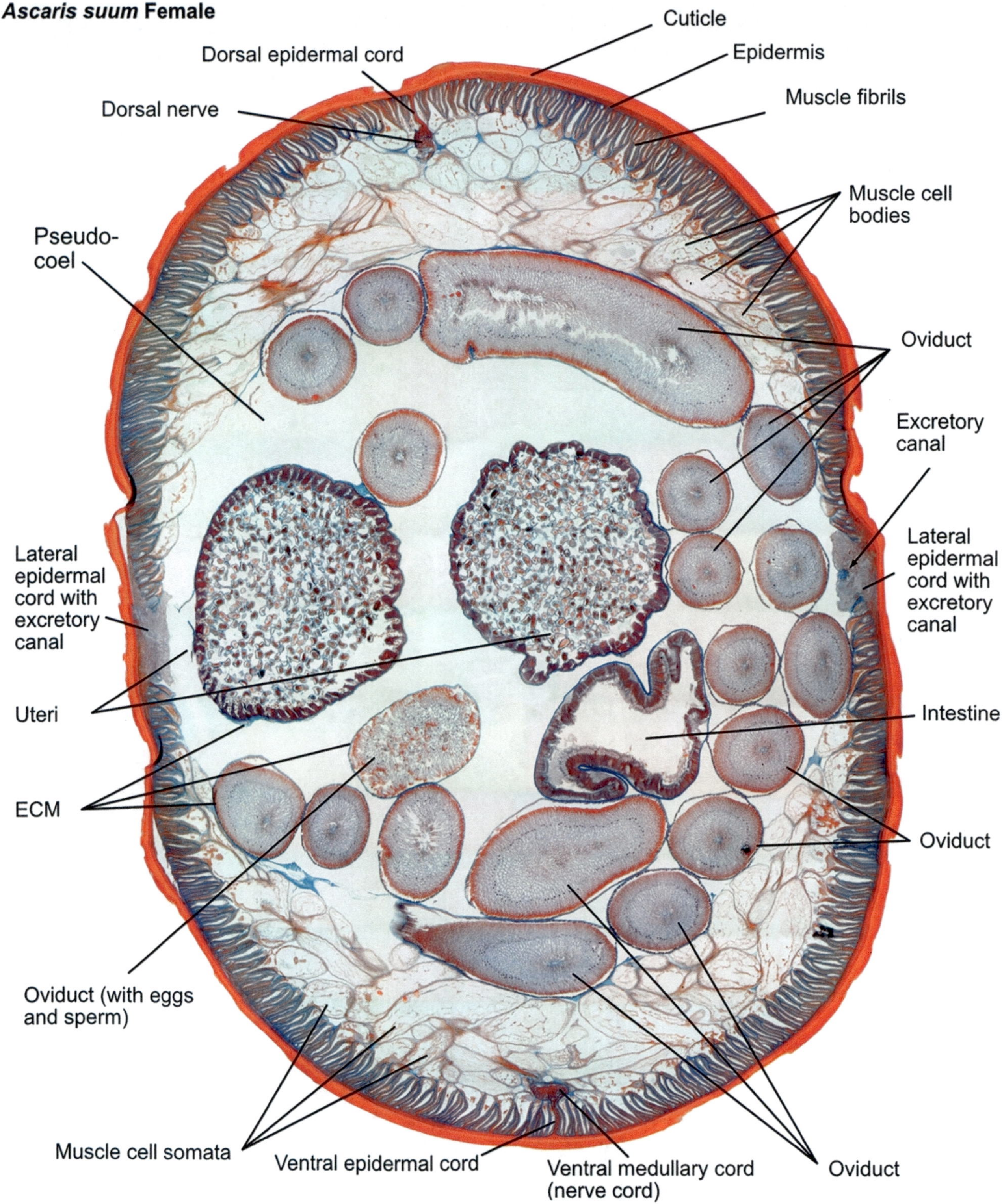

Fig. 7.4 _Ascaris suum._ Cross-section of a female, midbody region, hematoxylin/azan staining. Specimen from the Biological Collection of the University of Osnabrück

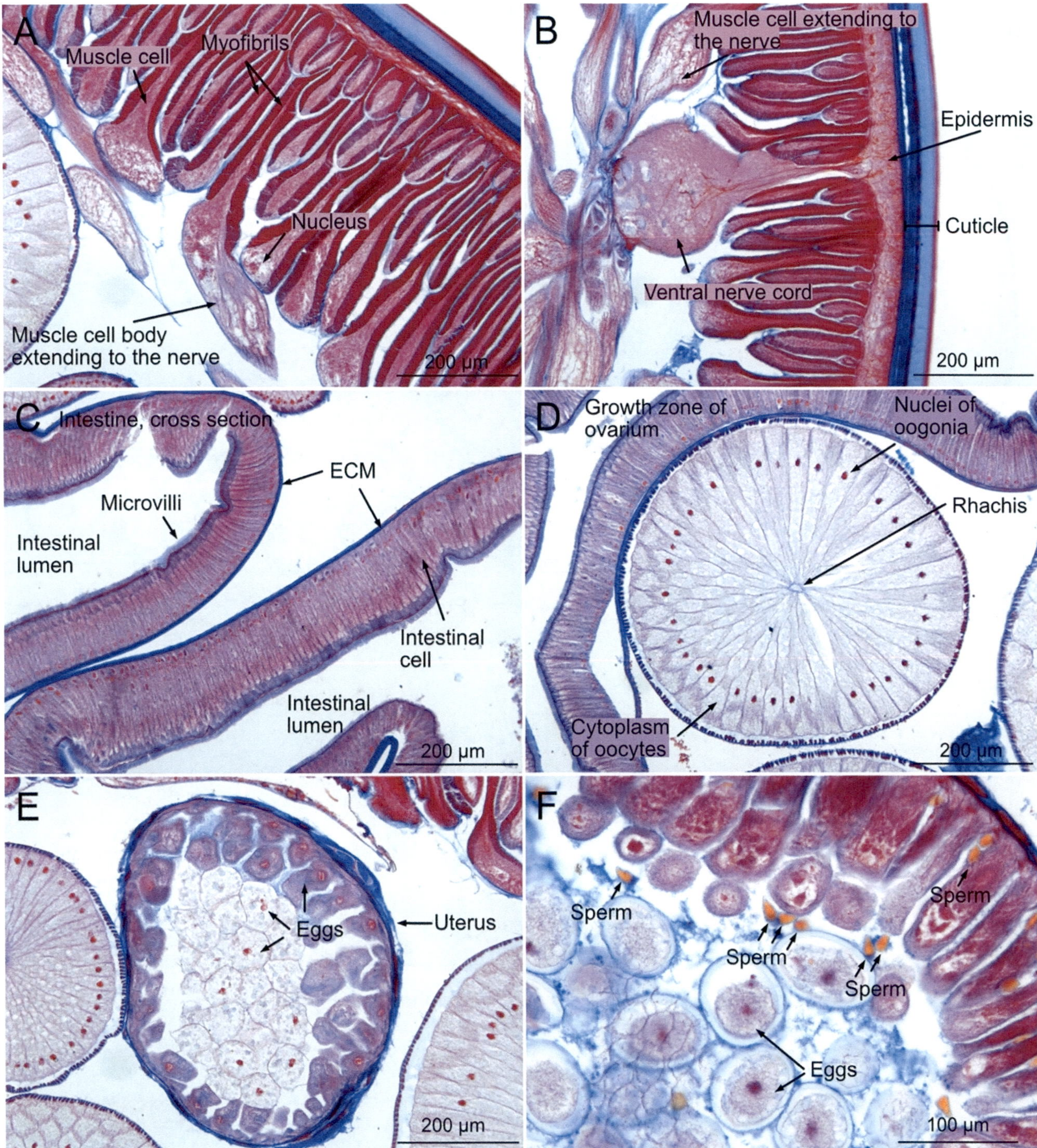

Fig. 7.5 *Ascaris suum*. Cross-section of a female, midbody region, hematoxylin/azan staining. (**a**) Cuticle, epidermis, musculature with myoneural synapses, extensions of the muscle cells, which project to the dorsal or ventral nerve cord. (**b**) Ventral nerve cord. (**c**) Intestine with immediate ECM coverage and microvilli-bearing intestinal cells. (**d**) Growth zone of the ovary with immature eggs. (**e**) Cross-section through an early uterine section with egg cells. (**f**) Uterine cross-section with differentiated sperm between eggs. Specimens from the Zoological Teaching Collection, Philipps University Marburg

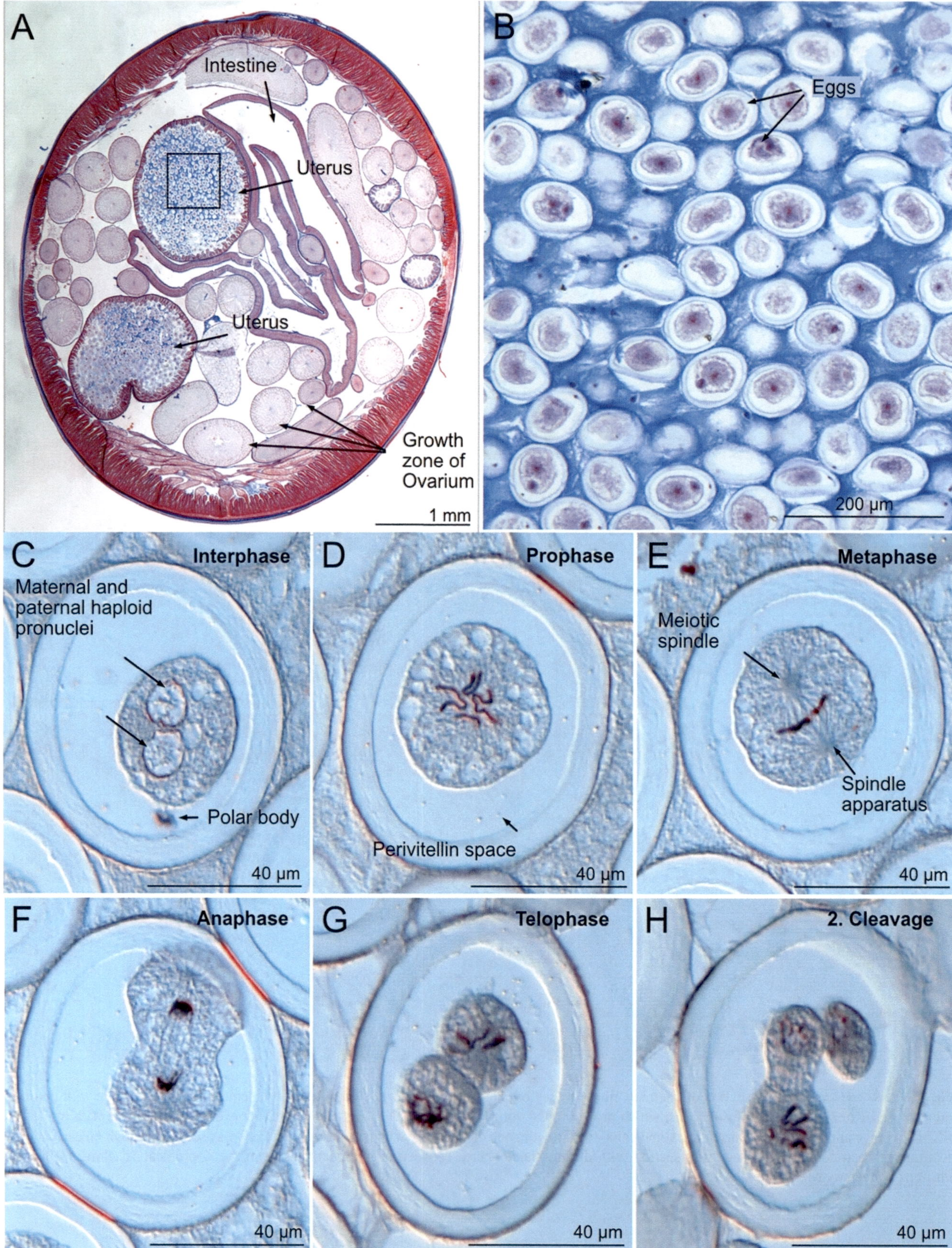

Fig. 7.6 *Ascaris suum*. Cross-section of a female. (**a**) Overview. (**b**) Uterus with eggs. (**c–h**) Meiotic stages, longitudinal section of the uterus. (**a**, **b**) Specimen from the Zoological Teaching Collection, Philipps University Marburg. (**c–h**) Specimens from the Biological Collection of the University of Osnabrück

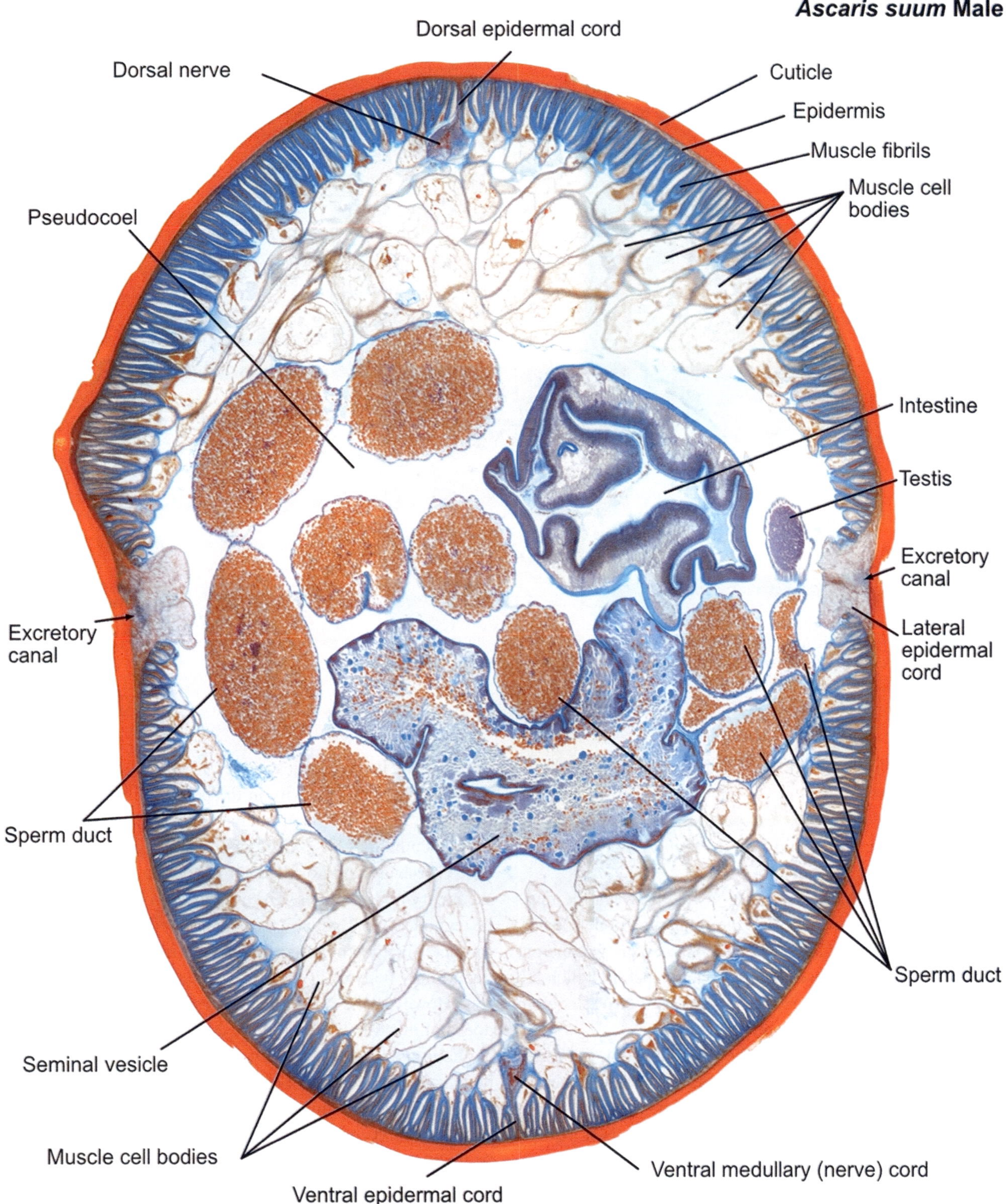

Fig. 7.7 *Ascaris suum.* Cross-section of a male, midbody region, hematoxylin/azan staining. Specimen from the Biological Collection of the University of Osnabrück

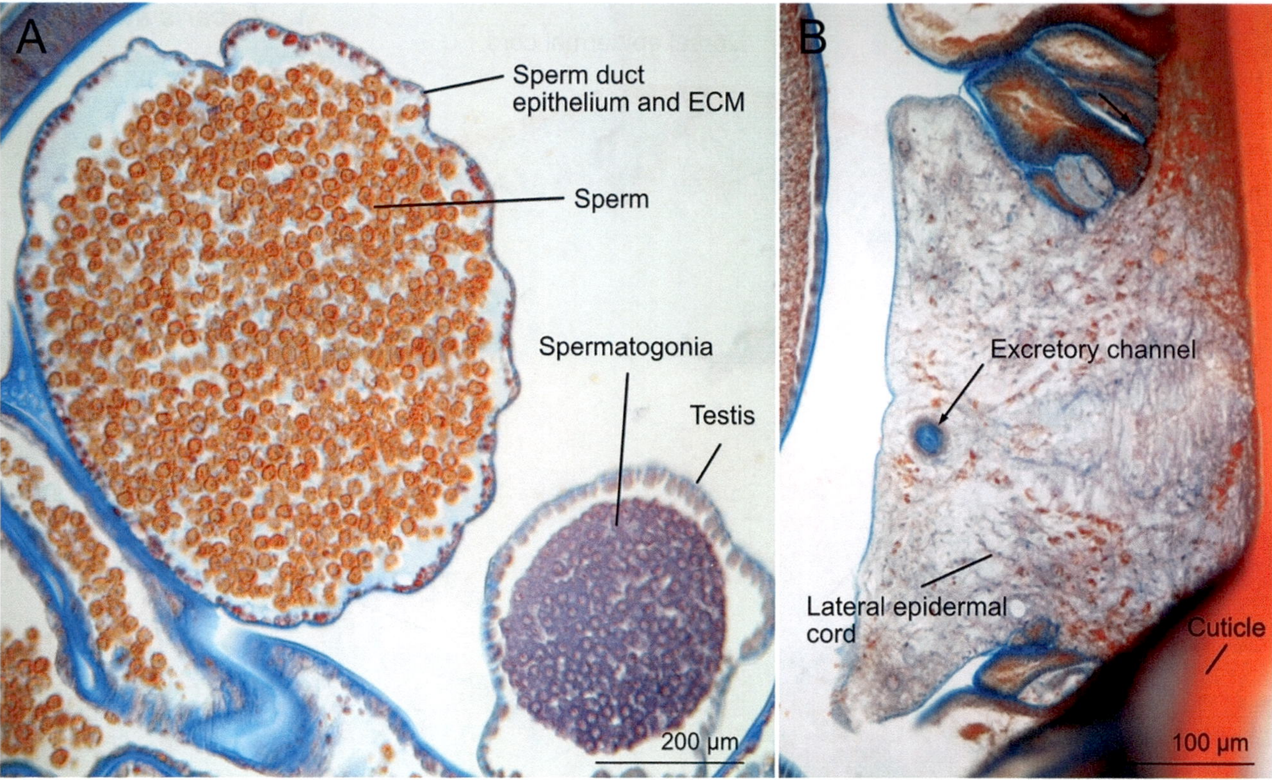

Fig. 7.8 *Ascaris suum*. Histology. (**a**) Cross-section of the vas deferens with sperm and of the testis with spermatogonia. (**b**) Lateral epidermal ridge with excretory canal. Specimen from the Biological Collection of the University of Osnabrück

and eel-like fish crawl on their belly, bending their bodies in the left-right axis.

The Pseudocoel: The Fluid-Filled Primary Body Cavity of Nematodes

There usually isn't any free space visible between the musculature and the gastrodermis in the small nematode species. This is different in the larger species. *Ascaris* has an apparent large body cavity, the pseudocoel or hemocoel, in which the internal organs are located. So, how does a "true" coelom differ from a pseudocoel found in *Ascaris*?

The secondary body cavity, the "true" coelom, is formed in all metazoans during embryonic development; either by folding from the dorsal primitive intestinal roof (in Deuterostomia) or as the result of the immigration of cells from the primordial mouth region into the primary body cavity, and the subsequent formation of solid cell bands that split into the mesodermal cavities (in Spiralia, a subgroup of Protostomia). Thus, a second new body cavity is lined by epithelia. During further ontogenesis, the mesoderm primarily gives rise to the musculature and parts of the circulatory system and, in vertebrates, the cartilaginous or bony skeleton. The coelom is a cavity in the body limited by mesodermal cells, either by flat epithelial cells, the coelothelium, or by myoepthelial cells. As a result, all coelomate metazoans have two separate fluid-filled compartments, the primary and secondary body cavities. The composition of their fluids can be regulated separately by the organisms due to the epithelial character of the coelomic lining. The remaining cavities of the primary body cavity become the lumina of the blood vessels. However, the fluid-filled body cavity of the nematodes does not meet the criteria for a secondary body cavity, as an epithelium does not enclose it. It corresponds to the primary body cavity, which is not compressed by musculature in the larger species. In nematodes, only the epidermis among the four epidermal cords is covered by longitudinal muscle cells. At the same time, the intestine, which runs freely in the body cavity, is only limited by an extracellular matrix (ECM) toward the body cavity (Figs. 7.3, 7.4 and 7.5). The same applies for the genital organs (Fig. 7.6). In nematodes, circular and longitudinal muscles usually attached to the intestinal wall in other metazoans are absent (Figs. 7.4 and 7.7). Also, the musculature is not organized in an epithelial manner, so the body cavity of the nematodes is only limited by ECM on all sides. Nematodes, therefore, do not have a coelom. Instead, they have a pseudocoel originating from the primary body cavity.

Longitudinal Muscle Cells, Nervous System, and Sensory Organs

In nematodes, the brain forms a ring that encircles the pharynx. A few nerves project anteriorly, connecting with various cephalic sensory organs. *Ascaris* has sensory papillae and sensillae at the mouth, with which nematodes perceive mechanical and chemical stimuli. Laterally, there are the so-called amphids or side organs, chemoreceptive organs, located in pit-shaped depressions of the epidermis. Only a very few free-living aquatic nematodes are known to have light sensory organs. They are situated as paired ocelli behind the amphids and above the pharynx or oral cavity.

Two large nerve cords are particularly well visible in histological preparations: the main longitudinal nerve cord on the ventral body side and the dorsal nerve cord. Both nerve cords project into the epidermal ridges and run from the anterior to the posterior end of the animal. Only the ventral nerve cord originates from the cerebral ganglion. This is not the case for the dorsal nerve cord, formed by neurites (nerve cell processes) of the ventral cord that extend laterally to the dorsal side. The ventral nerve cord bears somata along its entire length and is not separated into ganglia and connecting nerves; this is called a medullary cord. The ventral nerve cord includes motor and sensory fibers (Fig. 7.5b). In contrast, the dorsal nerve cord is purely motor, thus essentially coordinating the movement of the dorsal musculature. The entire nervous system is situated within the epidermis, called an intraepithelial nervous system.

Nematodes also possess stretch receptors, the so-called proprioceptive receptor cells found in the lateral epidermal ridges. They perceive the body curvature and, thus, control locomotion. Nematodes have only 250–302 neurons, regardless of their body size. All neurons are formed in a strictly defined manner during embryonic and post-embryonic development. Only a few metazoans have such simply structured nervous systems. Nematodes, therefore, represent a great model system and the nervous system of nematodes has been analyzed in great detail in terms of its structure and function.

The large, translucent longitudinal muscle cells, which are well visible under the stereomicroscope, lie on the inner side of the epidermis. They represent a peculiarity in Bilateria. Each muscle cell has a myofilament-containing contractile part anchored to the epidermis (Fig. 7.5a, b). An axon-like cell extension is in contact with the ventral or the dorsal medullary cord, respectively. This is very different from other neuromuscular systems in metazoans, with only a few more examples, like the one of the lancelet (*Branchiostoma lanceolatum*; see Chap. 10). Usually, neuromuscular connections are established by neurons that project to the muscle cells, not the other way around.

Digestive System

Ascaris, as a parasite, feeds on easily accessible nutrients found directly in the small intestines of its vertebrate hosts. The digestive tract is divided into a mouth and an oral cavity, a muscular pharynx, a long midgut, and a short rectum (Figs. 7.1, 7.2 and 7.3). The mouth is surrounded by one dorsal and two ventral lips (Fig. 7.1e). Depending on the diet of the species, the oral cavity is adapted accordingly and can be equipped with intricate tooth-like structures of various shapes. The pharynx is a single-layered tube of epitheliomuscular cells forming a triangular lumen. The myofilaments are arranged radially. Their contraction causes an expansion of the lumen, and thus pressure decreases and nutrients are sucked in. Flap-like valves are present between the pharynx and midgut and between the intestine and rectum. The midgut cells have a microvillar brush border for surface enlargement (Fig. 7.5c). Since the midgut lacks an attached visceral musculature—it lies in the pressurized body cavity without a mesodermal lining—, the intestinal contents cannot be moved forward by intrinsic peristaltic muscle contractions. The changing local pressure conditions in the body cavity, caused by the wriggling locomotion of the nematodes, are responsible for transporting the intestinal contents. But why is the intestine not squeezed and emptied by the high internal pressure? The reason is relatively simple: There is a flap valve under muscular control at the end of the pharynx and the end of the midgut. When the intestine is emptied, particular muscles open the posterior valve, and the entire body's musculature contracts simultaneously. The roundworms can then squirt their intestinal contents many centimeters away. The anus is located subterminally at the posterior end of the roundworm. In male animals, the vas deferens also ends here. When the intestine and reproductive organs share a common opening, this is called a cloaca.

Osmoregulation, Excretion, and Gas Exchange

Nematodes possess a unique organ for osmoregulation, which most likely is not homologous to nephridial structures in other Metazoa. In free-living nematodes, it consists of one or two glandular renette cells that open either into the pharynx or directly to the outside. The renette cells of many nematodes serve osmoregulation, rather than detoxification. Variations of this system are found in different nematodes. However, most nematodes excrete ammonia as ammonium ions (ammoniotely), similar to what is also known of other aquatic animals, such as many crustaceans, the larvae of amphibians, and bony fish. Nitrogenous waste compounds are initially released from the pseudocoel via the intestinal wall into the intestinal lumen and thus reach the external environment.

In *Ascaris*, the aforementioned renette cells are missing. Instead, a single cell forms an H-shaped tube system, which projects into the lateral epidermal cords. Anteriorly, just behind the mouth, the two lateral ducts of this tubular system are connected to each other. Therefore, the anatomy of the organ, seen from above, looks like a capital H and is referred to as the H-organ. From this cross-connection, an excretion canal projects ventrally to the excretory pore. The H-organ serves, according to our current knowledge, for excretion and osmoregulation. Sections of the H-organ can be seen in histological cross-sections as single ducts in the two lateral epidermal cords (Fig. 7.8b).

Nematodes do not have specialized organs for gas exchange. The adult roundworms live anaerobically in the small intestine of their host. However, in *Ascaris*'s juvenile stages, which not a aerobically living, oxygen uptake and carbon dioxide release occur directly over the entire body surface. A vascular system and a heart are absent in nematodes. Hemoglobin, as a respiratory protein, has been detected in the body cavity fluid. This suggests that the pseudocoelomic fluid transports oxygen as well.

Reproductive Organs

Like most nematodes, *Ascaris suum* and *Ascaris lumbricoides* are gonochoristic. Sexual dimorphism is visible regarding body size and posterior-located mating organs. Therefore, in an introductory zoology course, both sexes are prepared and demonstrated (Figs. 7.2 and 7.3).

The female reproductive system is paired. If the animals used for dissection are not too hardened, the female organs, which run through the entire animal in numerous loops and twists, can be laid out in the dissecting dish. Usually, this is much easier with fresh material. The tubes each start with a very thin, thread-like terminal section, the ovary. The production of oocytes (oogenesis) occurs in the ovaries' walls (Figs. 7.4 and 7.5d). During oogenesis, the mature oocytes are pushed further and further toward the central canal of the ovaries (rhachis) and released there. The oocytes then pass from the ovaries into the oviducts, and from there into the uteri (Figs. 7.5e, f and 7.6a, b). The oviducts and uteri almost completely fill the body cavity of the animals. The two uteri unite in a very short common vagina, which opens through the female gonopore on the ventral midline. The vagina should be sought and identified for orientation during preparation (Fig. 7.1). The fertilization of the oocytes takes place within the uterus, and the aflagellate sperm are transferred during mating, moving in an amoeboid manner in the uterus (Fig. 7.5f). There, the fertilization of the oocytes and the first steps in embryonic development also occur (Fig. 7.6). A single *Ascaris* female produces about 60 million eggs during the active reproductive phase, up to 200,000 per day. The

fertilized eggs are enclosed by a thick-walled, protective shell and are released into the environment with the host's feces. Therefore, in sexually mature animals, the uteri are always filled with countless eggs, which can be seen in histological preparations. Some mature eggs should be taken directly from the uteri regions close to the vagina for examination under a microscope.

The male reproductive system consists of a single, unpaired, thread-like testis. The germarium, where the spermatogonia are formed mitotically, is situated at the terminal section of the testis. The spermatogonia undergo meiosis, and they mature as haploid spermatids in the middle growth zone. After passing through the maturation zone, they enter the vas deferens as sperm (Figs. 7.7 and 7.8a). The vas deferens runs in long loops through the entire body cavity and opens posteriorly into a muscular ejaculation tube. From here, the sperm reach the cloaca. As copulatory organs, two so-called spicula, cuticular hooks, aid in the transfer of sperm into the female genital opening by widening the vagina (Fig. 7.1d). Free-living nematodes produce only a few gametes in both females and males. However, the egg size is independent of the body size and is approximately the same in all nematodes, usually only 50–100 μm × 20–50 μm. Since the probability that the eggs of the parasites will reach a favorable habitat again is relatively low, parasites must produce many more eggs than their free-living relatives. This also explains why parasites are usually significantly larger than their free-living relatives.

***Ascaris lumbricoides* and *Ascaris suum*. Is a New Species Emerging?**

The pig roundworm (*Ascaris suum*) and the human roundworm (*Ascaris lumbricoides*) are morphologically indistinguishable and still capable of interbreeding. *Ascaris lumbricoides* was first described in 1758, by the Swedish Naturalist Carl von Linné. Twenty-four years later, the German pastor and zoologist Johann August Ephraim Goeze discovered and described *Ascaris suum*. How are both species related? They could be two valid species that trace back to a common ancestor, which separated at the time of pig domestication, and thus represent a comparatively recent speciation event. If this hypothesis is valid, they would be sister species. Alternatively, *Ascaris lumbricoides* and *Ascaris suum* are the same species. To clarify this question, roundworms from pig populations and roundworms from infected human patients were collected, and their DNA sequences were analyzed for sequence similarities. It turned out that the genomic sequences of ascarids from pigs and humans are so similar that the authors of the study postulated one species. If these findings are confirmed, the name of the first description, *Ascaris lumbricoides*, applies, and *Ascaris suum* is a so-called junior synonym.

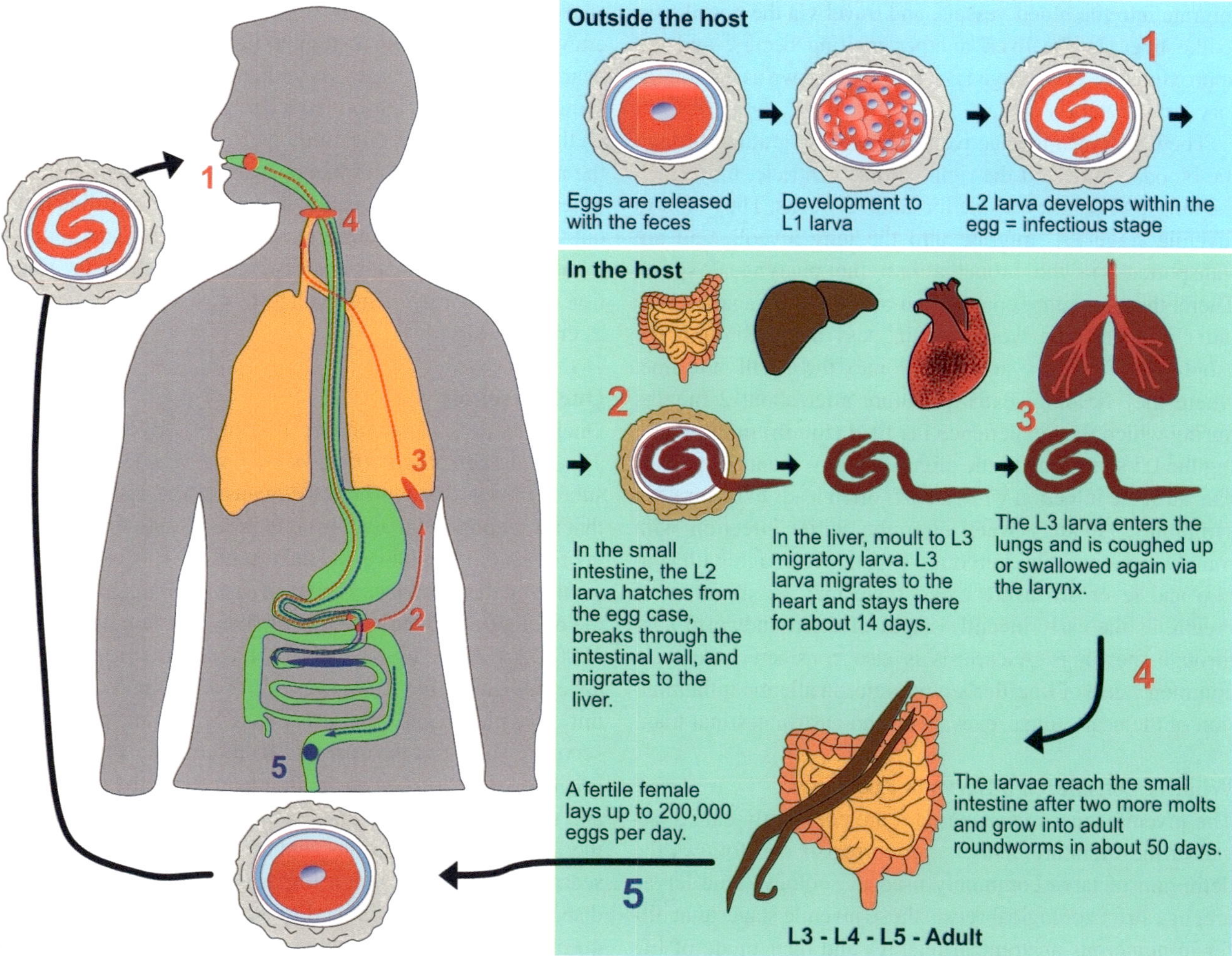

Fig. 7.9 Diagram of the life cycle of the human roundworm (*Ascaris lumbricoides*). According to various authors

Helminthiasis

Helminthiasis refers to all diseases caused by parasitic worms. The name is derived from the ancient Greek word for worm, *hélminthos*, and was introduced in zoology and medicine in previous centuries as a collective term for parasitic worms living in the intestine (Helminthes). A few clinically relevant diseases caused by parasitic nematodes will be presented below.

Ascariasis

The World Health Organization (WHO) estimates that about one billion people are infected with *Ascaris*. Ascariasis primarily occurs in warm, tropical regions and under poor hygiene conditions. The infectious eggs are constantly ingested orally. Contaminated drinking water is one of the leading causes of ascariasis transmission. High hygiene standards are essential for interrupting the infection pathways of these parasitic nematodes. Adult worms colonize the small intestines of the hosts and can cause various stomach and intestinal complaints, including nausea, vomiting, and diar-

rhea, depending on the intensity of the infestation. Health impairments often occur shortly after the initial infection. Serious complications are rare in adults, but in children, it can lead to intestinal obstruction (ileus). In cases of severe infestation, *Ascaris* occasionally migrates into the bile ducts and blocks them. According to the WHO, about 1% of cases result in the death of the patient.

Sexually mature females produce vast quantities of fertilized eggs daily in the hosts' intestines. The eggs then enter the free environment with the infected person's stool (Fig. 7.9). The L1 larva develops there. Still, within the thick-walled eggshell, the first molting and, thus, the formation of the L2 larva take place. The speed of the oxygen-dependent development described above depends to a large extent on the ambient temperature and other parameters. It can take between 8 and 50 days. The eggs with the L2 larvae are highly infectious and represent the crucial step in transmission. Humans become infected with L2 worm eggs upon drinking contaminated water. In the small intestine, the larvae push out of the eggshell, penetrate the intestinal wall, migrate into the blood vessels, and travel via the portal vein to the liver. In the liver, another molting occurs, and the approximately 2 mm large L3 larva, also known as the migratory larva, hatches.

The L3 larvae move actively, and after migrating through the blood vessels and the right cardiac ventricle, they reach the lungs, where they reside for about 2 weeks. They molt to juvenile stage L4, migrate into the lung alveoli, and are transported via the bronchia into the pharyngeal space. There, they are either coughed up or swallowed again. If the latter happens, the worms again pass down through the esophagus, stomach, and finally into the small intestine, where they become sexually mature after about 2 months during which they experience the final (fourth) molt, reaching the L5 state. Due to the migratory movements of the larvae, an acute infection with *Ascaris* often leads to respiratory symptoms. About 1 week after the initial infection with roundworm eggs, children often experience cough with sputum and fever. In severe cases, there may be shortness of breath, as the early juvenile stages of the roundworm pass through the lungs. Ascariasis is also considered the most common cause of Loeffler's syndrome, an allergic inflammation of the heart, lungs, eyes, skin, and gastrointestinal tract.

What Is Considered a Larva in Zoology?
The juvenile stages of nematodes are commonly called larvae in the literature. This is not correct if one follows the definition of larva commonly used in zoology. True larvae are, in a strict zoological sense, those juvenile stages that differ in numerous anatomical features and their mode of life from the adults. Moreover, genital organs have not yet developed. For example, the larvae of flies do not possess wings, eyes, or genital organs; these structures develop only during metamorphosis. For flies, the term larva is correct. In nematodes, the hatching worms have, in principle, a similar morphology and habitat to the adults, therefore, the term larva is incorrect in a strict zoological sense. However, in parasitology literature, the term larva is often used for the juvenile stages of nematodes; therefore, the term is retained here for traditional reasons.

Enterobiasis
In Western Europe, the "pinworm," *Enterobius vermicularis*, popularly also known as the threadworm (in the United Kingdom, Australia, New Zealand), the seatworm, or in Germany, even as the "kindergarten worm," is the most common, medically relevant parasite of humans. Exact numbers are not available, but estimates suggest that every fourth to fifth child in Western Europe becomes infected with pinworms at least once during kindergarten or another early school period. Around 50% of all people are affected at least once in their lives. Only a few millimeters long, pinworms live in the large intestine of their hosts and occasionally

cause inflammations there. The sexually mature females crawl out of the rectum at night to lay their eggs in skin folds near the anus. This causes very intensive itching, which subsequently leads to vigorous scratching. Usually, eggs remain on the fingers or under the fingernails, are transferred back to the mouth or onto other objects, and can lead to infection in other people. Often, eggs also remain in the bed linen, dry out there, and are later stirred up as "dust eggs" and possibly inhaled again. This leads to a new self-infection or the infection of other family members. The disease itself is referred to as enterobiasis.

Onchocerciasis
Onchocerciasis is commonly known as "river blindness." The adult nematodes (*Onchocerca volvulus*) settle in the subcutaneous tissue, where they cause large visible nodules that are an important diagnostic feature. Two to three females up to 50 cm long live in each nodule. The considerably smaller males, which can grow up to 4 cm long, migrate into the nodules and mate with the females. The females then lay 2000 eggs daily, after which the infectious juvenile stages develop. Black flies (*Simulium* sp.), which ingest and transmit juvenile stages of *Onchocerca* while sucking blood, serve as vectors. Once present in the human bloodstream, juvenile stages of *Onchocerca* spread throughout the body and may reach the eyes and colonize the orbit, the retina, or the optic nerve. When the nematods die, they cause tissue scarring, leading to vision impairment and blindness. The disease is primarily confined to Africa, where an estimated 40 million people are infected, many of whom are blind. In 2015, the Nobel Prize in Physiology or Medicine was awarded to William C. Campbell and Satoshi Omura, among others, for discovering "ivermectin" as a highly effective agent against all known clinically significant nematodes (antiparasitics). Ivermectin is an integral part of today's onchocerciasis treatment.

Elephantiasis (Elephant Disease)
An infection with *Wuchereria malayi* or *Wuchereria bancrofti* causes elephantiasis. *Wuchereria* is a filarial (arthropod-borne) nematode that is the primary cause of filariasis, a severe lymphatic disease. The females are viviparous in filariae, and the minute larvae I (microfilariae) are distributed in the blood. Blood-sucking insects (fleas, flies, and especially mosquitoes) are these parasites' intermediate hosts and vectors. *Wuchereria* microfilariae enter the lymphatic system through mosquito bites and cause inflammation with lymphatic congestion. In the case of *Wuchereria* sp., the entire development occurs inside the infected patient's body. Therefore, hygiene measures cannot protect against infestation. Patients in an advanced stage of the disease suffer from an extreme accumulation of lymph fluid in the tissue (Fig. 7.10). This leads, in the final stage, to a swelling of the

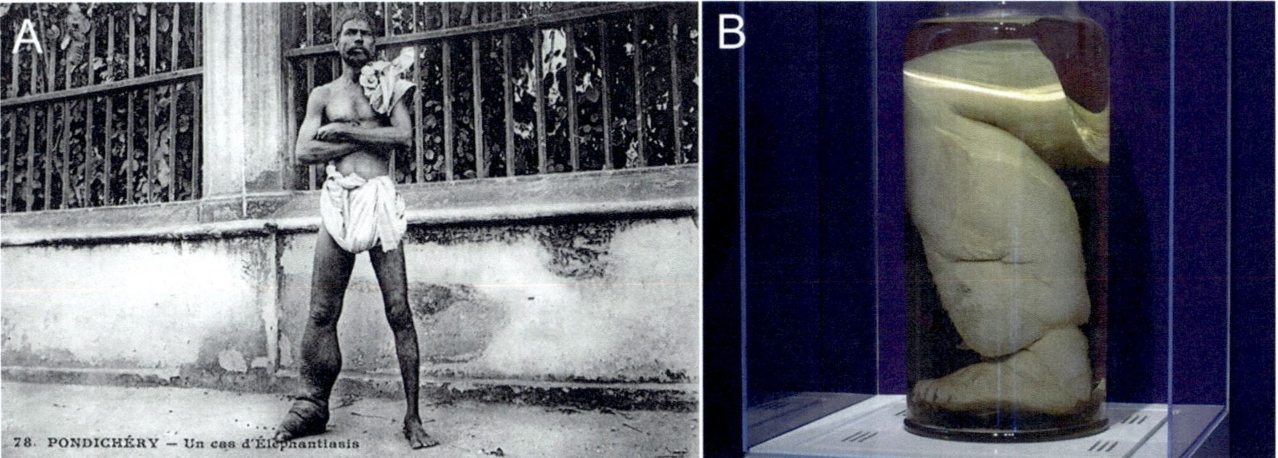

Fig. 7.10 Parasitic nematodes and disease patterns. (**a**) Postcard from Pondicherry, until 1954, the capital of French India. A man with elephantiasis in the right leg, caused by a filarial infection with *Wuchereria* *malayi* or *Wuchereria bancrofti*, is shown. (**b**) Elephantiasis, preparation of a complete human leg. National Museum of Health and Medicine, Silver Spring, Maryland, USA

tissue (edema) up to many times its original size. Arms and legs are frequently affected, as are the breasts in women and the scrotum in men. A real cure does not so far exist. Worldwide, about 120 million people in over 70 countries are infected.

Trichinellosis

Trichinellosis is a ubiquitous zoonosis caused by nematodes of the genus *Trichinella*. The adult nematodes colonize the intestines of carnivorous or omnivorous hosts. These include humans, monkeys, pigs, and bears. Females, up to 4 mm in size, are viviparous and produce 1000–2000 larvae daily. The larvae are released into the lymphatic or blood system, where they are transported via the blood to the cross-striated skeletal muscles or the myocardium. They develop further if they reach highly vascularized aerobic muscle tissue (e.g., diaphragm, mastication muscles). The larvae enter the muscle cells and encapsulate themselves in a calcified cyst (muscle trichinosis; Fig. 7.11). The presence of larvae often causes fever and edema in the face of the infected person due to inflammation of the muscles. If the cardiac muscles are severely affected, inflammatory reactions can also occur there that may lead to death. The encapsulated larvae are highly infectious and are transmitted to the next host through the eating of raw or insufficiently cooked meat. They continue to develop in the intestine through the second, third, and fourth larval stages until they reach adulthood. In the case of a *Trichinella* infection, each final host also becomes the intermediate host for the next generation. Thus, usually, only carnivorous animals can be infected. Adult trichinae typically cause abdominal pain and diarrhea in humans, often accompanied by dizziness. In many European countries, the occurrence of trichinae is notifiable. During slaughtering, the meat must always be checked for a possible infestation with trichinae; samples must be taken from the diaphragm and mastication musculature.

Fig. 7.11 Parasitic nematodes. (**a**) *Trichinella spiralis*, male and female, total specimens. Specimen of the Biological Collection of the University of Osnabrück. (**b**) *Trichinella spiralis*, capsular larvae in muscle tissue, histological section, stained. Specimen by Johannes Lieder, Ludwigsburg

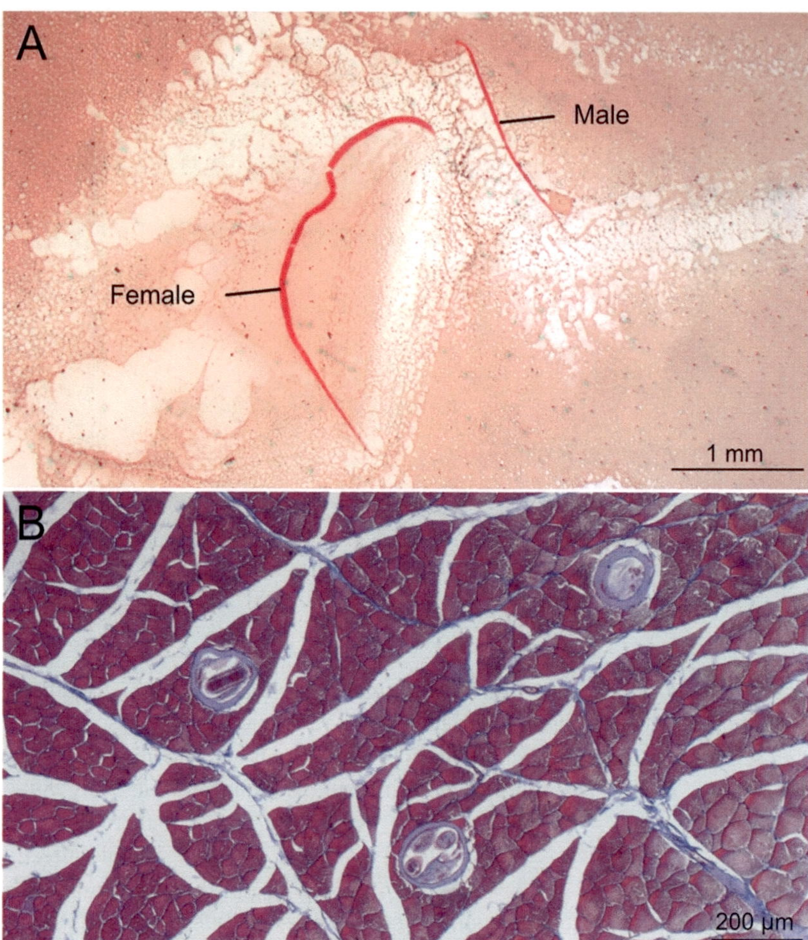

► **Summary** Arthropoda or arthropods are a large taxon of animals with more than 1 million identified species. They are the world's most diverse group of animals. Arthropods play a vital role in maintaining the balance of our global ecosystems. They help pollinate plants, feed birds, and keep our water systems clean. Some of them, such as bees, are essential for our economy. They are also a significant source of food for many animals, and insects, in particular, are being considered as a potential superfood for the future due to their high protein content. Arthropods have a segmented body consisting of separate sections, articulated appendages, and an exoskeleton made of chitinous cuticle and sclerotized protein.

The phylogenetic relationships within arthropods are still a topic of debate. Recent studies include molecular data, the fine structure of the nervous systems, and other features, in addition to classical anatomical characteristics. According to this, the arthropods can be classified into three major monophyletic groups:

- Chelicerata (jaw claw bearers, arachnids, spiders)
- Myriapoda (millipedes)
- Tetraconata (also called Pancrustacea with crustaceans and Hexapoda)

Chelicerata is the sister group of Myriapoda and Tetraconata, which are united as Mandibulata, referring to their characteristic mouthparts, called mandibles. Tetraconata include the paraphyletic crustaceans and Hexapoda (hexapods, meaning six legs). The Hexapoda include the insects and some closely related groups such as the Entognatha, which include, for example, the springtails (Collembola), the Diplura and the Protura. As with all Metazoa, also arthropods originate from the sea. Marine arthropods have repeatedly successfully conquered freshwater and terrestrial habitats, each conquest accompanied by corresponding transformations of their bodies and their physiology. Of the three major groups, we discuss here two representatives of the Tetraconata, an insect and a crustacean.

The epidermis of Arthropoda secretes a cuticle that sclerotizes and hardens. The cuticle effectively protects the internal organs from mechanical damage and in insects prevents the body from drying out. The hardened cuticle forms an exoskeleton, divided into rigid parts linked by soft, flexible, and hinge-like connections. Thus, the exoskeleton allows for protection and structural support, maintaining the body's shape and mobility. Arthropods have a segmented body, but it is still unclear whether these segments were originally of the same type or were already divided into functional units (tagmata; sing. tagma, greek, division) in their last common ancestor.

A unique feature in all arthropods is the formation of a head, where trunk segments are attached to the anterior end. This process is called cephalization. In the head of Myriapoda and Tetraconata, five appendage-bearing segments are fused, which carry the first and second antennae, the mandibles, and the first and second maxillae. However, it is only in crustaceans that all five head appendages are present. Millipedes and hexapods lack the second antennae. Initially, each body segment carries a pair of appendages. Each appendage consists of individual tubular sections connected by joints. This ensures a high degree of mobility. The individual appendages are moved by a system of antagonistic muscles consisting of flexors and extensors. If the latter are missing, the extension occurs passively by increasing the hemolymph pressure. Such appendages are called turgor appendages, as they are found in certain crustaceans, for example, in the well-known water fleas *Daphnia* spp. or the brine shrimp *Artemia salina*. It is very likely that the appendages were originally multifunctional and served for locomotion, respiration, and food uptake. In the course of evolution, a specialization of individual body sections occurred, the formation of tagmata. This was accompanied by a specialization of the appendages, which then took over only one function. In addition, the appendages of arthropods also serve sensory or reproductive purposes. The evolution of segmentation and the development of an exoskeleton were important evolutionary inventions responsible for the extraordinary evolutionary success of arthropods.

The nervous system of arthropods is a ventral, so-called rope-ladder-like nervous system in which a pair of ganglia (one per segment) is connected by longitudinally running connectives and transversely running commissures. In addition, there is usually a median longitudinal nerve. During cephalization, the ganglia of the foremost segments are combined into a complex brain (syncerebrum); the ganglia of the subsequent segments also merge into the composite subesophageal ganglion. In the brain, the so-called mushroom bodies represent important association centers. Further ganglia can merge depending on the group considered, so the ladder-like nature is not always visible.

Compound eyes are another characteristic of arthropods. Compound eyes are composed of many small individual eyes called ommatidia. Their number is extraordinarily variable and can reach up to 30,000 ommatidia in dragonflies. Although best developed in pancrustaceans (crustaceans and hexapods), they are also present in chelicerates and millipedes. However, in these latter two taxa, the compound eyes are usually either divided into groups of modified ommatidia or reduced; and are thus no longer recognizable as such. In addition to the paired compound or lateral eyes, arthropods possess median eyes or ocelli. Primarily, there are four of these ocelli, which may become reduced to two or three or fused to form a single unit. Other sensory organs are the sensilla distributed all over the body, often ending in cuticular bristle structures. They are usually differentiated as chemo- or mechanosensory structures.

The body cavity of arthropods is a so-called mixocoel, also called a hemocoel. It is formed by the fusion of the primary and the initially created secondary body cavity. As a result, the blood and the coelomic fluid form a uniform body fluid, which is now called hemolymph. Since these two body compartments are not separated, as in metazoans with a closed blood vascular system, we speak of a true open circulatory system in arthropods. The most crucial element is the dorsally located tubular heart, from which, primarily, an arterial system originates. This can be pretty extensive in primitive arthropods. In many arthropods, the heart is concentrated in the body region where the respiratory organs are also located. The hemolymph transports oxygen and carbon dioxide to the various organs, and from these to the primary respiratory organs, usually the gills. These may even be completely absent in smaller aquatic groups. In some terrestrial arthropods, decoupling of the respiratory and breathing system occurs. This is especially the case with the tubular tracheae in insects, most myriapods, and some arachnid groups. These tubular tracheae directly supply the organs with oxygen. The gills of aquatic arthropods serve not only for gas exchange but also for osmoregulation and excretion.

The original excretory organs of arthropods are nephridia, which are derived from metanephridia. They start with a small coelomic space, called a sacculus, followed by the proper nephridium. In the ground pattern, nephridia are laid out in each body segment. In recent arthropods, however, we always find nephridia in only a few anterior segments; in crustaceans, there are, at most, two pairs in the segments of the second antennae or second maxillae. Therefore, they are also referred to as antennal or maxillary glands. As a rule, however, most crustaceans have only one pair. During evolution, the development of alternative excretory organs, the Malpighian tubules, probably occurred several times in the course of invading terrestrial environments and because of the necessity to save water. Malpighian tubules are blind-ending tubes originating at the midgut-hindgut border from the intestinal tract and releasing their products via the intestine. They have the decisive advantage that excretion is coupled with a much lower water loss.

Arthropods are mostly gonochoristic, but other forms of reproduction occasionally occur. Hermaphroditism is found, for example, in remipedes. Protandry, individuals that first mature as males and then later become females, also exists in a few marine isopods. Parthenogenesis is known in several Hemiptera and Hymenoptera, ants, bees, and wasps. The list of reproductive modes can be easily extended. The paired gonads are of mesodermal origin and the gametes are discharged via more or less complex mesodermal ducts, which end with a short ectodermal part to the outside. It is assumed that arthropods originally released their gametes into the open water. Thus, the diverse reproductive modes are primarily a result of the conquest of terrestrial habitats—accompanied by a direct transfer of sperm to the females. Therefore, in terrestrial arthropods, the aquatic larval stages, which are important for the propagation of a species and still occur in crustaceans, become dispensable. After embryonic development, either a larva with a few segments, like the nauplius larva of the crustaceans, which only has three segments with appendages, or juvenile stages with an incomplete number of segments or with the number of segments of the adult animals will hatch. For zoologists, the term "larva" is only intended for those juvenile stages that differ in prominent morphological-anatomical features, as well as in their way of life, from the adult animal. For example, the larvae of dipteran insects have neither wings nor eyes, with both structures only developing during metamorphosis. Also, the sexual organs are still missing. Supplementary movies can be found at figshare: sn.pub/xr9qhi.

8.1 The European Crayfish (*Astacus astacus*, Crustacea)

Crustaceans (Crustacea) are a diverse group of predominantly aquatic species and play a prominent role in marine food webs. Crustaceans and hexapods have an exclusive last common ancestor, and are united as Pancrustacea or, due to the similar structure of the compound eyes, referred to as Tetraconata.

Most of the over 50,000 known crustacean species live in marine habitats, either as part of the plankton in open water or as benthic organisms on or within the seabed. Other crustaceans colonize similar freshwater habitats, like the native European crayfish (*Astacus astacus*), also known as the noble crayfish, which is introduced in this section (Fig. 8.1). Only a few crustaceans have entirely conquered terrestrial habitats, no longer being dependent on aquatic biotopes even for reproduction, for example, the common rough woodlouse (*Porcellio scaber*). The size of crustaceans ranges from microscopic planktonic forms such as the species-rich copepods (Copepoda), which are only about 0.2–2 mm in size, to the Japanese spider crab (*Macrocheira kaempferi*, Decapoda), which can measure up to 3.7 m from leg tip to leg tip when stretched out. Crustaceans play a dominant role in the food webs of the sea. Even more important are the numerous small representatives of the aforementioned marine Copepoda, which, as primary consumers of phytoplankton, form the second trophic level of many food webs and themselves mainly serve as food for many smaller fish. The Antarctic krill (*Euphausia superba*) forms huge swarms in the ocean and is the primary food source for certain baleen whales, such as blue and fin whales. An adult blue whale filters up to 4000 kg of krill daily from the ocean. The worldwide biomass of krill is estimated at over 500 billion tons and exceeds that of humans by more than double. Crabs are highly valued as a source of food by humans. The European lobster (*Homarus gammarus*, Fig. 8.2), the brown shrimp (*Crangon crangon*), and related species such as prawns are considered delicacies. However, we cannot go into the significant diversity within crustaceans here. Above all, the segmentation of their bodies, the number of segments, and the associated appendages are quite different among the various groups. Instead, the ten-footed crabs, Decapoda, are presented as an example. There are over 15,000 marine and freshwater species, including lobsters, shrimp, langoustines, the Japanese giant crab, and edible crabs. The native Noble or European Crayfish (*Astacus astacus*) also belongs to this group.

Mating and Reproduction of the European Crayfish, *Astacus astacus*

The European Crayfish (*Astacus astacus*, syn. *Astacus fluviatilis*) formerly inhabited numerous rivers throughout Central Europe. Its population is now severely threatened by

Fig. 8.1 *Astacus astacus*, European Crayfish, living individual

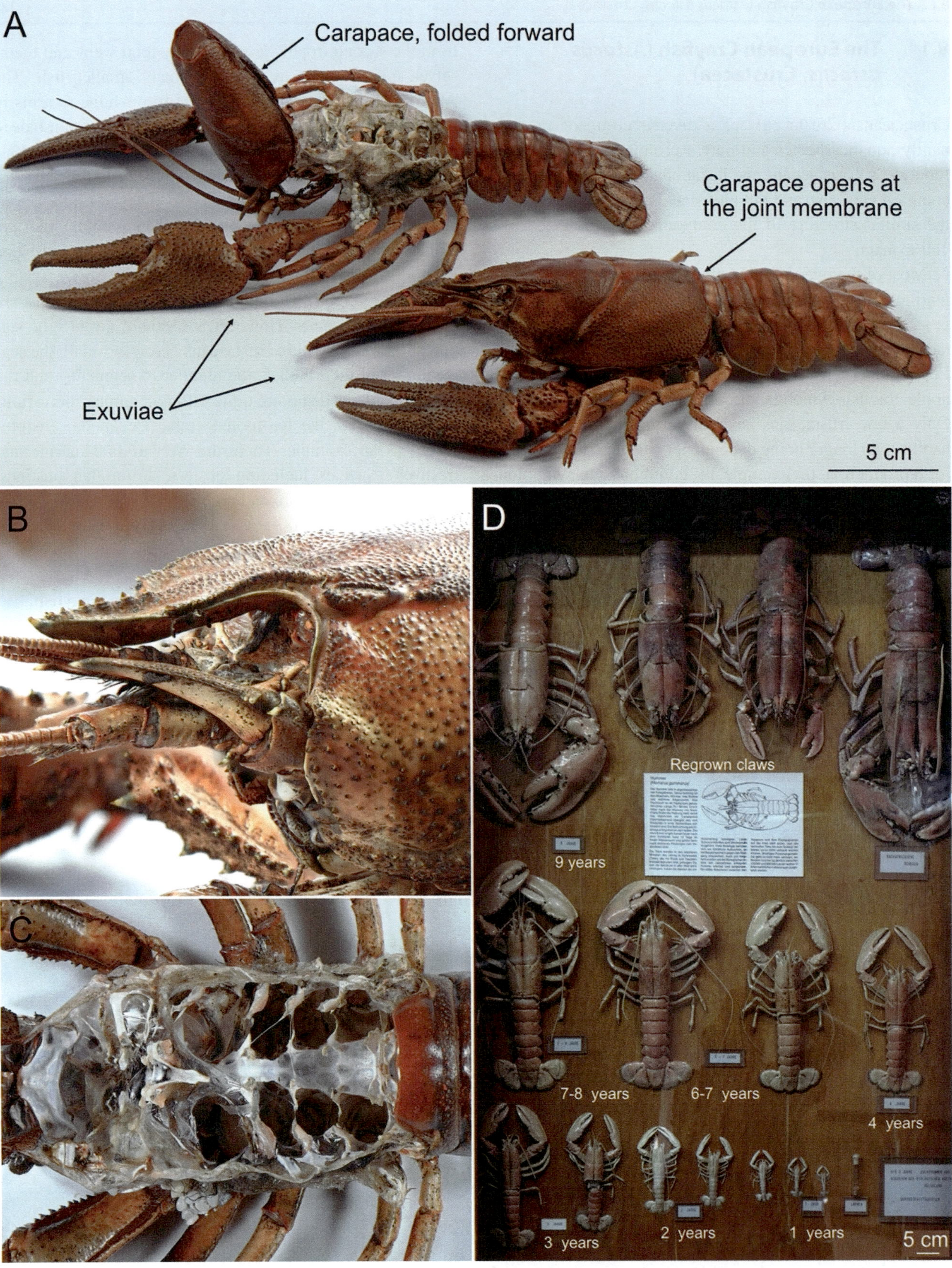

Fig. 8.2 *Astacus astacus*, European Crayfish. (**a**) Exuviae of two individuals. The elastic skin connection between the carapace and abdomen breaks open to allow the crayfish to hatch, still with a soft body. The animal pulls itself out of the old exoskeleton within 15–20 min. (**b**) The carapace of the left crayfish, still in a fresh state, has been folded forward to reveal the remaining internal structures. The exuviae of the right crayfish is as found in our aquarium, which we set up for our zoology course. (**c**) Close-up of the head region of an exuvia. The exuviae of freshly hatched crayfish are initially dark-brown or black, like the exoskeleton of a living crayfish, see Fig. 8.1, but turn red-brown upon drying in the air. (**d**) *Homarus gammarus*, European Lobster. The development from a free-swimming larva (bottom right) to a mature animal takes many years and numerous molts. Different stages of development are on display in the Heligoland Museum, Heligoland

habitat loss and fungal diseases (crayfish plague). Crayfish plague (*Aphanomyces astaci*) is a water mold that arrived in Italy roughly 160 years ago, with imported crayfish from North America or with ballast water from ships. Within a few years, the crayfish plague spread all over Europe. *Astacus astacus* dies within a few weeks of being infected. In addition, the European crayfish competes with some invasive non-European species, such as the Spinycheek Crayfish (*Fraxonius limosus*), which is resistant against the crayfish plague and therefore more successful.

During the day, crayfish tend to remain hidden under rocks or vegetation. In areas with softer sediments, they dig burrows or caves in the shallower bank area of streams, rivers, ponds, and lakes. Crayfish venture out at night to search for prey such as mussels, snails, worms, and insect larvae. The cave, which can reach up to 1 meter in length, is always occupied by a solitary animal. Crayfish become sexually mature in the second or third year of life. Then, in early fall, the gonads significantly increase in size. The males stroll around searching for females and turn all conspecifics whose paths they cross on their backs. If it is a female, mating occurs, during which a spermatophore is transferred. The mating appendages of the male, formed from the first two pairs of pleopods, shape the seminal fluid emerging from the sexual openings into 1 cm long spermatophores, which are attached to the female close to the genital opening. The fertilization of the eggs then only takes place in late fall, 1 or 2 months after mating. The fertilized eggs remain with the mother throughout the winter. They are covered with mucus and thus stick to the female pleopods like a string. In spring, the young crayfish far along in their development are released, being, at this stage, well adapted to the benthic lifestyle of the adults. Like the European Crayfish, many freshwater crayfish lack a free-swimming larval stage. There is a simple reason for this. During a free-swimming planktonic phase, the larvae would drift far downstream or even into the open sea, and they would have to run far up the river systems again after their metamorphosis to reach the biotopes of the parents. Unlike their freshwater relatives, most marine crustaceans have a biphasic life cycle consisting of a planktonic larva, often the nauplius larva, which is essential for the dispersal of the species, and an adult, which, in many forms, is benthic.

General Anatomy of Crayfish

The body of a crayfish can be divided into two parts—the cephalothorax and the pleon (Fig. 8.3). The cephalothorax is formed by the head and the first eight thoracic segments, which are fused. This body region is easily recognizable by the saddle-like carapace. The carapace is a dorsal covering formed by an epidermal fold, which protects the body's anterior part and provides a safe space for the gills. The pleon is the second part of the thorax and comprises the remaining six thoracic segments and the telson. The telson is a non-segmented body section that may carry two sensory attachments in some crustaceans, known as the caudal furca or tail fork. However, these are absent in crayfish and most malacostracans and are not homologous to segmental appendages.

The sclerotized cuticle of the crayfish is divided into sclerites connected by articular cuticular membranes. The dorsal sclerites are called tergites, and the ventral ones sternites. The abdominal tergites extend laterally into thin plates, which are referred to as pleurites. In decapods, the cephalothorax carries the sensory organs (Figs. 8.4 and 8.5) and the most important internal organs, the chelipeds or pincers, representing the fourth pair of thoracic appendages transformed into a large chela (Fig. 8.6a) and the walking legs. The muscular pleon is primarily used for locomotion through swimming (Figs. 8.3 and 8.7). In shrimps, lobsters, and the crayfish, the pleon can be flipped forward at lightning speed on the ventral side. This is an impressive fast mechanism, allowing the animals to quickly escape potential danger and find safety (Fig. 8.6b, c).

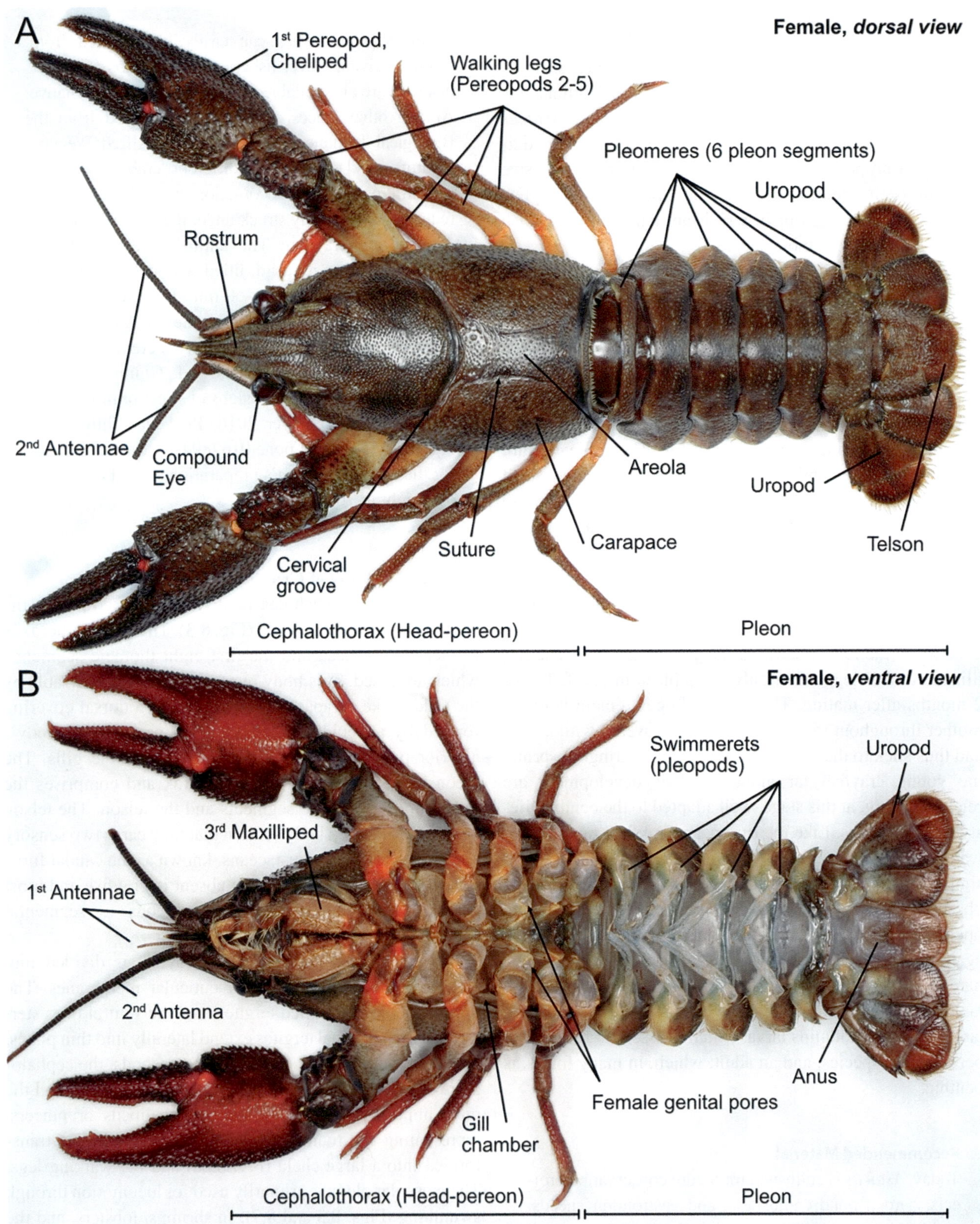

Fig. 8.3 *Astacus astacus*. Adult female in dorsal (**a**) and ventral (**b**) views

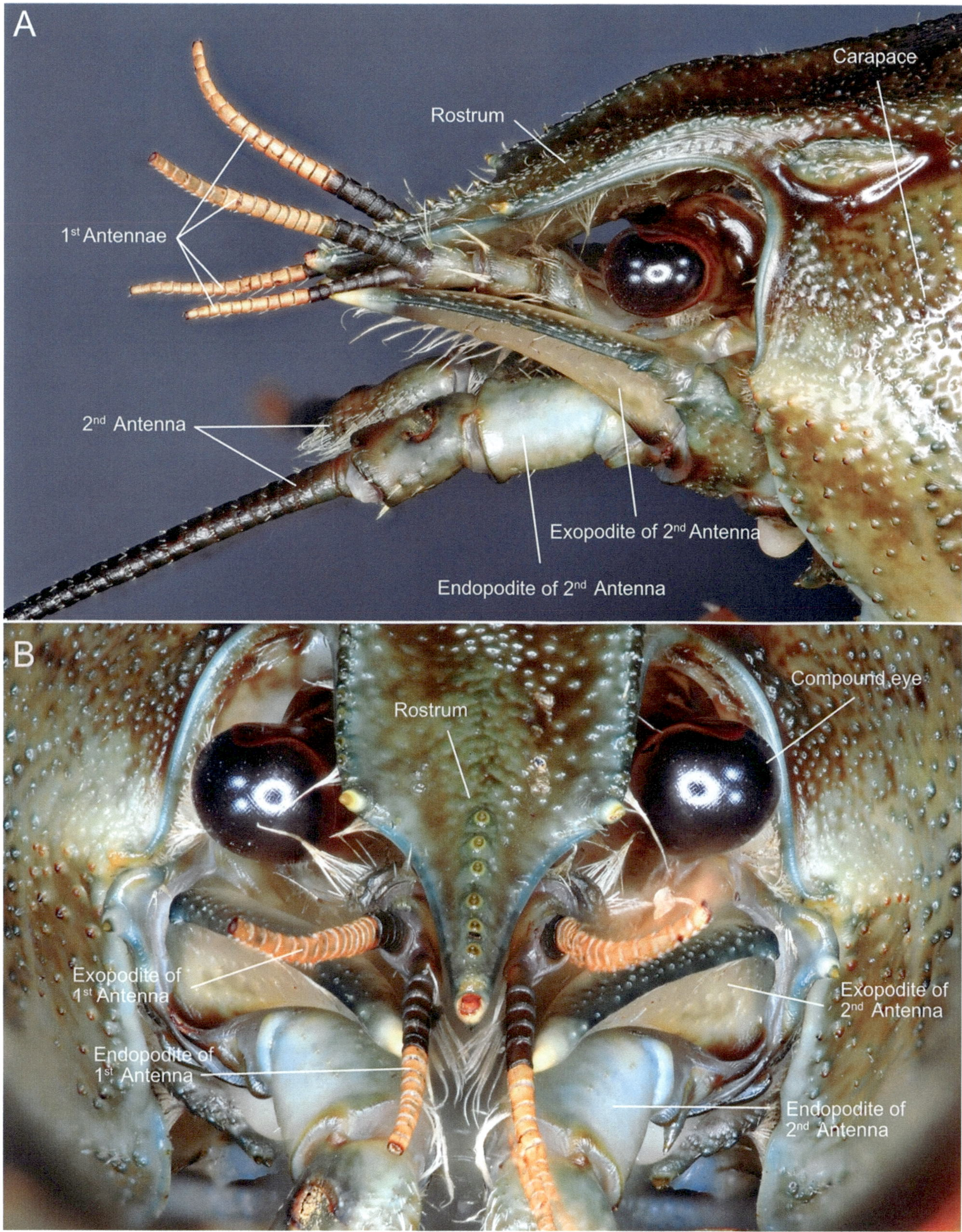

Fig. 8.4 *Astacus astacus*. (**a**) Lateral view and (**b**) Frontal view of the head

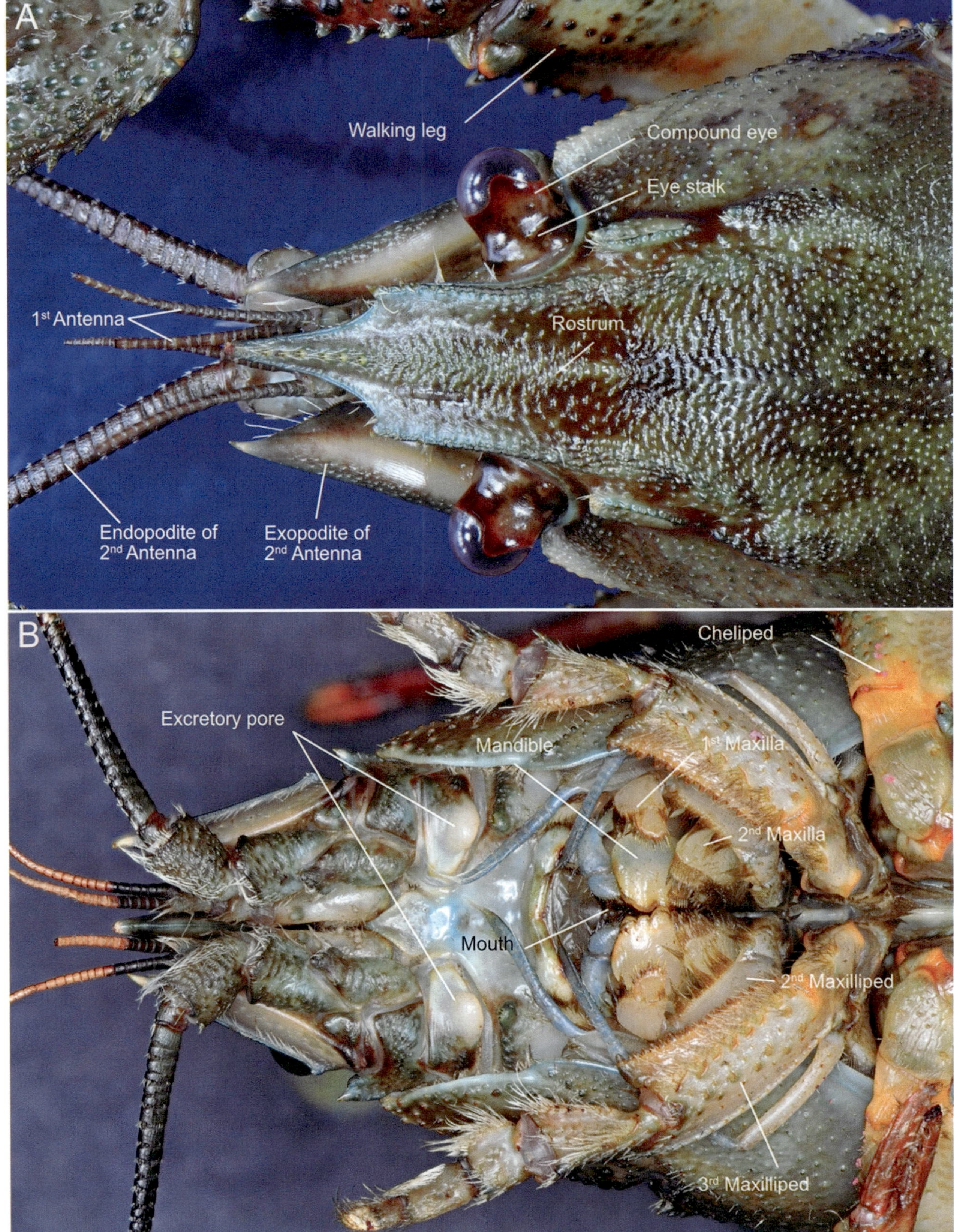

Fig. 8.5 *Astacus astacus*. (**a**) Dorsal view and (**b**) Ventral view of the head

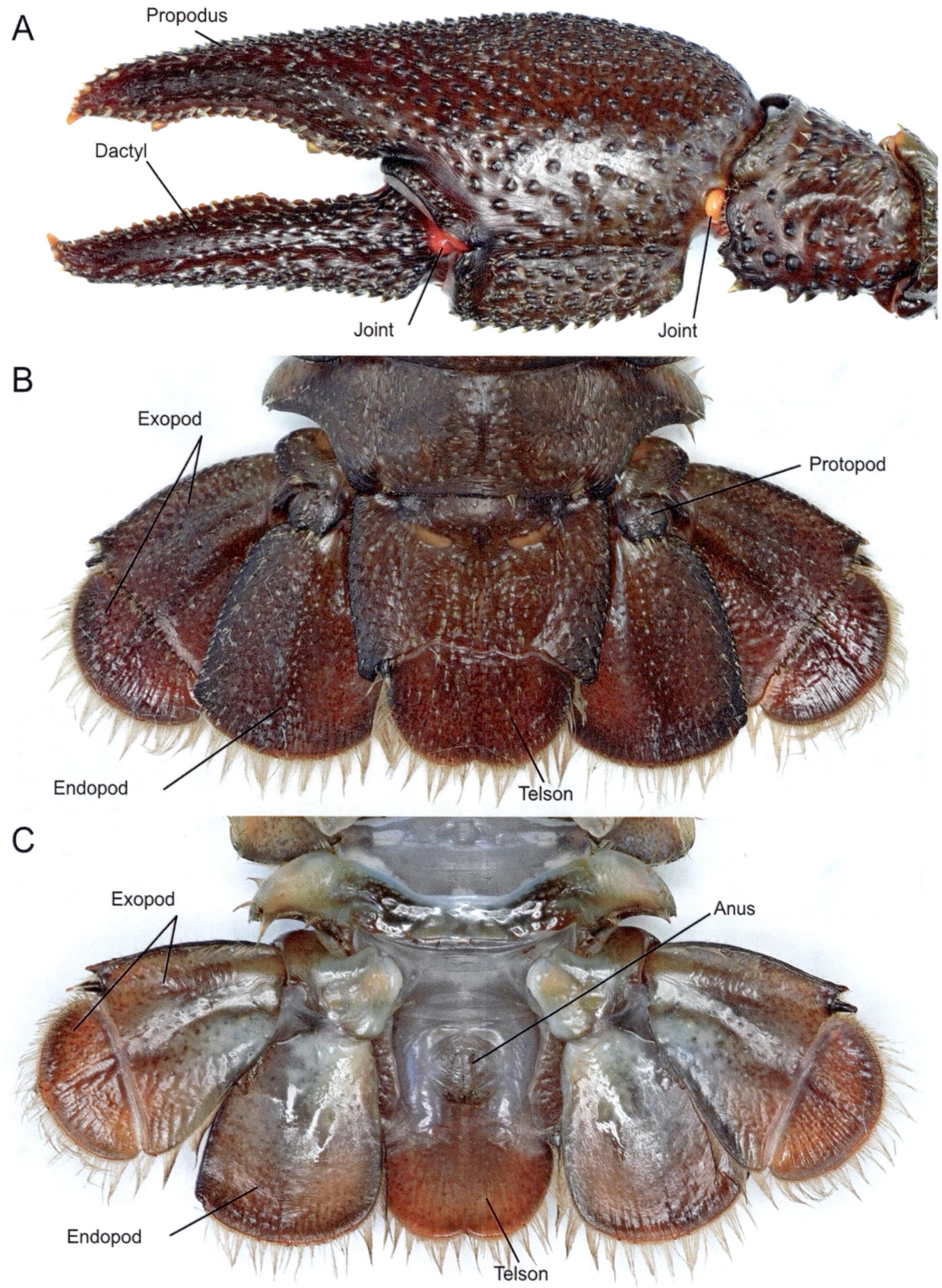

Fig. 8.6 *Astacus astacus*, female. (**a**) Large chela of the first walking leg (pereiopod). (**b**) Tail in dorsal and (**c**) ventral view

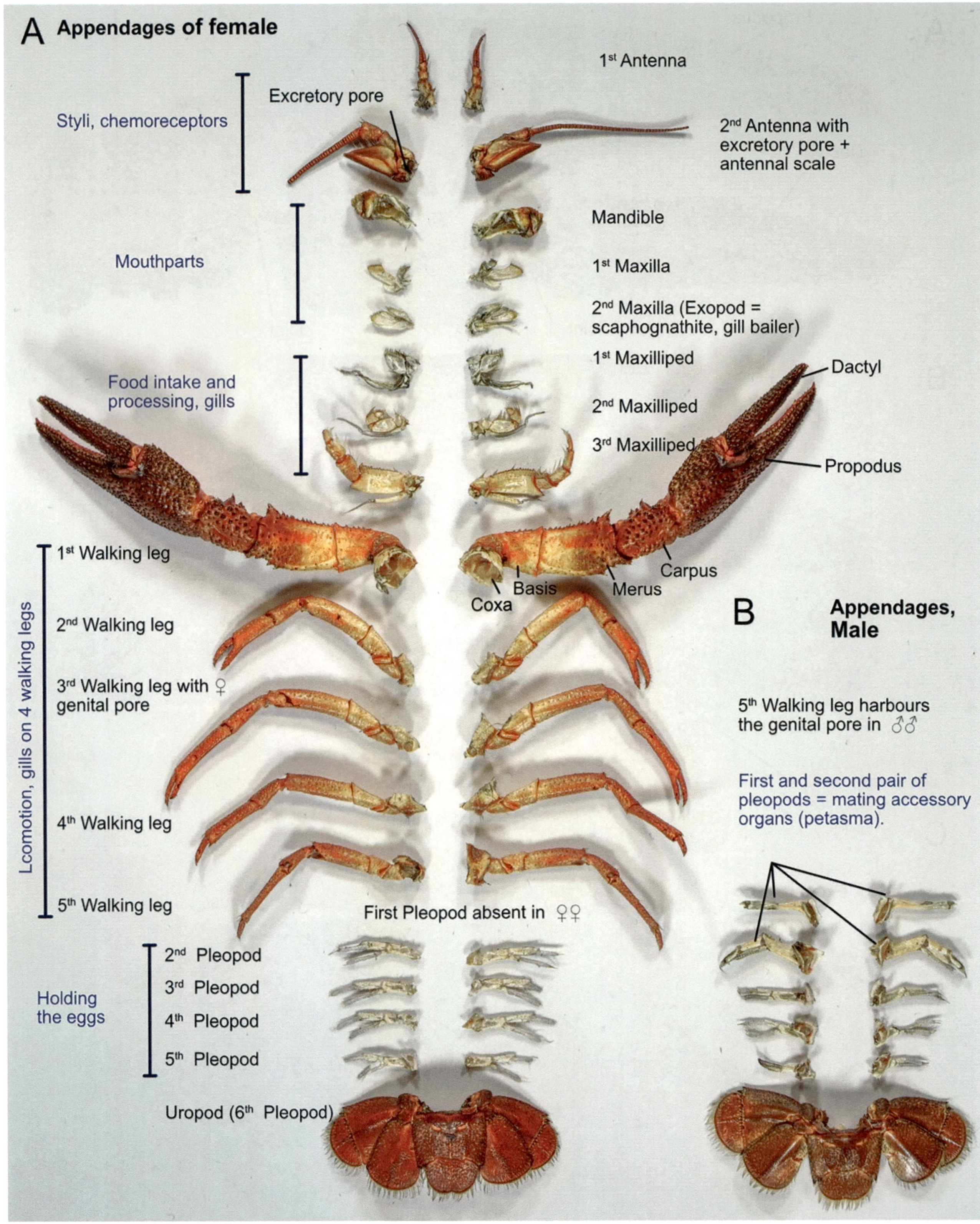

Fig. 8.7 *Astacus astacus*. Prepared appendages of a (**a**) female and a (**b**) male

The Many Functions of the Appendages

Before further dissection, observe the appendages in place. Note their position, movements, and the relationship to other appendages. Each body segment has a pair of appendages, which are built according to the principle of the biramous crustacean limb (Fig. 8.7). All appendages, except for the antennae, are primarily multifunctional and consist of three sections: A basal protopodite carries a larger outer branch (exopodite) and a larger inner branch (endopodite). The protopodite, divided into two sections in the crayfish, can carry additional appendages, which are referred to as exites and endites. Each section is assigned a specific function in the basic pattern: food collection, uptake and transport, respiration, and locomotion. The enormous variety of anatomical and functional adaptations of these appendages has undoubtedly contributed to the evolutionary success of this animal group. For example, the body appendages of the head are transformed into antennae that carry sensory organs or serve for food crushing and food intake (mouthparts). If you observe the crayfish from the ventral side (Fig. 8.3b), you can easily move all appendages back and forth and address them with a probe.

At the end of a course day, detach all appendages carefully and arrange them in sequence on a sheet of paper (Fig. 8.7). The functional diversity of the body appendages can be very well understood in this way. Let us first look at the appendages of the head. The smaller first antennae (antennulae) comprise three individual parts. They carry sensory bristles with chemoreceptors on the two flagella (Figs. 8.4 and 8.5). The statocysts (gravitational sense organs) are located in the first section of the first antenna. The large second antennae form the next paired body appendage. They serve as critical tactile organs (Figs. 8.4 and 8.5a). The first section of the second antenna, which is close to the body, is bipartite and corresponds to the protopodite. In the first part, the coxopodite, we discover the excretory pore of the antennal nephridium. We will discuss this organ, also called the green gland, later. To find the location of the nephridial opening, the coxopodite is searched under the stereomicroscope for a yellow elevation. In its center is the small nephridial pore, which is usually not directly visible (Fig. 8.5b). The outer branch of the second antenna has the shape of a scale and is called the scaphocerite, the antennal scale. It serves as an essential steering organ when swimming.

The following three pairs of appendages are the mandibles and the first and second maxillae, which have been transformed into mouth tools (Figs. 8.5b and 8.7). The first maxillae are also called maxillulae in crustaceans. The mandibles and first maxillae are primarily used to grind food. Part of the second maxilla is transformed into an auxiliary organ for respiration. The exopodite is spoon-shaped and is in permanent motion in the living animal. The exopodite, called the scaphognathite or gill bailer, generates a continuous flow of respiratory water in the gill cavity between the body wall and the carapace. The sensitive gills are well protected from damage by the carapace, but must be ventilated by the pumping activity of the scaphognathite. The respiratory water enters this cavity at the base of the appendages and from the back and exits at the front. The two pairs of antennae and the three pairs of mouthparts mark five appendage-bearing head segments.

Next are eight pairs of trunk appendages, with the first three pairs being auxiliary mouthparts, still serving for food intake and processing (Fig. 8.3b and 8.7). These are the first, second, and third maxillipeds. They are the first appendages of the thorax. The maxillipeds are formed as typical biramous crustacean appendages with proto-, exo-, and endopodites. The second and third maxillipeds each carry gills that project into the gill cavity formed by the carapace.

Four of the remaining five trunk appendages, which serve in decapod crustaceans as walking legs (pereopods), carry gills. They all lack an exopodite. The first pair of pereopods is called the chelipeds. These are the largest appendages, with huge chelae at the end. They serve mainly for grasping and tearing the prey (Figs. 8.3 and 8.6a). In territorial fights, these appendages are used as weapons. The two chelae are slightly different in size. The slimmer one is preferably used for gripping (tweezer chela), and the larger one for crushing (cracker chela). The remaining four pairs of pereopods are used primarily as walking legs. Only the first two of them carry small chelae and are called chelate. The chelae comprise the propodite and the dactylopodite; only the latter is moveable and is an opponent of the propodite. The pereopods also carry the genital openings of female and male crayfish. In males, the genital operculum is located at the base of the fifth pereopod, and in females, on the third pereopod (Fig. 8.8).

The pleon carries the final six pairs of appendages, called the pleopods or swimmerets. In several crustacean taxa, the pleopods serve as swimming legs, so they are called swimmerets. However, in crayfish, the first five of the six pleopods are no longer swimmerets and play no role in the locomotion of *Astacus* (Figs. 8.3b and 8.8). Instead, the pleopods have adopted new functions during evolution.

In males, the first two pairs of pleopods are modified and assist in transporting sperm to the female during copulation. These gonopods are also called petasma (Figs. 8.7 and 8.8). The sperm exits from the male genital opening on the fifth pereopod. Thus, within favorable reach of the tubular endopodites of gonopods (Fig. 8.8). They take up the sperm and form a spermatophore, which is then transferred to the female with the pencil-shaped endopodite of the second pleopod.

The first pair of pleopods is missing in females. Thus, they have only four rudimentary pairs of swimmerets,

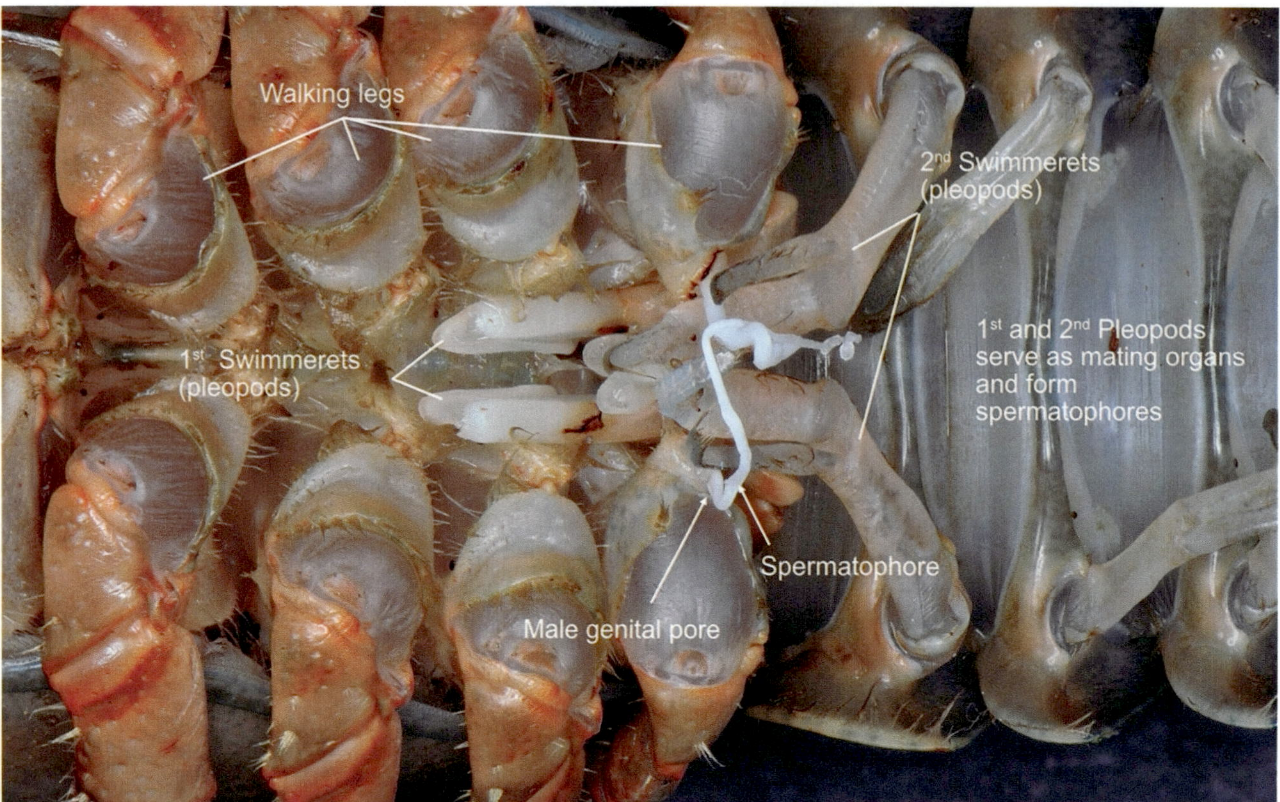

Fig. 8.8 *Astacus astacus*. Male, ventral view, sperm release is visible

another reliable feature for determining sex. The pleopods serve the female by holding the eggs, which are thus protected, perfectly ventilated, and supplied with fresh water.

The last segment of the pleon bears the posteriorly directed, flattened, paddle-like uropods, which, together with the telson, form the caudal fan (Fig. 8.6b, c). By folding the pleon towards the ventral side, the developing larvae are also protected from predators. When the body is straight, the powerful muscles of the pleon can fold the tail forward on the ventral side, producing a swift backward movement. Therefore, the tail fan is initially an escape device.

Crayfish Molt Their Sclerotized Cuticle as They Grow
The exoskeleton of the crayfish, along with all other arthropods, consists of a multilayered cuticle made up of chitin and tanned (sclerotized) proteins. The incorporation of calcium additionally increases the strength of the cuticle. The individual segmental sections of the exoskeleton are flexibly connected by thin membranous sections which are not sclerotized (see Fig. 8.3a and 8.10a). The carapace covers the cephalothorax, which also protects the gills outside the body like a shield. Anteriorly, the carapace forms a protruding tip, the rostrum, which laterally protects the two eyes. It also serves free-swimming crustaceans, for example, shrimps, in stabilizing their swimming movement (Figs. 8.3, 8.4, and 8.5a). However, animals with a firm and rigid exoskeleton

cannot grow continuously. Therefore, as with all arthropods, growth in crayfish is inevitably always associated with molting. Crayfish, which can live up to 20 years, molt about once or twice a year. During the molts, crayfish are unprotected, as they lack the hard exoskeleton for several days. Crayfish are called soft-shell crayfish during this approximately 1-week phase, and remain in their hiding places for the time it takes for the new cuticle to harden. The European lobster, which reaches an age of up to 50 years and grows up to 60 cm, has undoubtedly undergone several dozen molts (Fig. 8.2). Some time ago, animal rights activists found a particularly large American lobster in the tank of a New York restaurant and later released it on the Atlantic coast in Maine. Based on its weight of about 9 kg, the age of this lobster was estimated to be 140 years.

Crayfish Have Compound Eyes
The crayfish has two compound eyes, sitting on stalks to the left and right of the rostrum (Figs. 8.4a and 8.5a). They can be moved independently of each other using the muscular eye stalks and can be retracted into designated indentations, for example, in case of danger. Like insects, crustaceans have compound eyes, which comprise a multitude of individual minute eyes called ommatidia. The rectangular cornea of each ommatidium, typical for decapods, can be observed under the stereomicroscope (Fig. 8.9). The cornea is hexago-

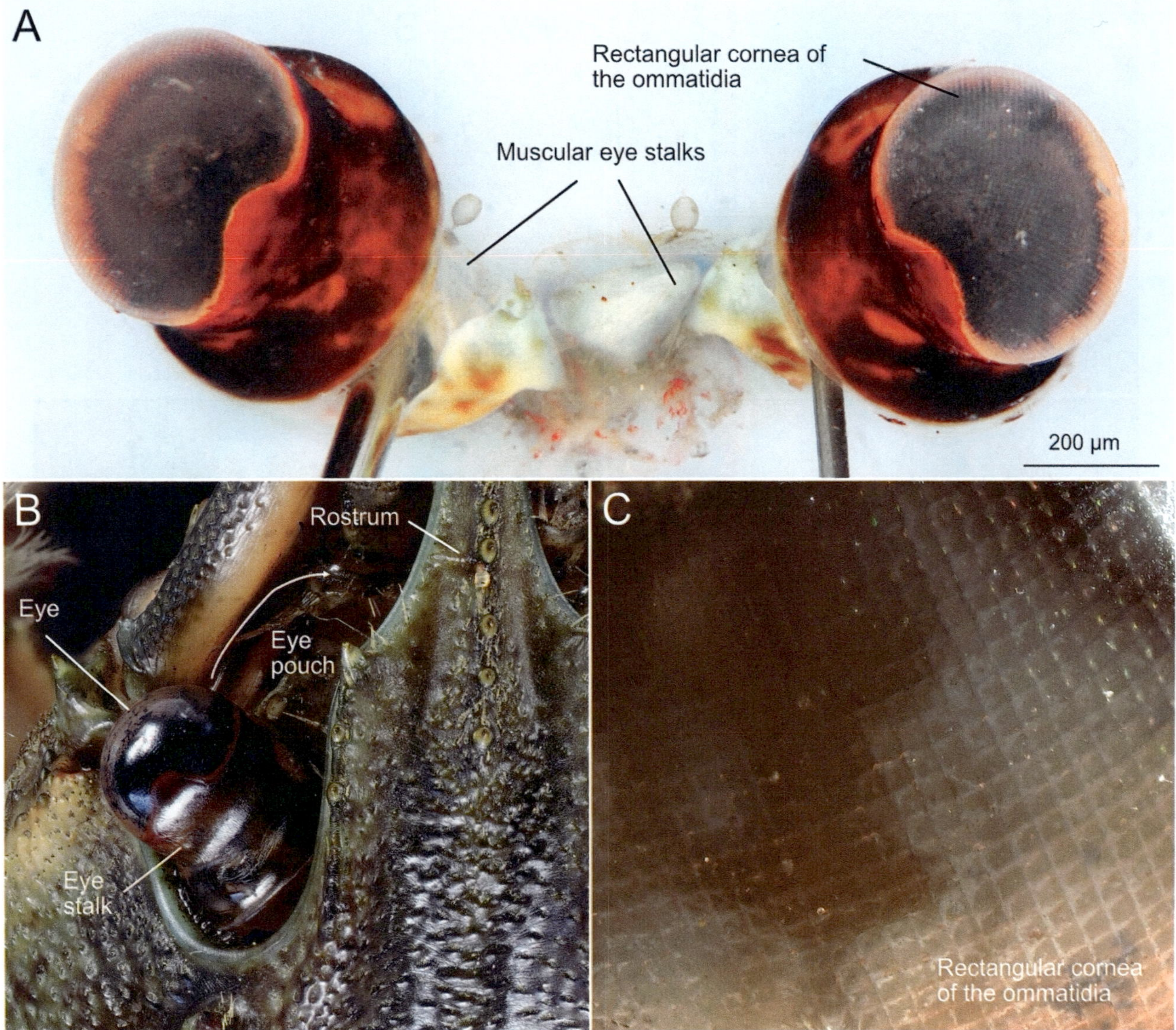

Fig. 8.9 *Astacus astacus*. (**a**) Dissected eyes with eye stalks. (**b**) Left eye of *Astacus astacus* in its natural position. Seen from above. The eye can be folded forward into an eye pouch for protection. (**c**) The surface of ommatidia of an *Astacus* eye at high magnification. Note the rectangular cross-section of the cornea

nal in the other crustacean groups, as well as in insects. Apart from that, the cellular structure of the ommatidia is completely identical in all crustaceans and hexapods. Their visual performance depends on the number of ommatidia, the optical or neuronal interconnections, and the quality of the optical system in the individual ommatidia. Crayfish have a so-called reflecting superposition eye, in which several ommatidia are interconnected by reflection. This mainly improves light sensitivity, but at the expense of the resolution compared to an apposition eye, in which each ommatidium works independently from the others. For a predominantly nocturnal animal, it is, however, important to have a good perception of light and dark.

Astacus: Preparation of the Internal Organs

(a) *Remove the carapace*: The dissection steps are shown in Fig. 8.10. Start at the posterior end of the carapace and cut the soft membrane, which connects the carapace with the pleon (Fig. 8.10a, cut 1, red line). Next, a rectangular window is cut from the posterior margin of the carapace across the cervical groove to near the eyes using a solid pair of scissors (Fig. 8.10a, cut 2, yellow line). When lifting the carapace, it is best to continu-

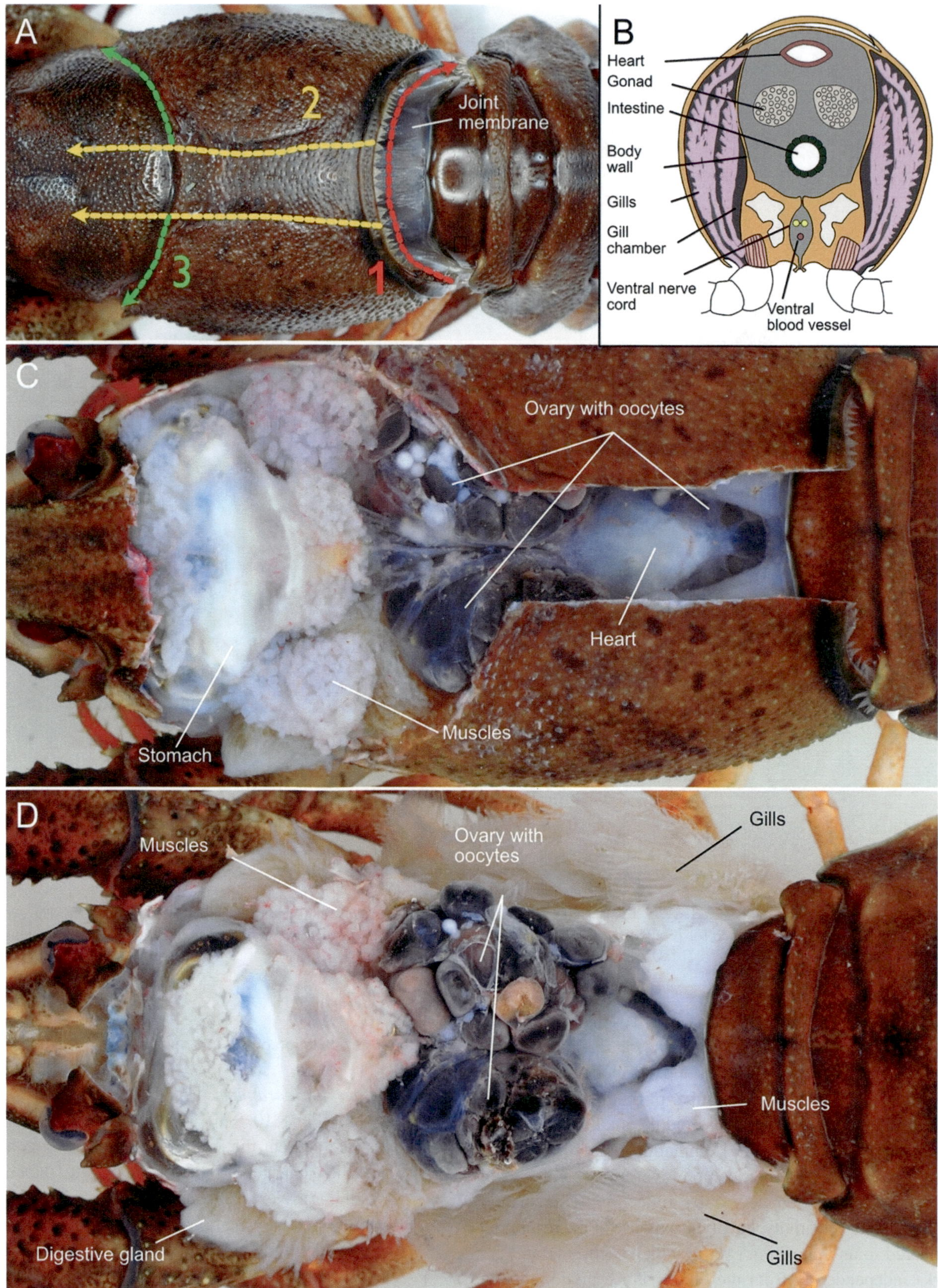

Fig. 8.10 *Astacus astacus*. Opening of the crayfish shell. (**a**) The first three cuts in the cephalothorax area. (**b**) Schematic representation of the location of the internal organs. According to various authors. (**c**) Opened cephalothorax with a view of the stomach, gonads, and claw muscles. (**d**) After removing the carapace, the gills become visible

ously separate tissue with a probe or scalpel to avoid damaging the organs. Finally, a cut is made from dorsal to ventral at the front, and along the cervical groove, after which a larger piece of the carapace is removed on both sides using a probe or scalpel (Fig. 8.10a, cut 3, green line). This provides a first view of the internal organs.

(b) *Internal organs*: If a female has been selected, the first things to be noticed are the ovaries filled with mature, violet eggs, the whitish transparent heart, and the sizeable white-blue stomach (Fig. 8.10c). The paired ovaries merge posteriorly, giving it a triangular appearance. This is best seen when the merged ovaries are isolated from the body and viewed under a stereomicroscope (Fig. 8.11). Now, the remaining carapace is removed using strong tweezers and scissors (Fig. 8.7d). This reveals the exposed gills on both sides of the animal (Fig. 8.10d). The stomach is prepared separately (Fig. 8.12) and opened to make visible the ectodermal cuticular teeth, which serve to grind the food (Fig. 8.13). Beneath the stomach, we recognize the digestive glands or ceca (Fig. 8.12b, c). Next, the dorsal region of the pleon is set free with two cuts. This reveals the strong musculature of the pleon (Fig. 8.12a), and the centrally running hindgut becomes visible. All internal tissues are carefully removed to get a clear view of the metameric nervous system (Fig. 8.15).

Ovary and Testes

The ovaries are immediately recognizable by their purple color after opening the cephalothorax of a female (Figs. 8.10c, d, and 8.11a). They are paired but fused in the posterior region. Only in sexually mature animals can the large, yolk-rich, purple-colored oocytes be seen. Usually, the ovaries contain eggs of different stages of maturity: oogonia and maturing oocytes, which often stand out due to a large cell nucleus (Fig. 8.11b). A single-layered, thin follicle epithelium covers each oocyte.

In the male, the long paired whitish vasa deferentia are the first structures to be noticed after opening the cephalothorax (Fig. 8.14a, b). They usually lie coiled behind the heart, and each vas deferens opens into the male genital opening on the coxa of the fifth pereiopod (Fig. 8.8). The paired testes are located behind the digestive gland and in front of and below the heart (Fig. 8.14a, b). They are partially fused, so their paired nature is not always easy to recognize. The testis itself consists of numerous testicular ducts in the wall, in which spermatogenesis takes place. The mature flagella-less sperm is later released into the interior of the testicular ducts (Fig. 8.14c, d).

Do Arthropods Have Blood?

The heart appears as an opaque, slightly translucent whitish organ near the dorsal midline (Fig. 8.10c). Anteriorly, three vessels branch off, usually being difficult to recognize without staining. They supply the eyes, as well as the brain (anterior aorta), the antennae, the stomach, and the excretory organs (lateral aortae), with blood. The abdominal artery (posterior aorta) runs backward, above, or laterally along the intestine and supplies tissue in the posterior part of the body. *Astacus,* like all arthropods, has an open circulatory system. The oxygen-poor blood collects, after being transported from the heart to the organs and tissues, in an open, ventral blood sinus, from where it flows to the gills. There, gas exchange occurs, and the blood is enriched with oxygen at the gills. It contains hemocyanin as a respiratory protein.

The whitish gills are located outside the body in the gill cavities (Fig. 8.10d). In total, there are 18 pairs of gills; they are located on the second and third maxillipeds, as well as the following four pairs of pereopods. The gills originate either from the trunk pleurites (pleurobranchiae), the coxae (podobranchiae), or at the joints between the coxae and trunk (arthrobranchiae). If the maxillipeds are gently moved back and forth with tweezers, the gills move with them. This is because the gills are directly connected to the legs. To examine the gills more closely, they are plucked off with tweezers and placed in a drop of water on a microscopic slide (Fig. 8.16). For surface enlargement, and thus improved gas exchange, the feather-like gills consist of a central shaft from which numerous pin-shaped appendages extend.

The term "blood" is often only used for animals with a closed circulation system, such as vertebrates. The coelom, the secondary body cavity and the blood vascular system form individual compartments and the fluid systems are completely separated from each other. In vertebrates, the blood contains respiratory proteins, as well as numerous other components, including cells of the immune system, carbohydrates, proteins (enzymes), dissolved gases, hormones, vitamins, and metabolic metabolites. In arthropods, the primary and secondary body cavities (coelom) are fused during embryonic development and form the so-called mixocoel. Coelomic spaces have largely disappeared and the blood has merged with the coelomic fluid to form the hemolymph. The hemolymph serves for, among other things, the storage and transport of lipids, collagens, signal molecules, hormones, and metabolites, and also to distribute the respiratory proteins. Thus, the hemolymph takes on tasks that, in animals with a closed blood circulation system, are fulfilled by two separate fluid systems. However, the open circulatory system of arthropods is always a low-pressure system, in which the circulation of the fluid, compared to a closed vessel system, is significantly slower. If the evolutionary nature of the body fluid in arthropods is to be expressed, the term hemolymph should be used. If the physiological aspect is in focus, the general term "blood" is also appropriate for cray-

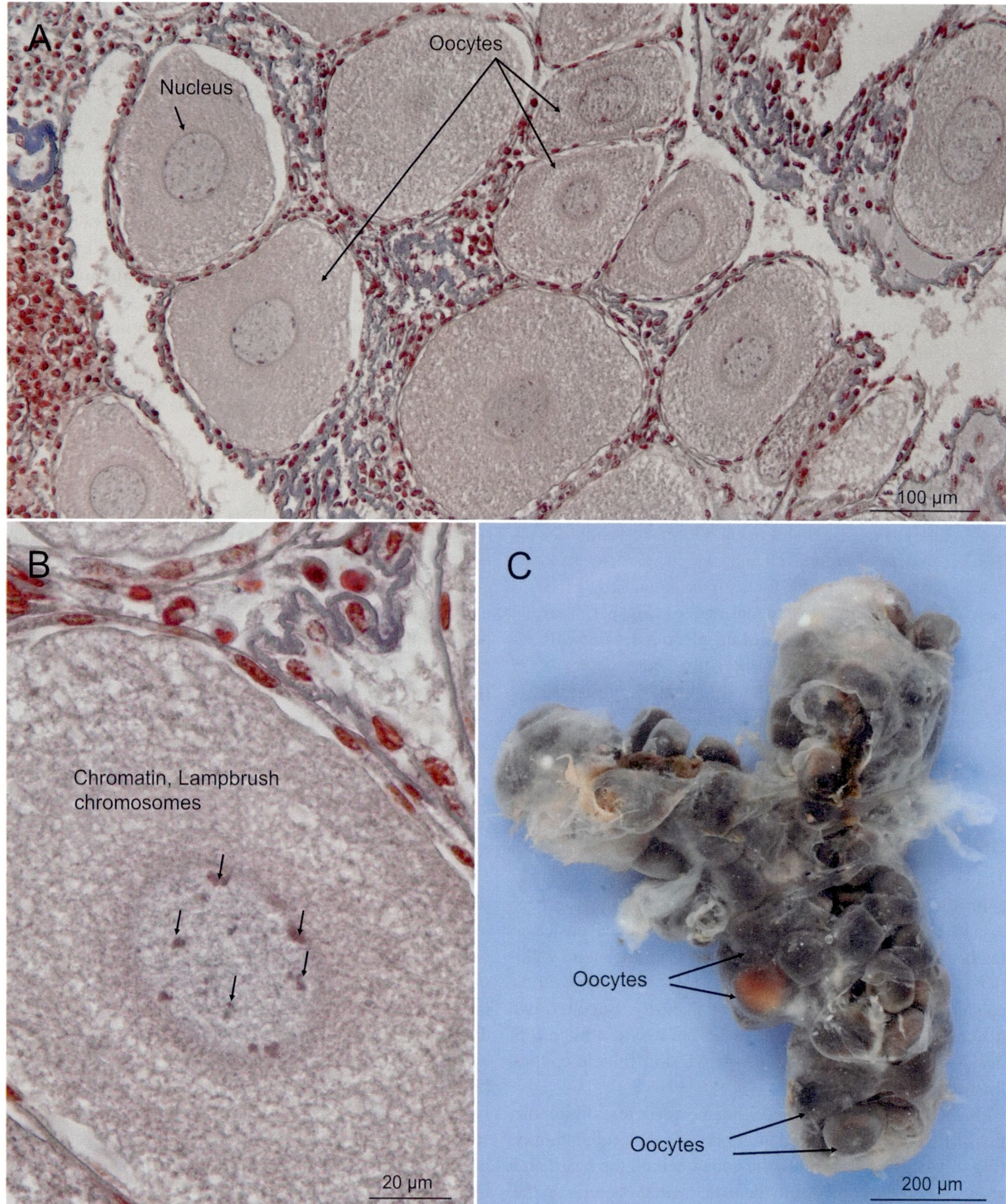

Fig. 8.11 *Astacus astacus*. (**a**) Cross-section of an ovary with various oocytes. (**b**) Histology of oocytes at higher magnification. (**c**) Exposed ovary with mature oocytes. Specimen from the Zoological Teaching Collection, University of Osnabrück

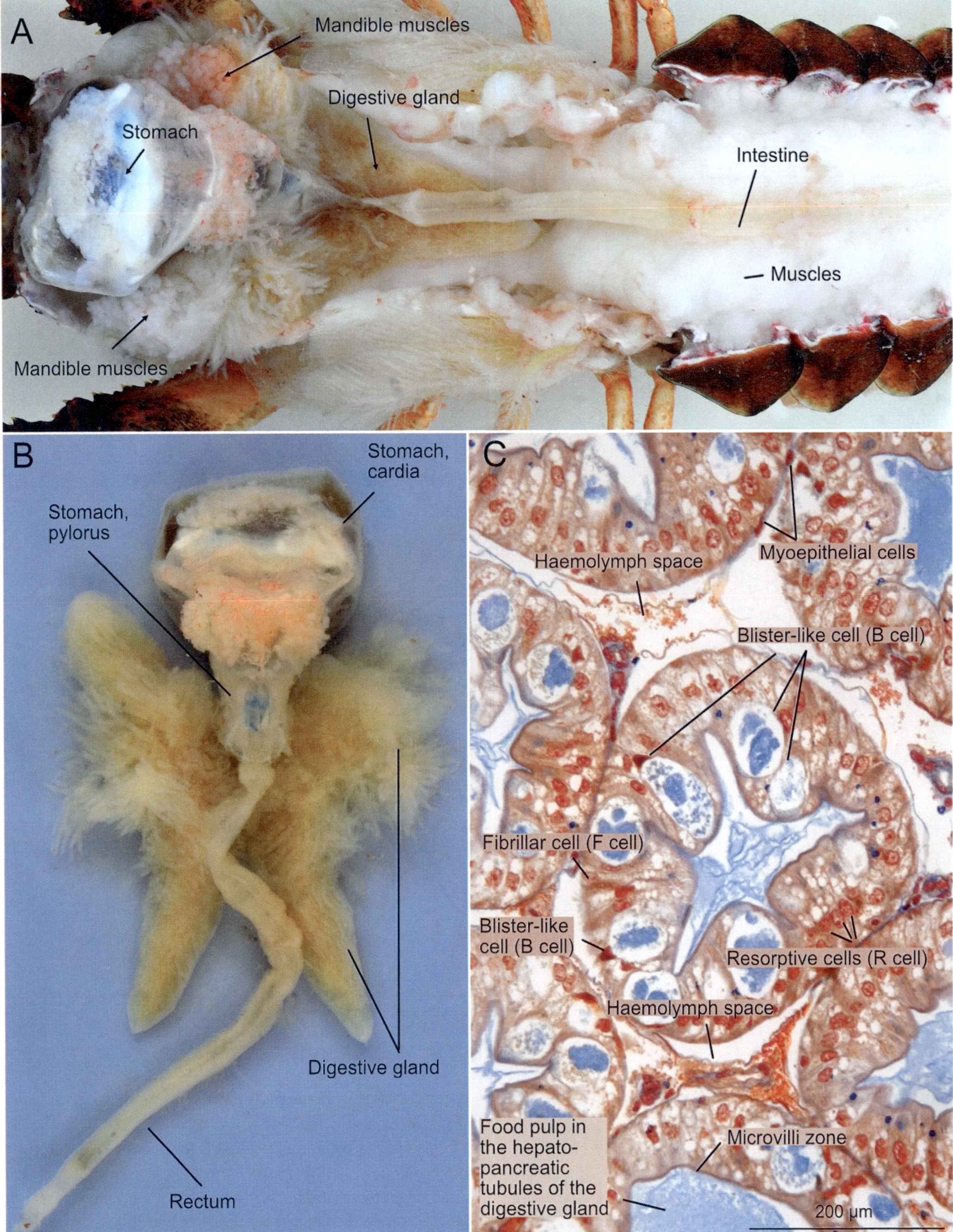

Fig. 8.12 *Astacus astacus*. Preparation steps for examining the digestive tract. (**a**) Exposed intestine with stomach and midgut gland. (**b**) Isolated stomach. (**c**) Cross-section through the midgut gland. Specimen from the Zoological Teaching Collection, Philipps University Marburg

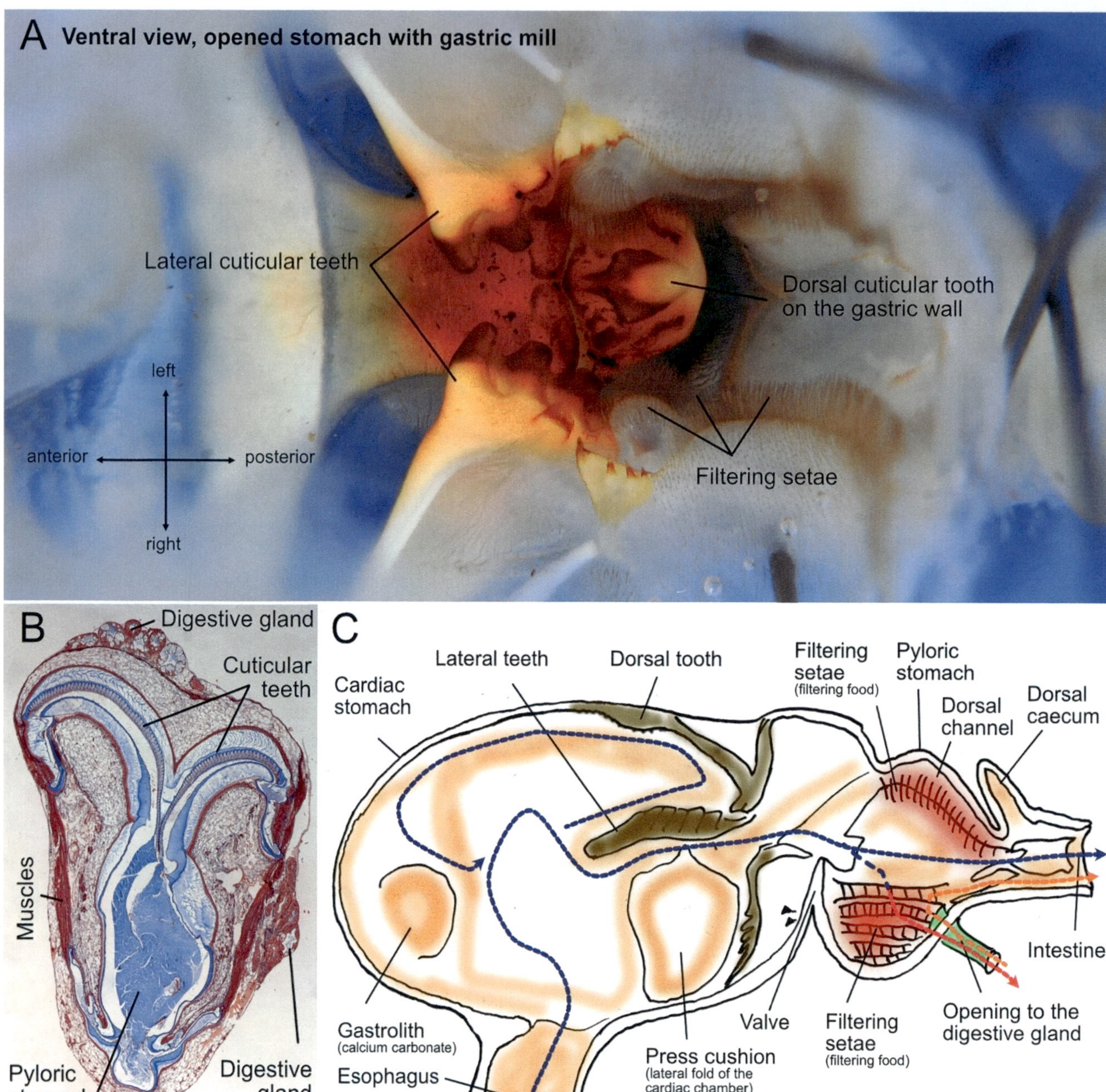

Fig. 8.13 *Astacus astacus*. (**a**) Open stomach (gastric mill). (**b**) Horizontal histological section through the stomach. Specimen from the Zoological Teaching Collection, Philipps University Marburg. (**c**) Schematic diagram of the decapod gastric mill, from various authors

Digestive Tract

Crayfish feed mainly on mussels, snails, insect larvae, annelids, and small crustaceans of all kinds. Young crayfish are herbivorous. The food is brought to the mouth by the maxillipeds, which are used for holding, while the mandibles are used for crushing and chewing (Figs. 8.4b and 8.5b). The digestive tract is examined in a dissection dish under water (Figs. 8.12 and 8.13). First, the rectum is cut at the back. Then, using a probe or a scalpel, the stomach is carefully lifted and detached from the surrounding tissue. This is best done when the animal is lying on its side and the stomach is pushed upward. Anteriorly, the esophagus should now be visible. It is separated from the stomach with scissors. With the help of probes, the stomach, with the attached digestive glands, is carefully removed from the animal. The digestive

fish, as it transports the respiratory proteins, like the vertebrate blood.

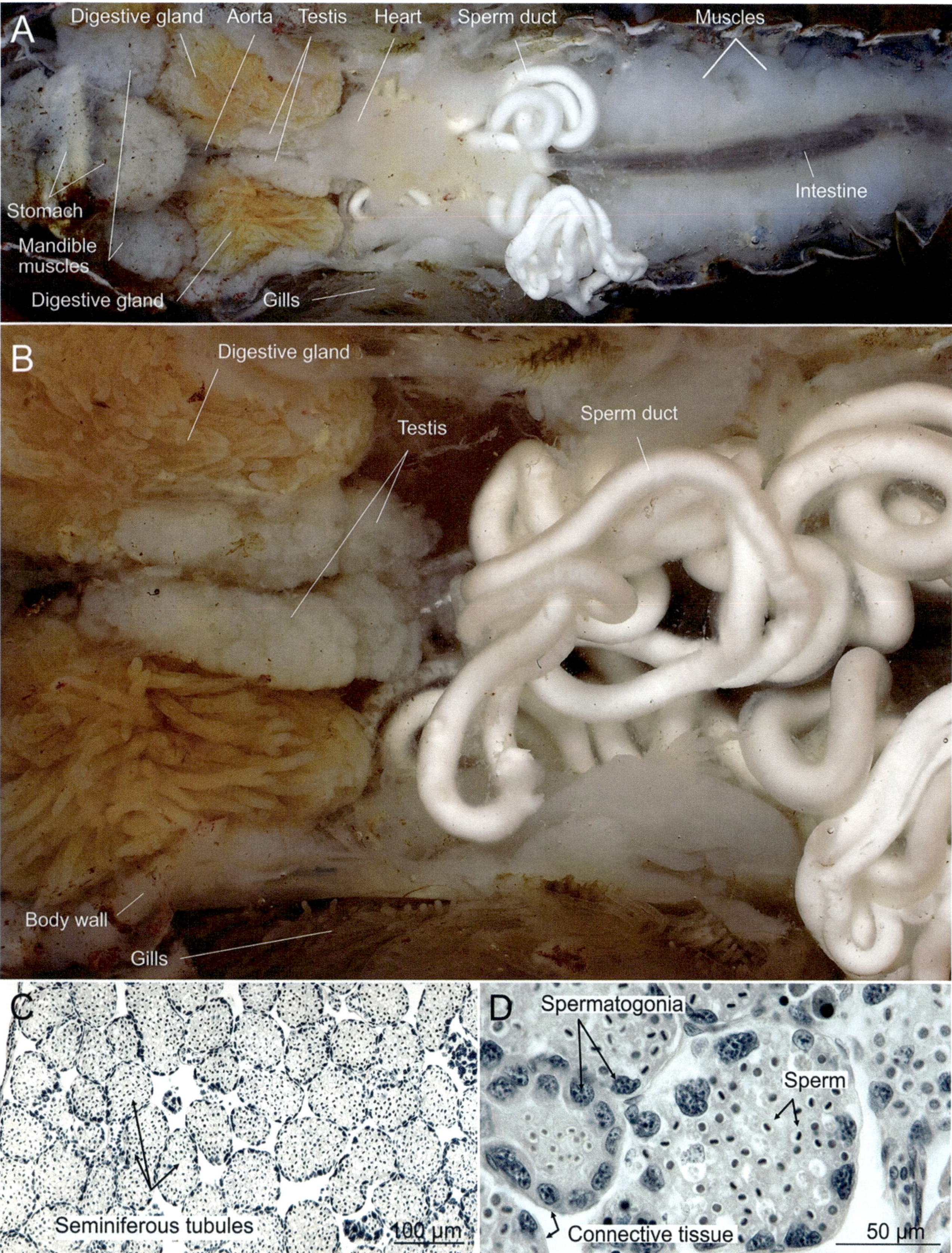

Fig. 8.14 *Astacus astacus.* (**a**) Male with digestive gland, stomach, testes, heart, and vas deferens. (**b**) Testes and vas deferens at higher magnification. (**c**) Section of testes with testicular tubules, overview, and (**d**) detailed view of some testicular tubules with spermatogonia and sperm. Specimen from the Zoological Teaching Collection, Heinrich Heine University Düsseldorf

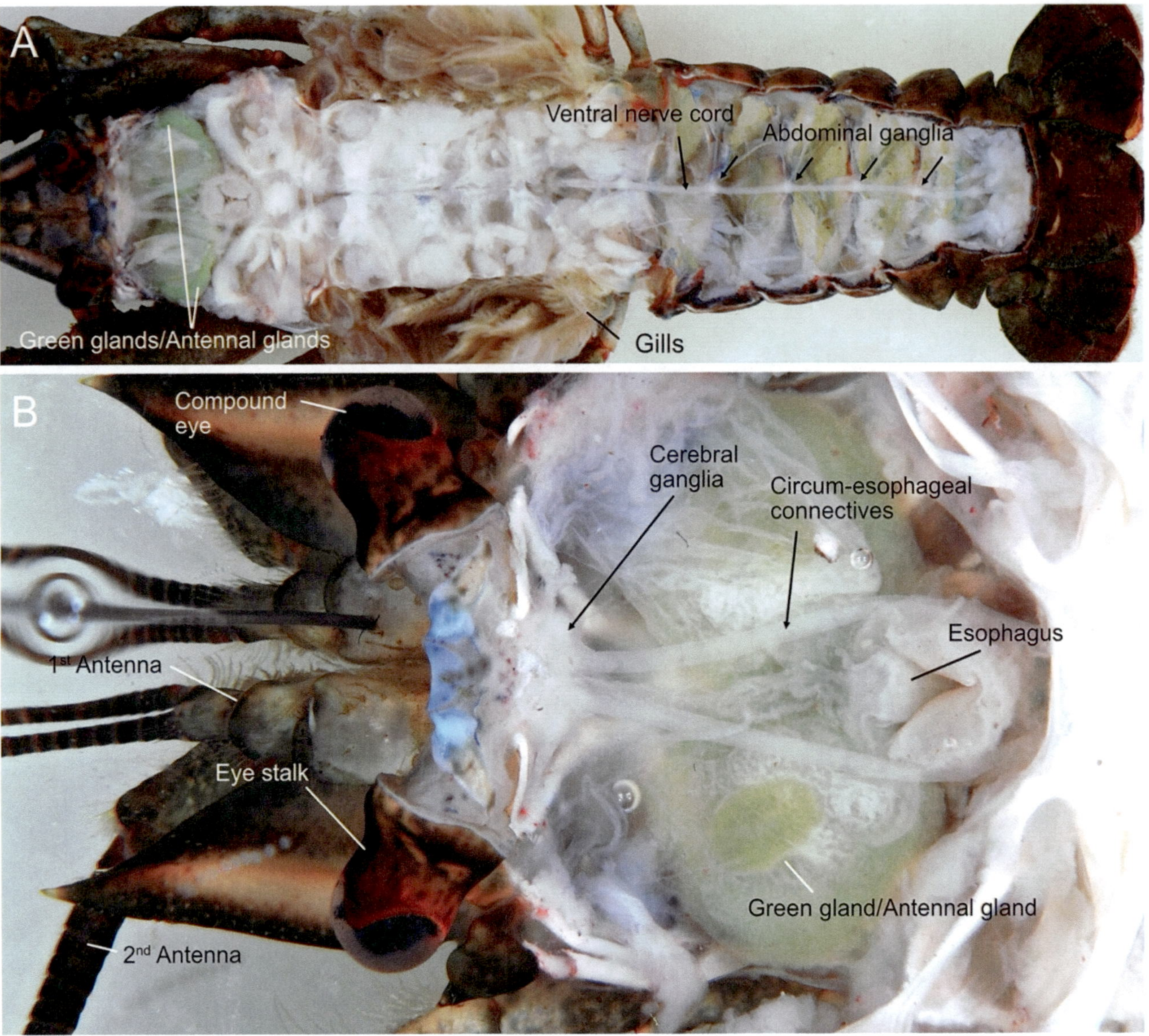

Fig. 8.15 *Astacus astacus*. Preparation steps for examining the nervous system and the antenna glands (= green glands). (**a**) Complete animal, opened dorsally, ventral nerve cord and abdominal ganglia are visible. (**b**) Exposed brain with esophageal connectives. Antenna glands are visible below the esophageal connectives

tract can be pinned down in its natural position with needles.

Via the esophagus, the food enters the bipartite stomach, which consists of the cardia (grinding stomach) and the pylorus (filter stomach) (Fig. 8.12b). At the transition from the pylorus to the hindgut is the short endodermal midgut with the openings of the digestive glands. The hindgut lies on the pleon musculature and extends as a thin tube to the anus. The digestive glands of crustaceans are not glands in the strict sense. Instead, they represent branched midgut diverticula. In the histological section of the digestive gland, several cell types are visible. Large bulbous bladder cells with a brush border of microvilli contain acidophilic secretory vacuoles. Filamentous cells and resorptive cells store lipids and glyco-

gen (Fig. 8.12c). However, before the food can be digested and absorbed in the diverticula of the digestive gland, it is mechanically comminuted. This is the task of the bipartite stomach, which is also called the gastric mill. As mentioned above, it consists of the grinding parts (cardia, gizzard) and the filtering part (pylorus). The digestion of food begins in the grinding stomach. Gastric denticles, teeth, assist in the crushing of food chunks. The teeth are arranged in rows, situated on three cuticle ridges (Fig. 8.13). Two rows are located laterally (one on the left and one on the right) and one median on the stomach surface. To make the teeth visible under the stereomicroscope, the stomach is placed on the dorsal side and freed from remnants of the esophagus. We now look into the gizzard. The opening is carefully widened and used for fur-

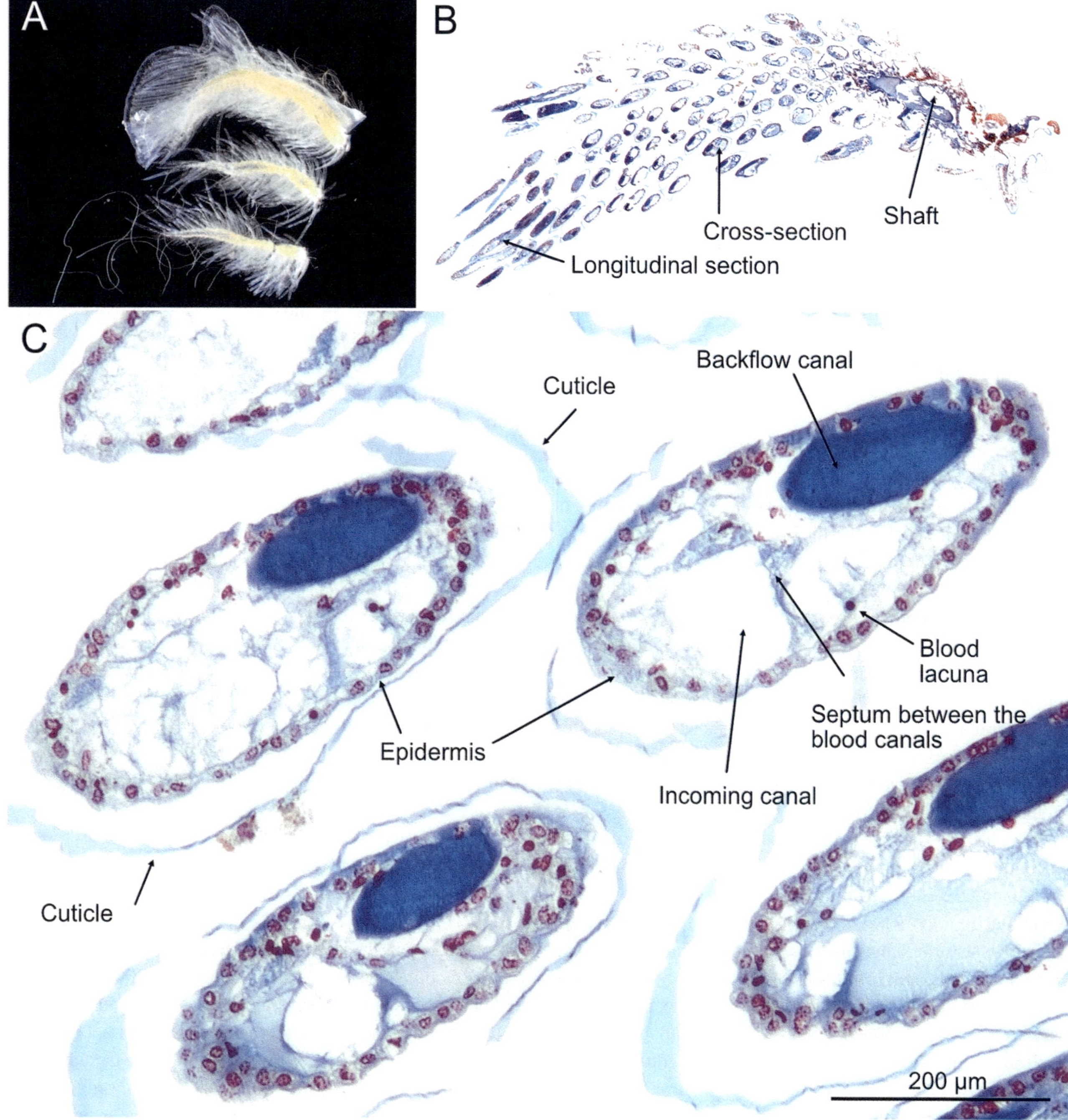

Fig. 8.16 *Astacus astacus*. (**a**) Gills. (**b**) Longitudinal section through a gill. (**c**) Cross-section through a gill branch with supplying and draining vessel. Azan staining. Specimen from the Zoological Teaching Collection, University of Osnabrück

ther sections until the stomach teeth can be seen (Fig. 8.13a). Digestion has already begun once the food is in the gizzard. This is possible because digestive secretions released further posteriorly from the digestive gland into the midgut are sucked into the stomach by muscle activity. However, at the end of the pre-digestive process, the food pulp is forwarded to the pylorus. The pylorus has a fold ventrally and medially that forms two channels separated by filtration bristles and leads to the digestive glands. The food pulp is pressed through the filtrating bristles into the midgut and the ducts of the digestive gland. Larger particles are held back and directly transported into the hindgut. The short midgut with the digestive glands is, by the way, the only endodermal part of the digestive tract; all other intestinal sections are ectodermal, visible thanks to, among other things, the chitin teeth in the grinding stomach. They, therefore, also renew with each molt. Sometimes a whitish gastrolith can be seen in the front part of the stomach (cardia). It serves as a calcium reservoir.

Musculature

The entire abdomen is filled with musculature. The muscles are white, indicating a low myoglobin content and a reliance on glycolytic enzymes. White muscles are less suited for sustained locomotion, but more so for rapid contraction. It allows the crayfish to make swift striking movements with the abdomen and the fan tail so that the animal can swim backward in the water or promptly pull back into the burrow as an escape response. Specific striking movements of the pereopods of crustaceans are among the fastest muscle reactions known in animals. The curling of the abdomen also plays an essential role in the brood care of the females and during copulation. The muscles responsible for the chewing movements of the mandibles are particularly strong (Fig. 8.10b, c). The histological structure and the myogenesis of the crustacean musculature resemble the muscle development in insects. As with many organisms with a hard skeleton, the musculature of the crustacean is cross-striated. Another important feature crustaceans, insects, and vertebrates share is the syncytial nature of muscle cells. In this process, the huge muscle cells are formed by the fusion of myoblasts.

Osmoregulation and Excretion

Toward the end of the crayfish dissection, attention will be turned to the excretory and nervous systems. First, carefully remove the remains of the stomach and the digestive glands. The excretory organs of decapod crustaceans are located behind the eyes and are usually clearly visible as larger greenish organs. Because of their coloration, they are called green glands (Fig. 8.15). These are modified and enlarged metanephridia. They each consist of a coelomic terminal sac (sacculus) that continues into a greenish, narrow-lumen, long, tangled nephridial duct. The primary urine is formed from the hemolymph via ultrafiltration into the sacculus. Then, it enters the labyrinth system (nephridial duct), which lacks a ciliated funnel. In this labyrinth system, the absorption of ions and the secretion of high molecular weight excretory substances occur (Fig. 8.17). From there, a longer nephridial canal extends, first to the urinary bladder and then to the base of the second antenna. There, the urine is released into the environment through the excretory pore (Fig. 8.5b). From the heart, a branch of the laterally projecting vessels (aorta lateralis) leads directly to the sacculus and transports

hemolymph to the green gland. The green glands in crayfish are relatively large, and more prominent than in their marine relatives, as crayfish live in a hypotonic medium, and thus constantly passively take up water through osmosis that must be released again via the nephridia. This problem does not exist for marine crustaceans, which are isoosmotic to the medium. Toxic nitrogen compounds (ammonia) are also released into the external water via the gills.

Nervous System

To expose the nervous system, all internal organs and the musculature in the pleon must be removed with solid forceps (Fig. 8.15). It is best to start posteriorly and work forward to the eyes (Fig. 8.15). The ventrally located nervous system is built on the principle of a rope ladder. In the pleon, we find the ventral nerve cord, with six pairs of ganglia, which can be assigned to the respective segments and their appendages (Fig. 8.15a). Anteriorly, the nerve cord can be traced to the subesophageal ganglion. The five thoracic pairs of ganglia each project to the walking legs, while the subesophageal ganglion can be assigned to the segments of the mandible up to the third maxillipeds. From the subesophageal ganglion, the paired circumesophageal connectives lead in an arc to the brain or supraesophageal ganglion (Fig. 8.15b). The brain is located between the bases of the eyestalks. As with all arthropods, it is multipartite. Nerves connecting to the two antennae and the compound eyes, among others, originate from here. Our movie clip collection, which includes movies showing living crayfish, can be found at figshare: sn.pub/xr9qhi.

8.2   The Argentine Cockroach (*Blaptica dubia*, Hexapoda, Insecta)

Insects represent the most diverse and ecologically successful group of animals on our planet, with over 1 million known species. Hexapods possess several characteristic morphological features (autapomorphies). Perhaps the most important key feature is their tripartite body, consisting of three tagmata (sing. tagma, Greek, division): the head (caput), the thorax, and the abdomen. These three tagmata comprise six, three, and eleven segments in the hexapod ground pattern. Each section is specialized for certain functions. The head carries the most important sensory organs, the brain, and the organs for feeding. The thorax serves primarily for locomotion and carries the six walking legs, as well as, in wing-bearing insects, the two or four wings. The abdomen bears the vegetative and generative organs. The body, originally composed of 20 segments, is anteriorly completed by an acron and posteriorly by a telson. These two sections are not counted as segments, lacking the repetitive internal structure. The telson carries the anus in insects.

The body structure of hexapods is generally subject to an enormous range of variation, often accompanied by a fusion or reduction of body segments. Various, sometimes fascinating anatomical adaptations have enabled insects to colonize

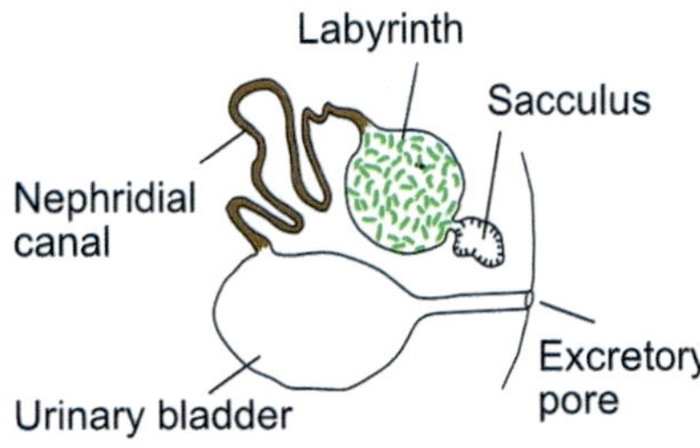

Fig. 8.17 Schematic representation of the antenna gland, according to various authors

widely different habitats in almost all climatic zones. For introducing the basic anatomy of insects, cockroaches (Blattodea) are often prepared in zoological courses. They show many characteristics of the insect ground pattern, and they are also available all year round in pet shops. Most of the approximately 5000 known cockroach species inhabit tropical or subtropical ecosystems. However, a few species are also found in the European Mediterranean region. For example, the approximately 1.5 cm large German cockroach (*Blattella germanica*) is native to large parts of Central Europe and prefers to colonize humid-warm habitats with good food supply, for example, large kitchens, hospitals, and similar buildings.

According to current knowledge, all hexapods go back to a single stem species. They are, therefore, monophyletic. Many laymen believe all insects can fly, as bees, wasps, flies, and butterflies can. However, this is not the case. The well-known silverfish (Zygentoma) and the rock jumpers (Archaeognatha) are primarily wingless and, therefore, unable to fly. Wings appeared later, during the course of evolution. The wing-bearing insects are grouped into a separate taxon called Pterygota. The cockroaches presented in this chapter also belong to pterygotes. These have two pairs of wings, one on the second and one on the third thoracic segment. However, in Diptera, the second pair of wings is reduced. Instead, the so-called halteres are formed that perceive rotational movements as mechanosensory organs and enable the dipterans, for example, flies, to perform acrobatic flight maneuvers. In Hymenoptera, which include bees, wasps, and many other insects, the second pair of wings, if not completely regressed, is usually very small and mechanically coupled to the first pair. Functionally, these insects also fly with only two wings. In addition to the primarily wingless insects mentioned above, it should be mentioned that there are also secondarily wingless insects, for example, fleas. The insects known today are currently divided into about 30 higher taxa, traditionally called orders.

External Examination of *Blaptica dubia*

Male and female specimens of the Argentine cockroach (*Blaptica dubia*) are very easy to distinguish (Fig. 8.18). The males possess two large and strongly chitinized, sclerotized forewings (tegmina) covering the entire abdomen and the underlying flight wings. Females have considerably shortened tegmina that virtually do not cover the abdomen at all (Figs. 8.18 and 8.19). The three tagmata can be recognized in cockroaches in both sexes. The head is equipped with antennae and mouthparts (Fig. 8.10), namely, the paired antennae, mandibles, maxillae, and the labium. The latter consists of the fused second maxillae, whereas the segment of the second antenna does not bear appendages in hexapods. As in crustaceans, the head segments are invisible as such. They are externally defined by the arrangement of the antennae and mouthparts and internally by the corresponding ganglia in the central nervous system. The wing- and leg-bearing thorax consists of three recognizable segments. These thoracic segments are called pro-, meso-, and metathorax. This is followed by the abdomen, originally consisting of 11 segments. In *Blaptica dubia*, due to reductions, fusions, and the interlocking of segments, usually only seven to eight segments are visible. Also, in both sexes, we find paired mechanosensory abdominal appendages, called cerci, which sense the environment (Fig. 8.18). The internal organs of each body segment are protected by chitinous plates (sclerites): a dorsal tergum, a ventral sternum, and two lateral pleurites (Figs. 8.18 and 8.21). These are flexibly connected by weakly sclerotized areas. Under the stereomicroscope, laterally arranged pairs of stigmata can be seen. Stigmata form the openings of the respiratory system (tracheae) to the outside.

Wings

Insect wings evolved from outgrowths of the epidermis. They originate in all Pterygota from the second and third thoracic segments. Wings are the most critical innovation evolved in insects. Wings allow access to the third dimension and, in this way, the rapid colonization of new habitats, meeting potential mates, and are also highly effective when it comes to escaping from a potential predator. This is the only way to explain that more than 99% of all insect species belong to Pterygota. Among the winged insects, there is often an apparent functional separation between juvenile and larval stages on the one hand and the adult stages or imagines on the other: The flight-capable adult stages often contribute exclusively to the reproduction and dispersal of the species, while growth and molting is reserved for the larval stages. Very likely, the reason for this lies in the delicate structure of the wings which might not be easy to molt during growth without damage. In fact, there is only one wing-bearing group of insects in which a winged-stage moltes once, namely the mayflies, Ephemeroptera. In many species, the imagines no longer consume any food, as they often live only a few days or weeks and die shortly after a successful mating. This is most impressively demonstrated by the mayflies (Ephemeroptera) that live only a few hours as adults. Many insect wings carry a series of hair sensilla, often on the leading margin of the wing. These are mechanoreceptors that respond to the flow of air over the wings in flight, so as to, for example, serve to improve flight ability. However, chemoreceptors and mechanoreceptors that respond to mechanical stimuli such as touch are also common. As with many other cockroach species, only the males of *Blaptica dubia* can fly. Only they possess a functional second pair of hindwings or flight wings (Figs. 8.18 and 8.19), which is protected by the forewings. Since the forewings of cockroaches and, for example, stick insects (Phasmatodea) have functional tracheae, they are different from the hardened forewings of beetles, the elytra, which lack these elements. Therefore, a different term for the protective wings in cockroaches has been coined, tegmina. The second pair of wings, the hindwings, of the females of *Blaptica dubia* is greatly reduced and does not allow for active flight.

Fig. 8.18 *Blaptica dubia*. Adult female (right) and adult male (left). (**a**) dorsal view. (**b**) ventral view

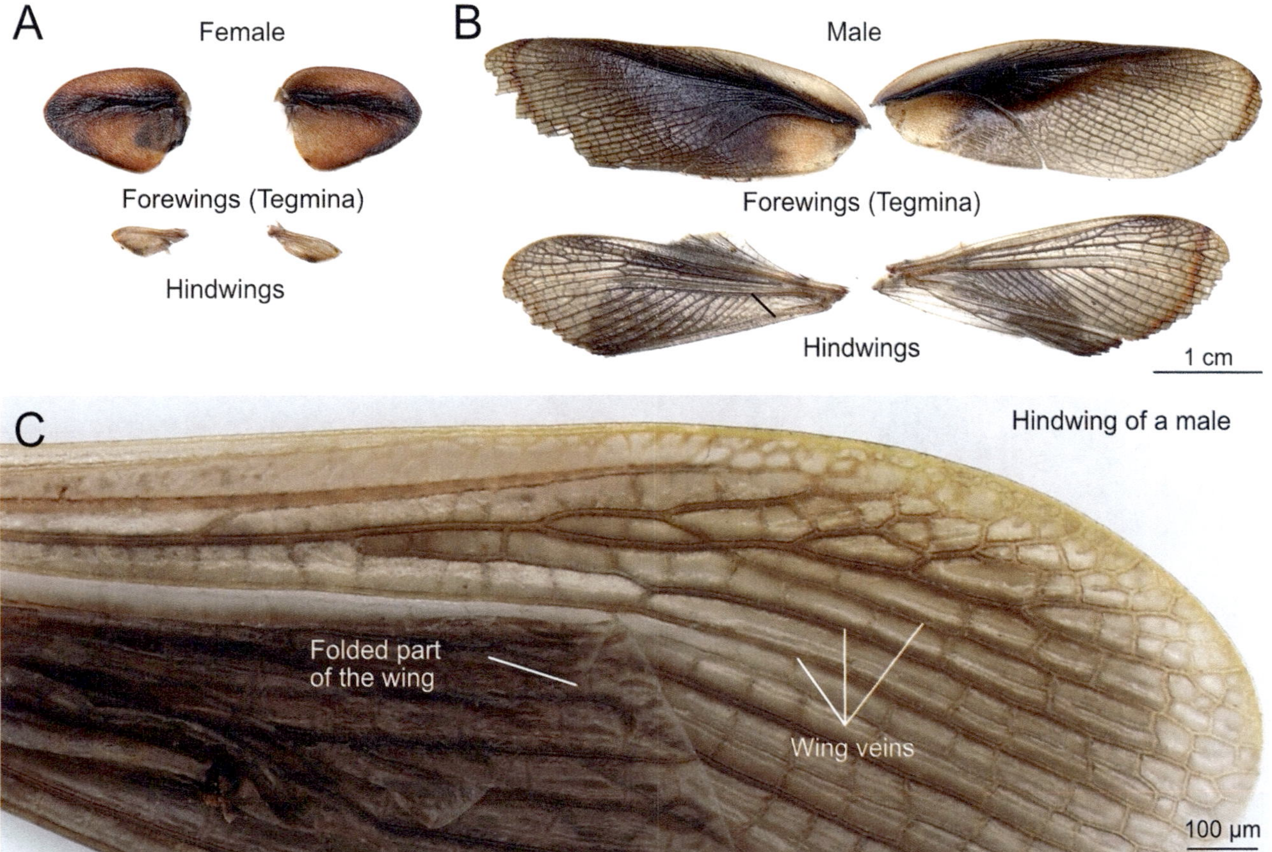

Fig. 8.19 *Blaptica dubia*. (**a**) Tegmina (forewings) and flight wings (second pair of wings) of female (**b**) animals. (**c**) Flight wing of a male in detail

Legs

The three pairs of legs of the cockroaches are typical walking legs (Figs. 8.18 and 8.21). They consist of the individual sections: coxa, trochanter, femur, tibia, and the five-part tarsus ending in a pretarsus and claws. On the tarsal segments are the whitish-appearing pulvilli or plantula (Fig. 8.21). These are smooth pads that serve as attachment devices. They allow for close contact of the foot with the ground so that Van der Waals and capillary forces can act. These allow many insects to walk on very smooth and vertical surfaces. The film of moisture on the surface required for capillary forces is produced by the insects themselves by releasing fluid on the tarsi. Two large claws are noticeable on the pretarsus, with which the animals can move safely on rough terrain.

Compound Eyes

The large, laterally arranged paired insect eyes (Fig. 8.20a, b) are compound eyes, as they are composed of numerous individual eyes, the ommatidia. The number of ommatidia per eye varies between 10 and 12 in silverfish, 600 and 1000 in most flies, and up to 30,000 in dragonflies, and naturally is closely related to their individual lifestyles.

Dragonflies, for example, are diurnal hunters that hunt other insects in flight. They, therefore, have high-resolution compound eyes. Each ommatidium has a chitinous lens formed from the cuticle, through which light enters the ommatidium proper. There, the light falls through the crystal cone into the rhabdom, which is formed by the closely apposed microvilli of the rhabdomeric photoreceptor cells. The rhabdom is the light-sensitive part of the ommatidia. The visual performance of compound eyes correlates with the structure and the optical or neuronal interconnection of the ommatidia. In the apposition eye of mostly diurnal insects, the individual ommatidia are optically separated from each other. This allows for a relatively sharp imaging performance, but the light sensitivity of the apposition eye is quite low. In the optical superposition eye of many nocturnal insects, such as that of moths, light that falls into an ommatidium also reaches adjacent rhabdomes. This leads to a more efficient light yield. Optical superposition eyes are, therefore, very light-sensitive, with a reduced spatial resolution. In the neuronal superposition eyes of fast-flying insects, for example, flies, the rhabdomeres of several ommatidia are neuronally interconnected. This type of eye is characterized by a very high

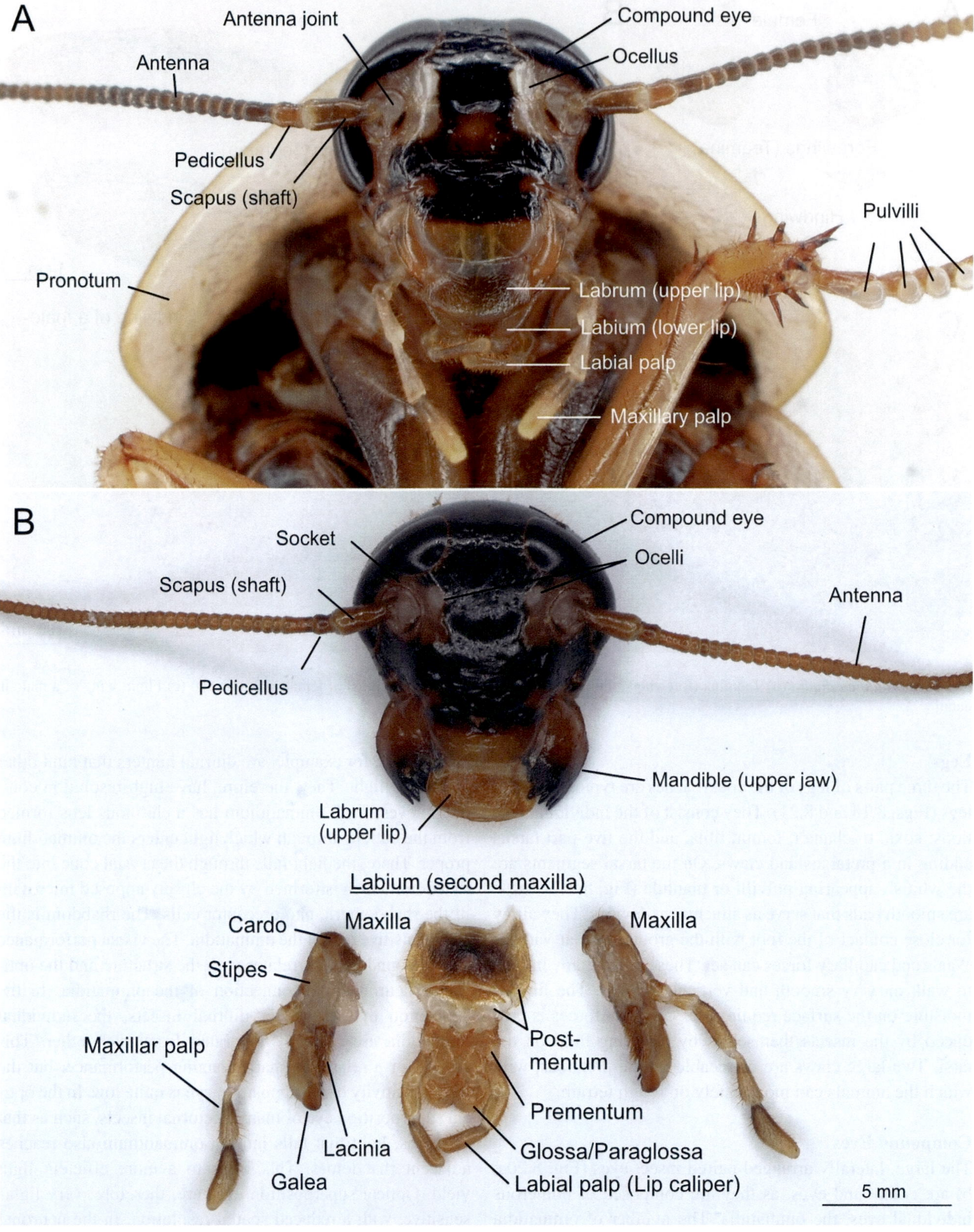

Fig. 8.20 *Blaptica dubia*. (**a**) Mouthparts in a natural position. (**b**) Prepared mouthparts

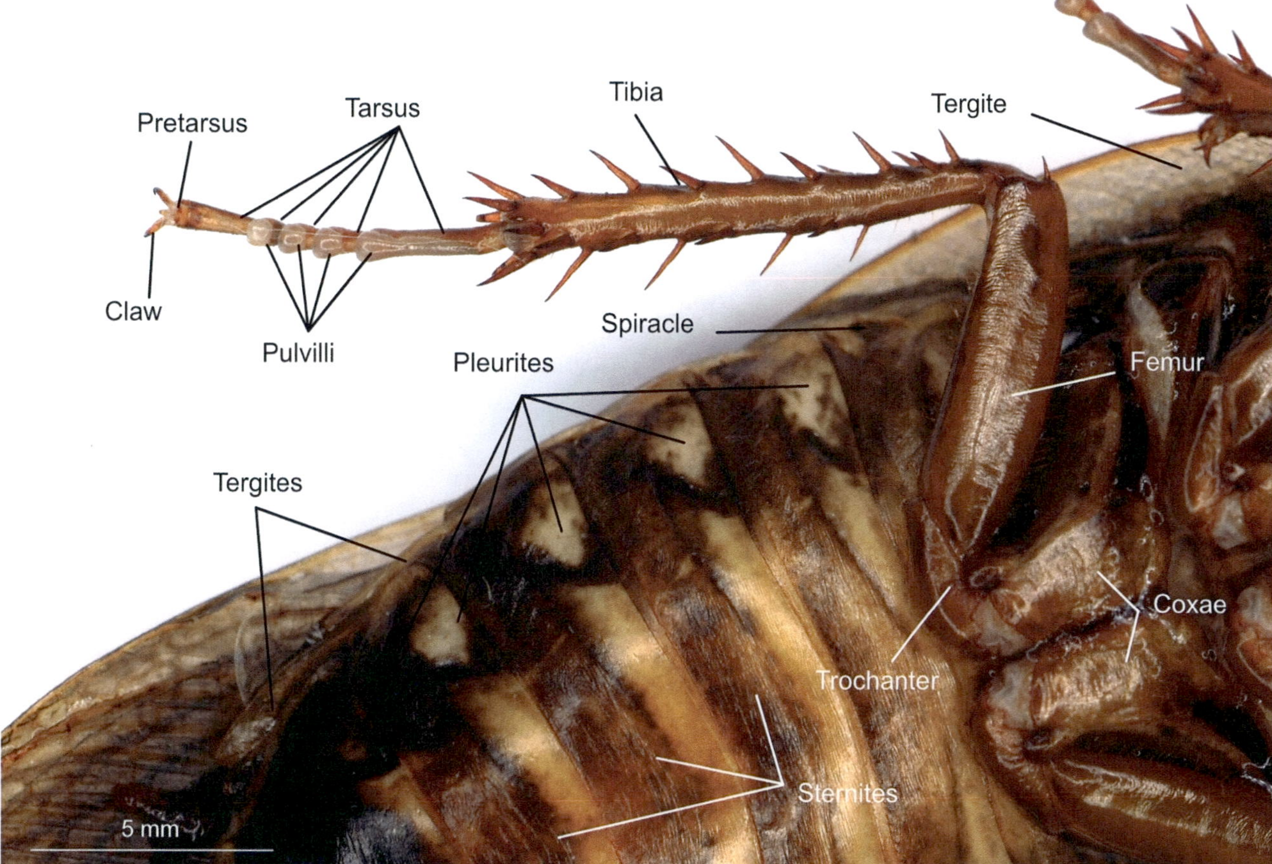

Fig. 8.21 *Blaptica dubia*. Leg and segmented body structure

temporal resolution and relatively sharp spatial imaging performance. Thus, a housefly (*Musca domestica*) can resolve about 300 images per second, about ten times more than a human.

In addition to the large compound eyes, hexapods also have other important light sensory organs, the median eyes (ocelli). Their number has been reduced during evolution from the original four in the arthropod stem species to three in hexapods. They enable directional vision and light perception, and are also part of the ground pattern of Arthropoda. Meanwhile, *Blaptica dubia* has only two ocelli, which are located frontally next to the antennal base under whitish structures, the so-called windows (Fig. 8.20). A neuronal linkage of the photoreceptor cells and the presence of shading pigment cells give the median eyes directional vision. For the *Blaptica dubia*'s close relative, the American cockroach (*Periplaneta americana*), it was recently shown that the ocelli are involved in the animal's optomotor performance, especially at dusk. In many cases, the exact function of the ocelli is still unclear. In dragonflies, for example, it is suspected that the median eyes serve to perceive the horizon, and thus contribute to the control of flight movements and horizontal flight position.

Are there also hexapods without eyes? Yes, the 0.8–1.3 mm small Protura, sometimes named coneheads, and the 4–12 mm large Diplura, the two-pronged bristletails, lack eyes. Both groups live in the upper soil layers, under stones, or in small spaces in the ground, i.e., in biotopes without light.

Antennae

The antennae, originating from the second head segment of cockroaches, are homologous to the first antennae of crustaceans. They are characterized by the presence of numerous sensilla, which primarily serve for the perception of odor, and thus harbor numerous chemoreceptors (Fig. 8.20a, b). The muscles responsible for the movement of the antennae are located exclusively in the shaft of the antenna, but not in the subsequent distal antennomeres. Antennae of this type are referred to as flagellate antennae. All hexapods with external mouthparts (Ectognatha = Insecta) always have flagellate antennae representing one morphological apomorphy. The shaft (scapus) connects the antenna to the head. The muscles in the shaft allow for the movement of the antennae via a joint. The second section of the antenna (pedicellus) contains the Johnston's organ, a so-called scolopidial organ (chordotonal organ), whose sensilla perceive movements of the antenna relative to the pedicellus. This organ is extremely sensitive and registers vibrations, such as sound, pressure changes, and air currents caused by movement. It allows cockroaches a very early escape reaction, for example, when a human enters the room. The actual antenna follows the

Fig. 8.22 *Blaptica dubia*. Developmental stages, nymphs

pedicellus. In *Blaptica dubia*, it consists of dozens of individual sections, which together carry more than 250,000 chemosensory sensilla. The second pair of antennae present in crustaceans is absent in hexapods, so the third head segment is without appendages. Statocysts are not found in *Blaptica dubia*. Springtails (Collembola) and diplurans (Diplura) have so-called segmented antennae, which are muscularized in all sections except the terminal one, allowing each section of the antenna to be moved individually. However, these hexapods are not discussed further here.

Mouthparts

Hexapods possess many differently shaped mouthparts (Fig. 8.20), sometimes assigned to the head segments. Typically, they consist of the paired mandibles (upper jaws), the first maxillae (lower jaws), and the labium (lower lip), which corresponds to the basally fused second maxillae as present in crustaceans. An unpaired labrum (upper lip) covers the mouthparts from the front. Hexapods have produced a variety of highly specialized mouthparts in the course of evolution as adaptations to different food sources. Thus, depending on the type of insect, the differently differentiated mouthparts allow for a variety of feeding methods, for example, chewing-biting (grasshoppers), sucking (butterflies, flies, proboscis), licking-sucking (bees), or piercing-sucking (female mosquitoes, bugs). During an excursion to a summer wetland, the variation of insect mouthparts can be experienced firsthand. In that instance, you could be bitten and sucked by female horseflies, stable flies (flies), horseflies (flies), or mosquitoes or simply bitten by wart-biters (grasshoppers).

Development

In hexapods, the individual development of an animal follows two different basic patterns. In hemimetabolous insects like cockroaches, the development is always gradual. With each molt, the juvenile stages, called nymphs, become more and more similar to the adults. Only during the last molt to the adult animal, the imago, are the wings finally formed (Fig. 8.22). In cockroaches, as well as other hemimetabolous insects, the lifestyles of nymphs and adults are often very similar. Insects with a holometabolous development, such as beetles, butterflies, and flies, always go through several true larval stages before a complete transformation to the imago occurs. This transformation is referred to as metamorphosis and occurs during pupal dormancy. The term "larva" is used in a strict sense only for the juvenile stages of holometabolous insects, to make clear that these juvenile stages differ greatly in their anatomy and lifestyle from the imagines. For example, adult holometabolous insects have compound eyes, reproductive organs and wings, which are always missing in larvae. The larvae of holometabolous insects usually also inhabit different habitats and show different food preferences than the adult animals. For the larvae of holometabolous insects, terms are often used that are unique to their morphology. Thus, the larvae of butterflies are called caterpillars (because they have real legs), those of beetles are called grubs (real legs and head), and those of flies, bees, and ants are called maggots (because they are legless). The holometabolous life cycle is ecologically very successful, the vast majority of insect species belong to the holometabolans, which, by the way, go back to a unique common ancestor.

The development of cockroaches begins with the internal fertilization of the eggs. Sperm are transferred from the male

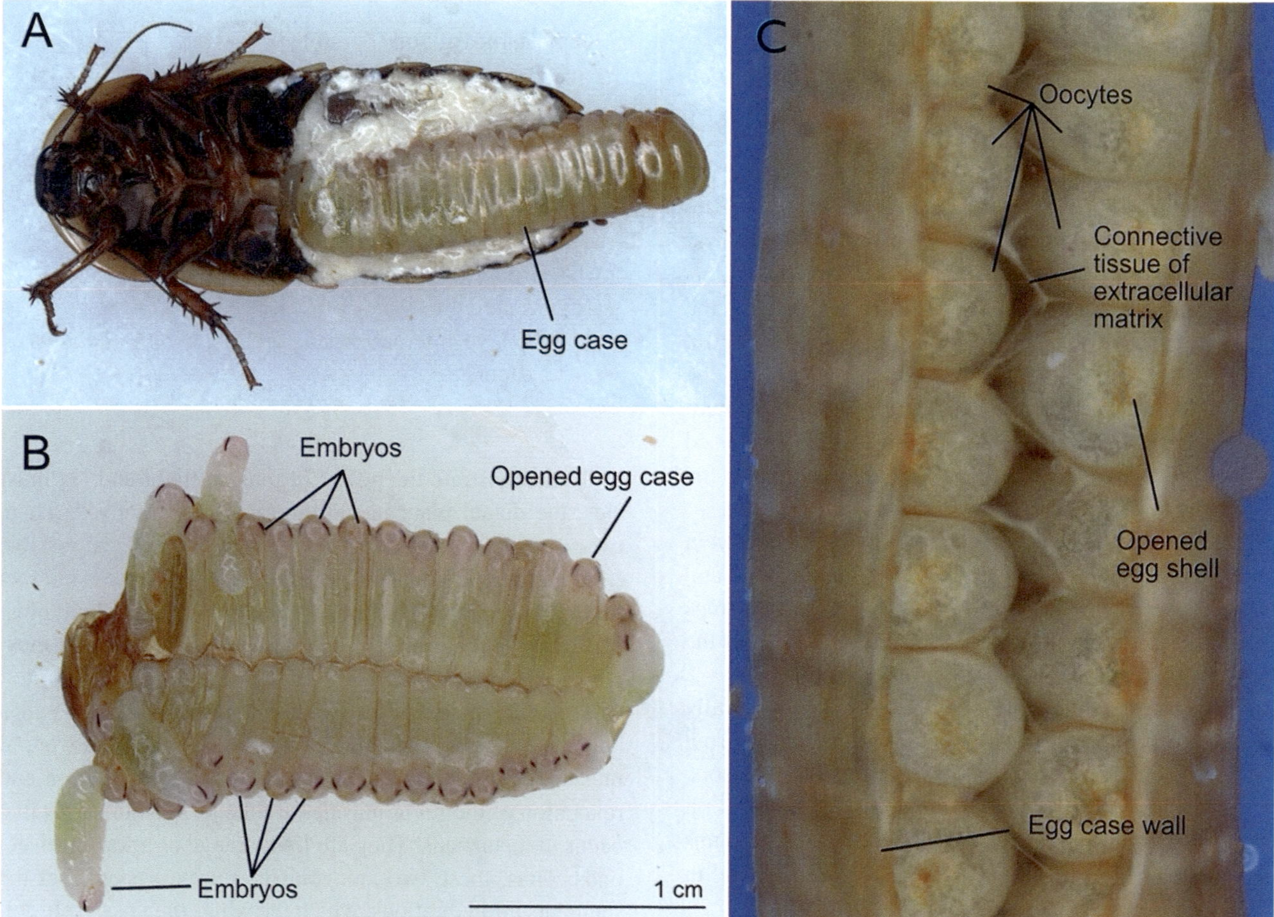

Fig. 8.23 *Blaptica dubia*. (**a–c**) Exposed ootheca

to the female by spermatophores, an encapsulated package of spermatozoa. The sperm are stored in the receptaculum seminis and later used for fertilization. The fertilized eggs in cockroaches develop within an ootheca (egg chamber) (Fig. 8.23), which, in some species, remains in the body until the first nymphs hatch. The females of *Blaptica dubia* carry their ootheca around for 3–4 weeks before it is then laid in a protected place for the hatching of the nymphs. Other species, for example, *Blatta orientalis*, lay their ootheca as soon as 1 day after fertilization of the eggs. The embryonic development then lasts about 30 days, depending on the temperature, and ends with the hatching of the young animals. Numerous molts characterize further development. In some cockroach species, up to 13 nymphal stages have been counted; for *Blaptica dubia*, 7 nymphal stages are usually described. In many species, the number of nymphal stages is variable and depends on the food conditions and environmental temperatures. In addition to bisexual sexual reproduction, as is the case for *Blaptica dubia*, parthenogenetic reproduction also occurs in some cockroach species. Movies showing nymphs and adults are shown at figshare: sn.pub/xr9qhi.

Hemimetabolous Insects

Mayflies (Ephemeroptera), dragonflies (Odonata), cockroaches (Blattodea), mantises (Mantodea), termites (Isoptera), stoneflies (Plecoptera), webspinners (Embioptera), grasshoppers (Orthoptera), stick insects (Phasmatodea), gladiators (Mantophasmatodea), booklice (Psocoptera), lice (Phthiraptera), true bugs (Hemiptera), and thrips (Thysanoptera).

Holometabolous Insects

Sawflies, wasps, bees, ants, and others (Hymenoptera), lacewings, antlions, alderflies, dobsonflies, snakeflies, and more (Neuropteroida), beetles (Coleoptera), stylops or wisted-wing parasites (Strepsiptera), caddisflies (Trichoptera), butterflies (Lepidoptera), scorpionflies (Mecoptera), true flies (Diptera), and fleas (Siphonaptera).

Internal Anatomy and Organs
Dissection of the abdomen and thorax of *Blaptica dubia*

(a) The cockroach can be dissected either from the dorsal or the ventral side (Figs. 8.24, 8.25, 8.26, 8.27, and 8.28). For the display of the nervous system, however, beginning the dissection at the dorsal side is preferable, while to demonstrate the heart, the dissection should start from the ventral side. The cockroach is first held between the thumb and index finger. Starting from the rectum or the genital pouch, a fine pair of scissors is used to cut sideways along the lateral edge up to the head.

(b) The animal is pinned with the ventral side down in a small dissection tray with insect needles and covered with tap water. All further dissection steps are now carried out at low magnification under the stereomicroscope.

(c) Now, the dorsal plate can be lifted very carefully with a pair of tweezers and slowly separated from the fatty tissue using a scalpel or a pair of scissors. Either the dorsal plate is completely removed, or it is folded over to one side. Often, the heart remains attached to the dorsal plate and can be viewed in this way.

(d) To see the large flight and leg muscles, the thoracic dorsal plate must also be carefully removed using scissors and a scalpel.

The organs that are already exposed are examined more closely

Fat Body and Musculature

After the dorsal plate has been removed and set aside for later examination, the whitish fat body, which extends throughout the abdominal cavity, is immediately visible (Figs. 8.24a and 8.25). The fat body of insects consists of fat cells and adipocytes, which, in the cockroach, are organized into numerous finger-like strands. The fat body serves as an important storage and synthesis organ and is traversed by large tracheal branches and fine tracheoles, which ensure gas exchange. Lipids, proteins, and sugars stored in the adipocytes are mobilized during longer periods of starvation and supplied to the metabolic system. In addition, adipocytes also produce and secrete many important proteins, such as collagen, which is essential for constructing the extracellular matrix.

Remove the Fat Body Carefully

(a) The abdomen is filled with adipocytes, and the fat body covers several organs, like the gonads or parts of the intestine. Therefore, during preparation, the fat body must be removed very carefully by plucking or suction without damaging other organs.

(b) Expose the intestine.

(c) Prepare the ovary or testis.

(d) Remove internal organs and expose the nervous system.

To demonstrate the powerful pinkish flight and leg muscles, the dorsal plate in the thorax is first removed. Then, tweezers are used to remove as much fat body as possible (Fig. 8.24a). Like most insects, cockroaches possess indirect flight musculature. This means that dorsal longitudinal muscles insert at the joints of the tergites and dorsoventral muscles insert at the surfaces of the tergites and sternites. Contraction of the dorsal longitudinal muscles leads to a bulging of the tergites and contraction of the dorsoventral muscles to a flattening. The times between contraction and relaxation of the flight muscles are the shortest intervals that occur in metazoans. They are in the range of a few milliseconds. Thus, these two antagonistic muscle systems set the segments into vibrations, which the wings then passively follow. Some insects, such as dragonflies, which catch their prey in flight and can perform very sophisticated flight maneuvers, possess direct flight musculature. In cockroaches, the downward movement of the wings is supported by direct musculature. The somatic muscles of insects, like those of vertebrates, are cross-striated and multinucleated, and arise from the fusion of individual myoblasts. In addition to the somatic muscles, the midgut is covered by mononucleated visceral muscle cells. These cells enable the contraction peristalsis of the intestine.

Nutrition, Digestive Tract, Excretion

As with all insects, the intestine of cockroaches is tripartite. It consists of the foregut, midgut, and hindgut (Fig. 8.24b). While the midgut ontogenetically arises from the endoderm, the foregut and hindgut are ectodermal derivatives. Their ectodermal origin is evident in the cuticular lining of the foregut and hindgut. *Blaptica dubia*, like the majority of known cockroaches, is omnivorous. The food is crushed with the mouthparts and first passes through the pharynx and esophagus into the crop, where it is collected. Paired salivary glands (Fig. 8.24c) secrete digestive secretions into the oral cavity, from where they pass through the esophagus into the crop and drive the digestion of the food. In the subsequent

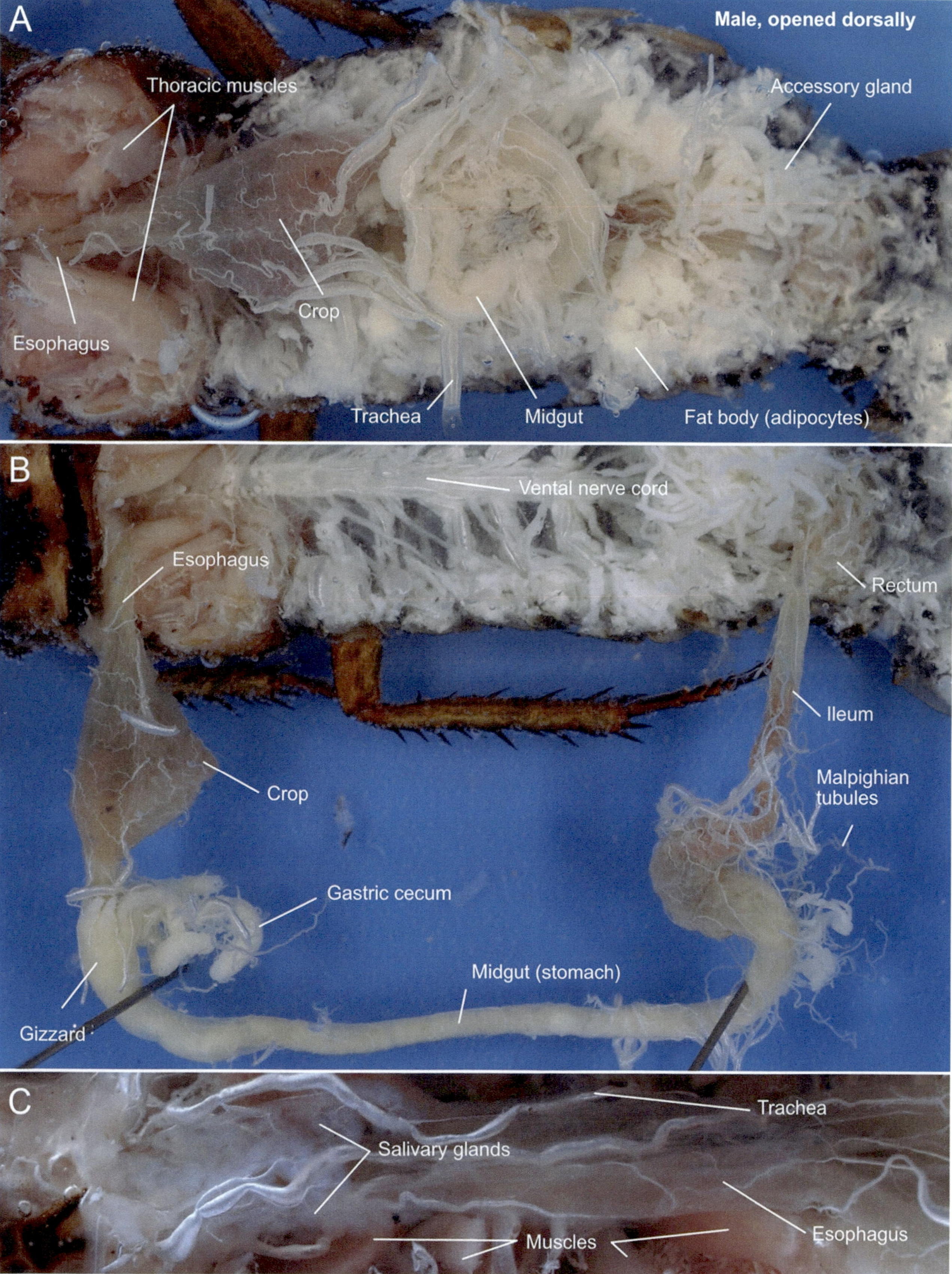

Fig. 8.24 *Blaptica dubia*. (**a**) Male prepared from the dorsal side. After removal of the dorsal plate, the internal organs are exposed. (**b**) After removing the fat body, the intestine can be prepared and the ventral nerve cord can be exposed. (**c**) The milky-appearing salivary glands become visible only after removing the thoracic muscles and the thoracic fat body

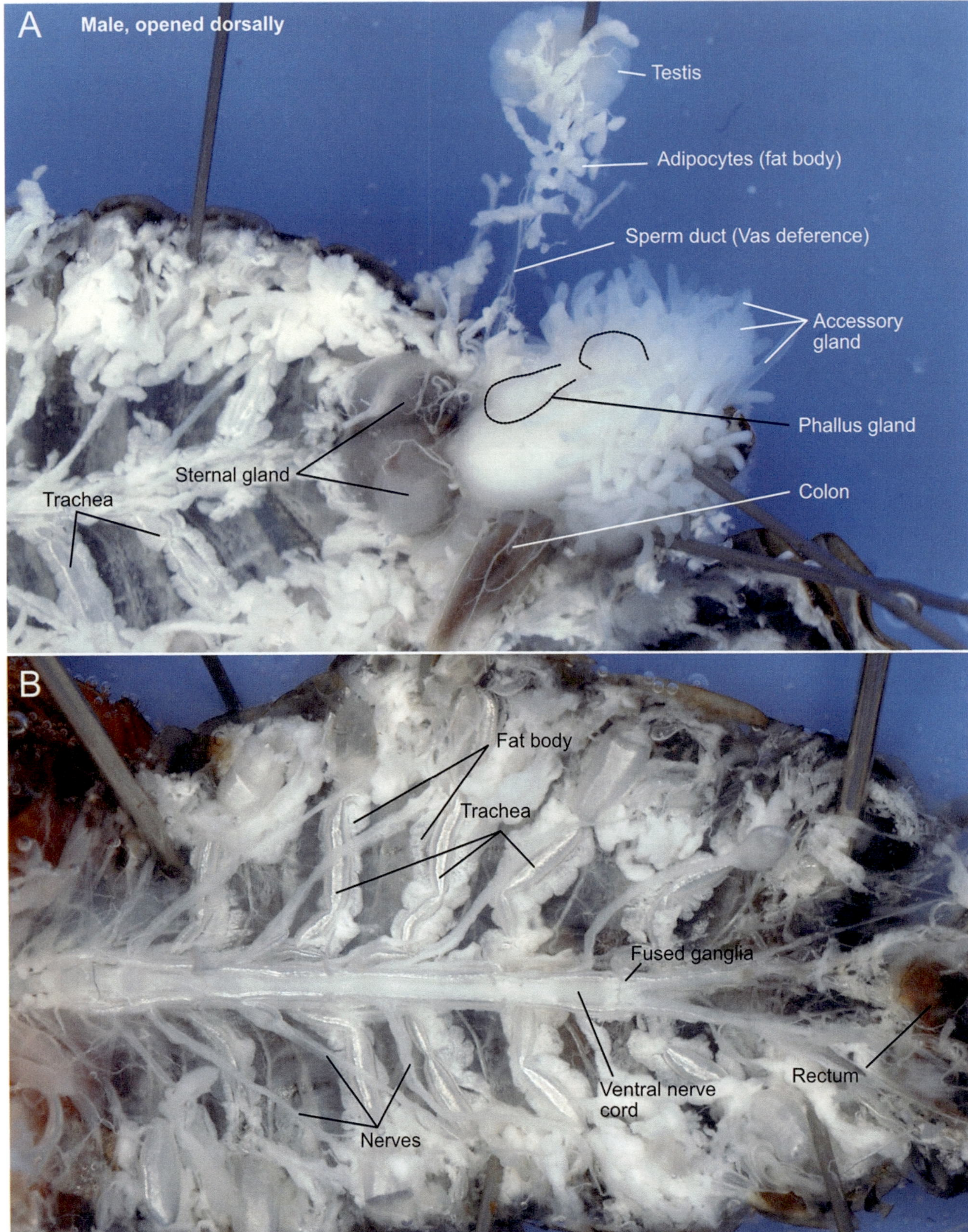

Fig. 8.25 *Blaptica dubia*. Male prepared from the dorsal. (**a**) Reproductive organs of a male (**b**) Exposure of the ventral nerve cord

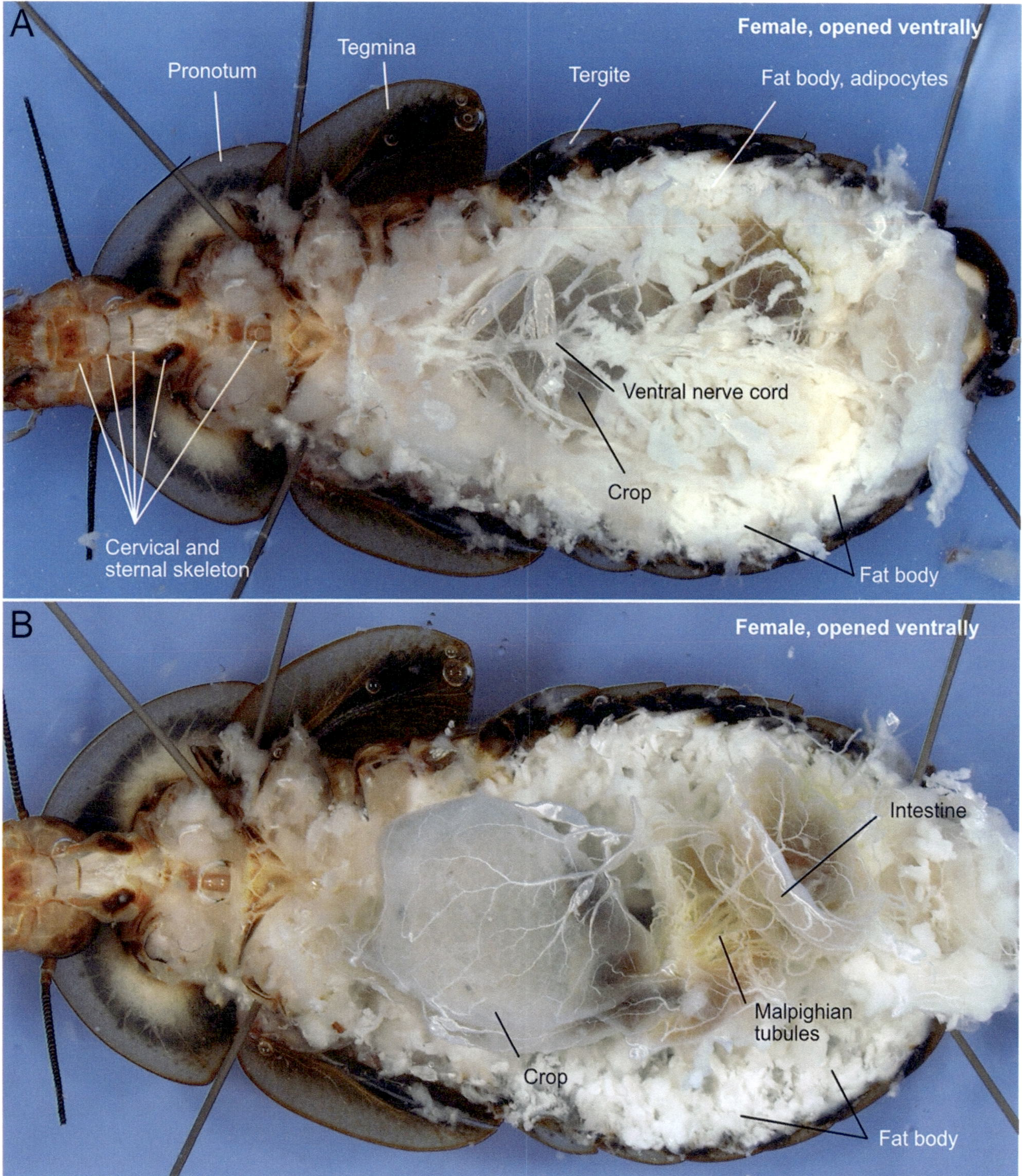

Fig. 8.26 *Blaptica dubia*. (**a**) Female, prepared from the ventral. After the removal of the dorsal plate, the internal organs are exposed. (**b**) After removal of the ventral cord and parts of the fat body, the intestine is exposed

gizzard, the food is further mechanically crushed with the help of chitin teeth. The oral cavity, pharynx, esophagus, crop, and gizzard belong to the foregut, followed by the midgut. The cells of the midgut secrete digestive enzymes and absorb nutrients. Special epithelial cells at the beginning of this intestinal section form a porous peritrophic membrane made of proteins and chitin. It envelops the food and protects the midgut epithelium from mechanical damage. The peritrophic membrane is constantly renewed and allows for the passage of dissolved food substances, but prevents the pas-

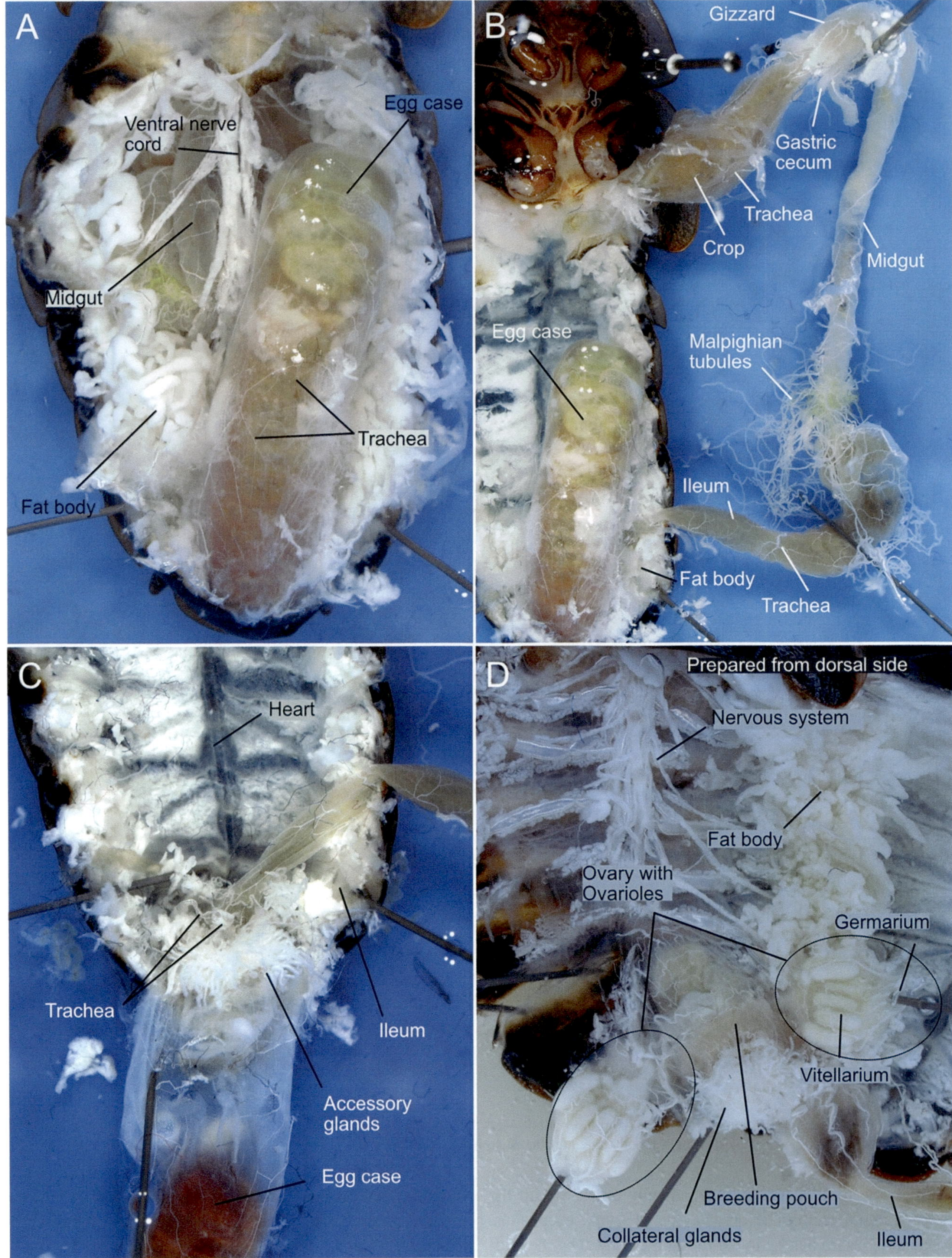

Fig. 8.27 *Blaptica dubia*. (**a**, **b**) Female with mature ootheca prepared from the ventral. (**c**) After removing the internal organs, the heart becomes visible. (**d**) Reproductive organs of a female cockroach, prepared from the dorsal

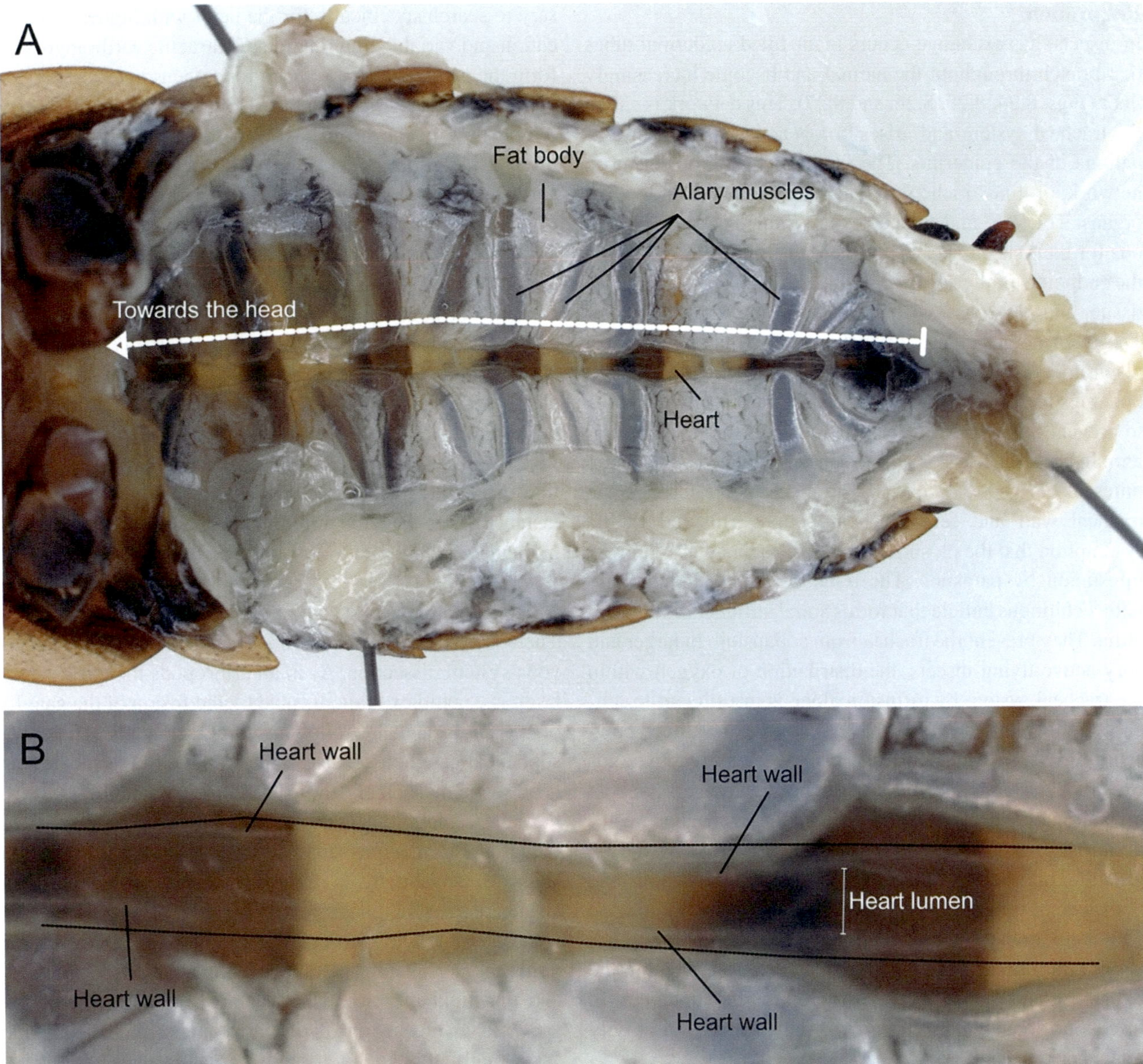

Fig. 8.28 *Blaptica dubia*. The heart, prepared from the ventral. (**a**) Abdomen opened and viscera removed. Along the midline, the heart is visible. (**b**) Details

sage of larger particles or parasites. Immediately in the first section of the midgut, finger-shaped pouches that open into the midgut are noticeable. These are the caeca, which can harbor bacterial symbionts and support the digestion of the food. The food is transported from the midgut to the hindgut by peristaltic contractions of the intestinal musculature.

At the transition from the midgut to the hindgut, the Malpighian tubules open into the hindgut (Figs. 8.24b and 8.26b). The Malpighian tubules, together with specialized individual cells, the nephrocytes, form the excretory and osmoregulatory organs of insects. They functionally replace the nephridia present in crustaceans. In the cockroach, the Malpighian tubules are formed by a few dozen thin, whitish, or slightly yellowish tubes. They arise from ectodermal and mesodermal tissue. Malpighian tubules are secreting and reabsorbing, blind beginning tubes that release their urine into the intestine. Most insects excrete their nitrogenous compounds as uric acid to save water. Further reabsorption of water takes place in the rectal bladder. As mentioned above, these formations can be interpreted as adaptations to the terrestrial lifestyle, by keeping water loss to a minimum while regulating the internal environment.

Respiration

In insects, gas exchange occurs in air-filled epidermal tubes that branch throughout the animal and become increasingly finer (Figs. 8.24, 8.25, 8.26, and 8.27). This network is called the tracheal system and arises ontogenetically from invaginations of the epidermis. The large tracheae branch into a network of fine tracheoles that project to the individual organs. Tracheae and tracheoles are immediately conspicuous for their sheen because of their air filling (Fig. 8.24). At the endpoints of the network, at the organs, oxygen enters the tissue, and CO_2 passes from the tissue into the tracheoles. The tracheae are connected to the external atmosphere via respiratory openings (stigmata, singular: stigma). These openings are often protected by a sieve or closure apparatus to prevent the entry of foreign bodies into the tracheal system. In bees, the larger tracheae can be colonized by parasitic mites of the genus *Acarapis*. The mites pierce through the tracheal wall and feed on hemolymph. This led to the assumption that the closure of the tracheae may also prevent infestation by parasites. The tracheae are lined internally with a chitinous cuticle that forms spiral stiffeners called taenidia. They prevent the trachea from collapsing. In larger and very active flying insects, the distribution of oxygen within the tracheal system by diffusion alone is insufficient to reliably supply all organs. They have, therefore, developed powerful air sacs. Many insects can accelerate active gas exchange through muscle contractions and compression and relaxation of the tracheal walls. Since the tracheae are ectodermal and carry a cuticle, they must be molted during growth. This sets narrower limits to the maximum achievable body size of the animals and is probably a significant reason why there are no "giants" among insects. In addition, a successful molting of those tracheae that are located within the wings is hardly conceivable. Therefore, the imagines in insects no longer molt and grow—quite unlike the closely related crustaceans.

Holometababolan insects use intracellular hemoglobin to bind oxygen. In particular, the cells of the fat body (adipocytes) and the trachea produce hemoglobin. Another well-known respiratory protein in insects is hemocyanin. Hemocyanins transport oxygen in the hemolymph of stoneflies, but also in Entognatha and most hemimetabolan taxa. Apparently, hemocyanin has been lost in Holometabola. Together with gas exchange via tracheae, the intracellular respiratory proteins play an essential role in tissue oxygenation.

Heart

To perfectly visualize the heart, also called the dorsal vessel, start the preparation from the ventral side (Figs. 8.27c and 8.28). The heart is a very fine transparent tube running from anterior to posterior in the dorsal midline. It might be necessary to search specifically for the heart with higher magnification and variable illumination. Contractile cardiomyocytes form the cardiac tube. The heart tube opens anteriorly and is closed posteriorly in hexapods. However, retrograde hemolymph flow in the heart has been described for numerous insects, including cockroaches. The posterior, wide-lumen part of the dorsal vessel is referred to as the actual heart and is separated from the narrow-lumen anterior part, referred to as the aorta, by a cardiac valve. The body fluid of insects (hemolymph) is drawn into the heart chamber through segmentally arranged inflow openings (ostia) and transported anteriorly toward the head. The driving force is an interplay between the contractile cardiomyocytes and the alary muscles, which span and stretch the heart tube laterally in the body. From the head, the hemolymph, which transports a variety of metabolites, hormones, and peptides, can freely flow back into the body cavity and bathe all organs. For instance, more than 500 different proteins have been detected in the hemolymph of the fruit fly *Drosophila melanogaster*.

Nervous System

There is usually insufficient time for a detailed brain and nervous system dissection. As in all arthropods the cockroach's brain is a complex brain, formed by the fusion of the ganglia of the acron and the following three segments. The subesophageal ganglion consists of the ganglia of the subsequent head segments, namely, those carrying the mandibles, maxillae, and labium. The ventral nerve cord, built in a rope-ladder-like form, is easy to expose in the thorax and abdomen (Figs. 8.25b and 8.27a). The structure with segmentally arranged pairs of ganglia and segmental lateral nerves is clearly visible. Pay particular attention to the three slightly larger thoracic ganglia pairs, which innervate the leg muscles and the flight muscles. The last large abdominal ganglion arises from the fusion of the last four segmental ganglia; accordingly, this ganglion sends nerves to the entire posterior body (Fig. 8.25b). The autonomic (visceral) nervous system consists of a posterior part, which originates from the abdominal ganglion, and an anterior part (stomatogastric nervous system), which originates from the posterior part of the brain. The visceral nervous system innervates the mouth and the esophagus, the stigmata, the openings of the tracheal system to the outside, the intestine, and the gonads.

Reproductive Organs

Like almost all hexapods, cockroaches are gonochoristic, in other words, they have separate sexes. Among the few exceptions is, for example, the cottony cushion scale (*Icerya purchasi*), in which the frequent hermaphrodites mate with each other or with the rarely occurring males. Only in the latter case do new males arise again.

Demonstrating the Reproductive Organs

To expose the paired reproductive organs in cockroaches, as much fat body as possible is first removed. Particular care must be taken in the posterior region of the animal, as the testes and the vas deferens are closely surrounded by adipocytes (Fig. 8.25a). It is helpful to cut and remove the intestine in the rectal region. The tracheae should be cut and not torn off.

In male animals (Fig. 8.25a), a vas deferens leads from each paired testis to the ductus ejaculatorius, which is to be found in the accessory glands. The vasa deferentia are usually hidden by the fat body and are torn off very easily. In the testes, spermatogonia develop into spermatocytes and sperm during meiosis. The numerous tubes of the accessory glands produce the secretions for forming spermatophores (a capsule containing a mass of spermatozoa). Spermatophores transfer the sperm into the female genital tract, thus enabling internal fertilization. Another small gland, the phallus gland, is probably used to produce adhesive compounds, which keep the spermatophore adhesive in the female.

The female's preparation is done the same way, starting dorsally (Figs. 8.26 and 8.27). After removing the adipose tissue and the intestine, the pair of ovaries becomes visible (Fig. 8.27d). Each ovary holds eight ovarioles (egg tubes); in which the oocytes mature in a linear arrangement. The ends of the ovarioles are combined into a filament attached to the body wall. Each ovariole consists of the distal germarium, in which the mitotically produced oogonia (stem cells, primordial germ cells) arise. From the oogonia, the oocytes differentiate. Each egg migrates into the second section, the vitellarium, initiates meiosis and develops, surrounded by follicle cells, into the mature yolk-containing egg. This type of ovary is referred to as a panoistic ovary. The ovaries are connected via oviducts to the genital pouch, into which the bifurcated receptaculum seminis also opens. The egg chambers (oothecae) are formed in the genital pocket after fertilization of the oocytes. The larger branch of the receptaculum seminis contains the sperm, and the smaller one represents an accessory gland. The fertilization occurs when the oocytes pass the opening of the receptaculum seminis. The sperm enters the eggshell through fine openings, the micropyles. Two asymmetrically large accessory glands (collateral glands) secrete the proteins of the cocoon shell into the genital pouch. As a rule, eight eggs come from each ovariole, so the egg chamber usually contains 16 eggs. When preparing cockroaches, one will always find animals with a filled egg chamber (Figs. 8.23 and 8.27a, b). Our movie clip collection, which includes movies showing living cockroaches, can be found at figshare: sn.pub/xr9qhi.

Recommended Material

(a) *Blaptica dubia* (Argentinian wood roach), *Periplaneta americana* (American cockroach, water bug, kakerlac, ship cockroach), *Blaberus craniifer* (death's head cockroach) or similar large species can be obtained from local pet shops, where they are often sold as live food for terrarium animals. It is best to ask the supplier to include some nymph stages for demonstration (Fig. 8.22). It is said that *Blaptica dubia* rarely, if ever, climbs on smooth surfaces, so the animals can be kept escape-proof in a glass aquarium for the duration of the zoological course. Additionally, the upper part of the glass aquarium is finely coated with vaseline on the inside. This measure is also intended to prevent the animals from escaping.

(b) Course preparation: A cotton ball soaked in diethyl ether or chloroform is placed in a 500-ml glass vessel with a tight lid. The cockroaches intended for dissection, about 10–20 animals per glass, die in about 10 min.

▶ **Summary Echinoderms have highly unusual body shapes but are descendants of bilaterally symmetrical ancestors. They show a deuterostome development, have a biphasic life cycle with a planktonic bilateral symmetric larva and possess, as adult animals, a pentameric ("five-rayed"), radially symmetrical body, and live as adults on or in the seabed. Echinoderms are exclusively marine.**

From the echinoderms (Echinodermata), we discuss here the sea stars (Asteroidea, Figs. 9.1, 9.2, 9.3, 9.4, 9.5, 9.6, 9.7, 9.8, 9.9, 9.10, and 9.11) and the sea urchins (Echinoidea, Figs. 9.12, 9.13, 9.14, and 9.15). In addition to these two groups, the feather stars and sea lilies (Crinoidea), the brittle stars (Ophiuroidea), and the sea cucumbers (Holothuroidea) are among the recent representatives of this group. All echinoderms exhibit several common characteristic key features. These include:

- A mesodermal endoskeleton located beneath the epidermis that is composed of many ossicles, small calcareous plates or pieces, embedded in the collagen-rich extracellular matrix.
- A water-vascular (ambulacral) system that is unique in the animal kingdom. It is formed from the coelom. With its numerous channels, it serves primarily for locomotion and food uptake.
- A metamorphosis, in which a bilaterally symmetrical larva is transformed into an adult animal with primarily pentameric, radial symmetry (Fig. 9.1). The bilateral body symmetry of some sea urchins ("Irregularia") and holothurians evolved secondarily from this pentameric symmetry. These groups also descend from radially symmetrical ancestors.

According to recent molecular studies, echinoderms (Echinodermata) and chordates (Chordata) are closely related. Both groups have in common that embryos arise from the so-called radial cleavages and that the blastopore gives rise to the anus. The mouth is always newly formed. This is called a deuterostome development. What are radial cleavages? When the fertilized echinoderm egg develops, the mitotic spindles are parallel or perpendicular to the main axis, so that the blastomeres form a radially symmetrical pattern to the main axis of the egg. Throughout metazoans, this type of cleavage pattern leads to deuterostome development of the later embryo. The type of cleavage and the formation of the mouth opening are characteristic for Echinodermata, Hemichordata (acorn worms and others), and Chordata, to which Cephalochordata (Acrania, no skull, for example, *Branchiostoma*), Urochordata (e.g., tunicates, for example, *Ciona*), and Craniota (Vertebrata, vertebrates, for example *Oncorhynchus*, *Rattus*) belong,

Echinoderms are exclusively marine. The majority of the 6000–7000 known echinoderm species colonize the benthos. They inhabit all habitats from the intertidal down to the deep sea. In local areas of the deep sea, sea cucumbers, in particular, can occur in very high numbers and constitute up to 90% of the benthic biomass. Like sea stars and sea urchins, they constantly roam around in search of food, although they move slowly. The body sizes of echinoderms vary considerably and can range from a few millimeters to 2–3 m: The smallest sea star, *Asterina phylactica*, has a maximum diameter of 1.5 cm and feeds on bacteria and diatoms (algae); one of the largest, a deep-sea species, *Midgardia xandaros*, reaches a diameter of 1.40 m. The smallest brittle star has a disc diameter of only 0.5 mm with an arm length of 3 mm each. The largest representative of Echinodermata is the snake sea cucumber, *Synapta maculata,* This species, found in the Red Sea, reaches a body length of up to 2.5 m with a body diameter of around 5 cm. One of the smallest sea cucumbers, only 5 mm long, is a resident of the interstitial meiofauna in the North Sea, *Leptosynapta minuta.*

© The Author(s), under exclusive license to Springer-Verlag GmbH, DE, part of Springer Nature 2025
A. Paululat, G. Purschke, *Metazoa – Morphology and Evolution of Animals*, https://doi.org/10.1007/978-3-662-69904-1_9

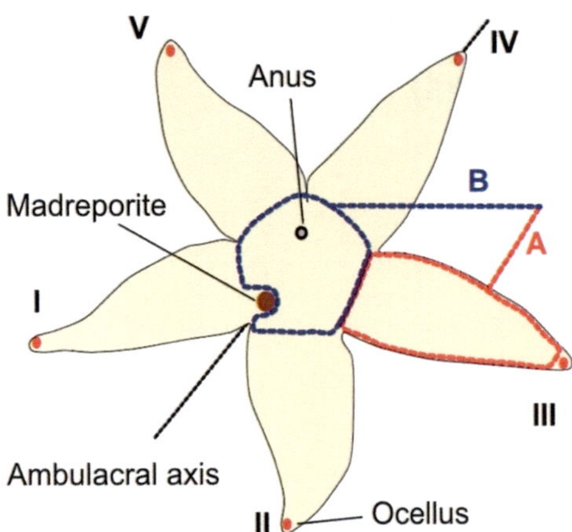

Fig. 9.1 Schematic representation of a sea star. The first two preparation steps are marked (**a, b**), I-V Arms

Echinoderms are either microphagous, taking their food directly from the seawater as microphagous suspension feeders (sea lilies and feather stars), omnivorous (sea urchins), grazing hard floors and feeding on algae, crustose red algae, cyanobacteria, sponges, hydrozoans, anthozoans or bryozoans (most sea urchins), or carnivorous (sea stars), preying mainly on mussels, but also other small animals. All sea cucumbers are microphagous as well: some sea cucumber species swallow soil substrate and digest the attached organic components (detritus processors); other sea cucumbers are also suspension feeders. Only a few species live sessile on a solid substrate, such as the Gorgonocephalidae, a family of ophiuroids known as basket stars. They have characteristic highly branched arms.

Pentameric Symmetry and Locomotion

As mentioned above, the pentameric echinoderms arise from bilaterally symmetrical larvae. The transformation of the body symmetry occurs during metamorphosis. What are the advantages associated with a pentameric symmetry? A current hypothesis suggests that the recent free-moving echinoderms are very likely descendants of a group of sessile suspension feeders. Evidence for this is provided by the fossil records of Echinodermata. There are many marine bilaterian species with a biphasic life cycle, beginning with a planktonic larva (distribution stage) and ending with a more or less sessile adult. Usually, such animals are microphagous suspension feeders. These often have a circularly arranged filtering device, in the form of a tentacle apparatus. The huge advantage of having a radial tentacle apparatus is the ability to catch prey from all sides. A current hypothesis is that this type of feeding supports radial symmetry of the body and, secondarily, the loss of the bilateral symmetry. Thus, representatives of other animal taxa also show a radially symmetrical tentacular crown. For example, many bryozoans carry a radial arrangement of the tentacles, which can be derived from a horseshoe-shaped original form. Likewise, Sabellidae and Serpulidae, feather duster worms, flowers of the sea,

annelids, possess such a crown of radially arranged tentacles. Corresponding tentacle apparatuses can also be found in peanut worms (Sipuncula) and entoprocts (Kamptozoa). Even within the Deuterostomia, this feature occurs in some hemichordates, example given, in some *Cephalodiscus* species, which belong to Pterobranchia, some worm-shaped animals living in secreted tubes on the ocean floor. Lastly, the polyps in Cnidaria also show a radial arrangement of the tentacles. Some examples of radially arranged tentacles crowns can be found in the supplementary movie clip database at figshare: sn.pub/xr9qhi.

Did the free-moving, benthic Echinodermata evolve from sessile suspension feeders? Due to their calcified skeleton, there is an exceptionally extensive fossil record of echinoderms that can be traced back about 550 million years to the early Cambrian. In addition to pentameric representatives, there are also asymmetric and bilaterally symmetrical forms. Representatives of the recent groups can also be traced back to the Paleozoic. The earliest fossils are indeed interpreted as suspension feeders, as they are sessile forms with upward-facing (substrate facing away) ambulacral grooves and feet. These were already pentameric, and some are classified into the stem line of Crinoidea, which, in all phylogenetic analyses, are the sister group of all other Echinodermata and are referred to as Eleutherozoa. Only Eleutherozoa have an opposite orientation of the body: their oral side faces toward the substrate. Crinoidea are semi-mobile suspension feeders that catch floating food particles in the water with their tentacle crown. The ambulacral system is thus primarily an adaptation for ingesting suspended particles and only secondarily harnessed for locomotion and other functions. Interestingly, the feather stars are stalked and sessile during their juvenile phase, like the sea lilies; they only detach later to become mobile animals. A movie showing a feather star in motion can be found at figshare: sn.pub/xr9qhi. The pentameric symmetry resulting from this lifestyle is optimally adapted to suspension feeding. After a loss of bilateral symmetry combined with a comprehensive reorganization of body structures, a return to the bilateral symmetry of the ancestors was practically impossible, despite the regaining of motility, convergent in some Crinoidea and Eleutherozoa. However, this does not apply to all recent echinoderms, because the sea cucumbers (Holothuroidea) have secondarily regained bilateral symmetry, albeit based on pentamery, and thus completely different from all other Bilateria, meaning that their echinoderm relationship is still evident. This hypothesis is also further supported by the absence of complex sensory organs and the lack of a brain, as evolution of sessility also goes hand in hand with a simplification of the sensory and nervous system.

As exclusively crawling bottom dwellers, Echinodermata have very limited individual action radii. For example, some sea cucumbers only cover about 30–40 cm in an hour, a little more than two or three times their body length. For some "fast"

Fig. 9.2 *Asterias rubens*, fixed specimen. (**a**) Top view of the aboral side with madreporite, anus, and central disc; this side is facing away from the substrate in life. (**b**) View of the oral side with mouth opening and ambulacral grooves

Fig. 9.3 *Asterias rubens*, fixed, preparation steps. (**a**) Arms are opened by cutting laterally along the arms, see red line. (**b**) Exposing the central disc. (**c**) The most prominent organs in all five arms are the pyloric diverticulae

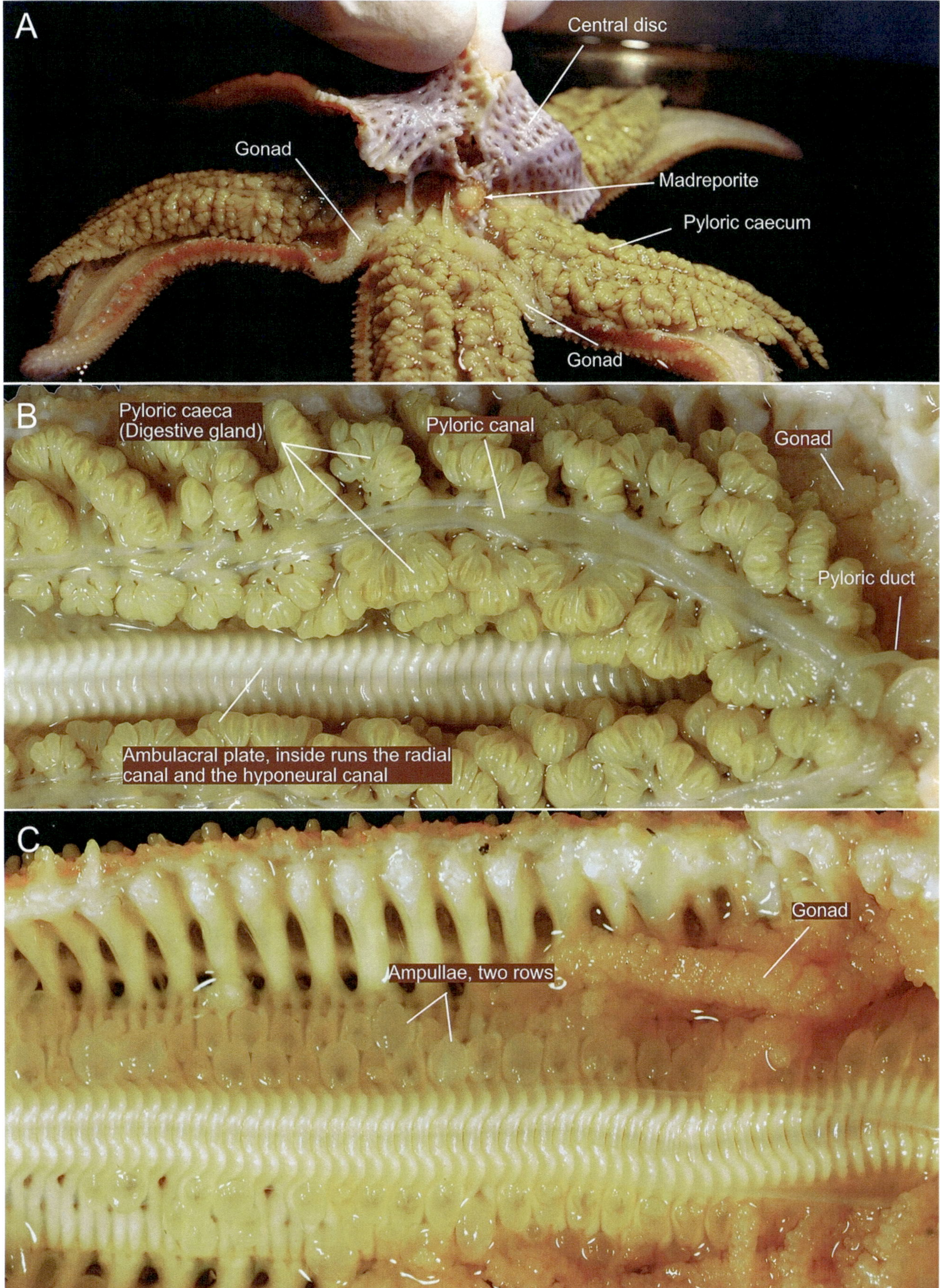

Fig. 9.4 *Asterias rubens*, fixed, preparation steps. (**a**) Removal of the aboral central disc. (**b**) Pyloric diverticula (pyloric glands), pyloric canal, and gonads at higher magnification. (**c**) Ampullae at higher magnification

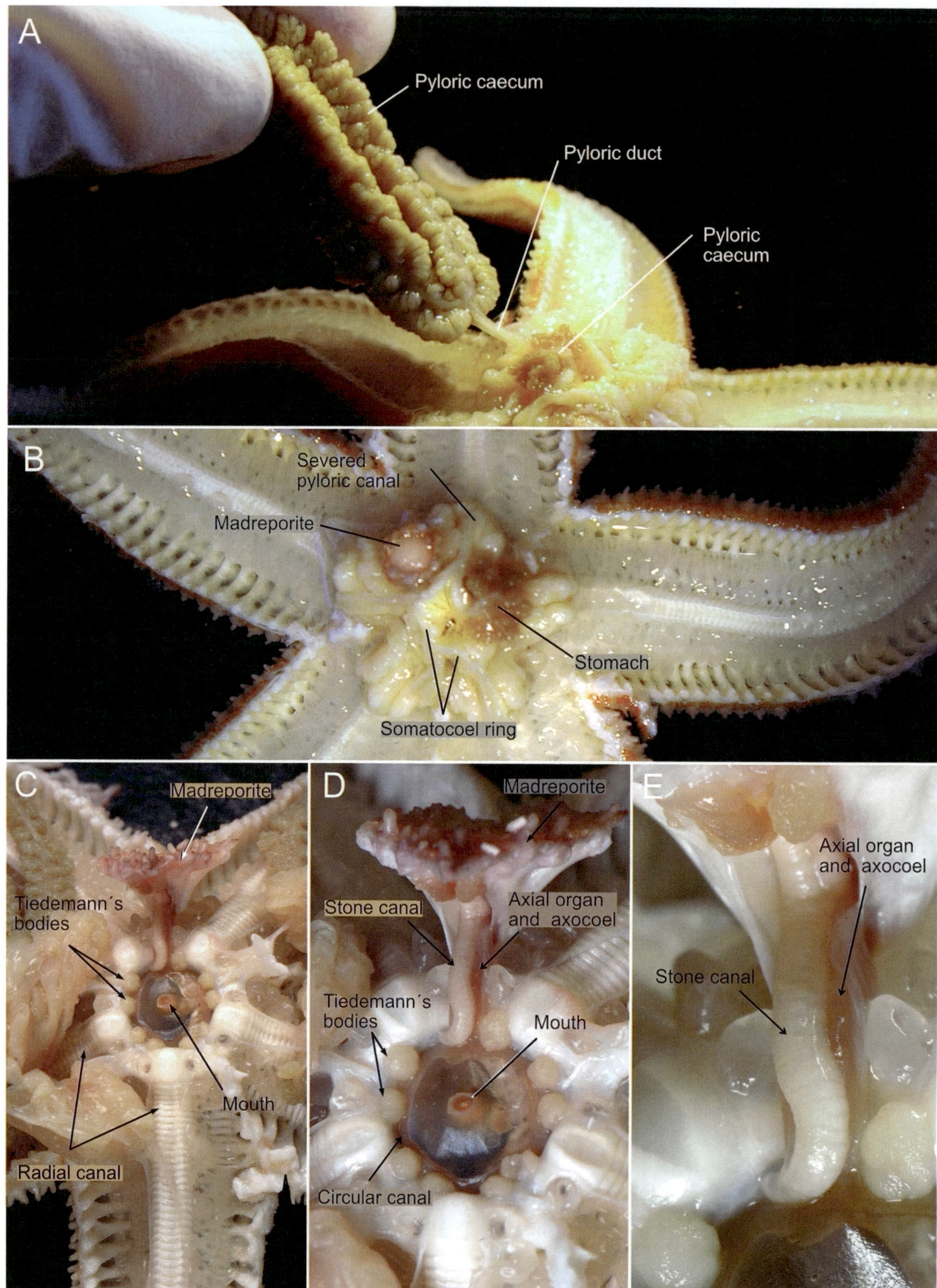

Fig. 9.5 *Asterias rubens*, fixed. (**a**) Demonstrating the pylorus duct of one arm. (**b**) Removal of the pyloric caeca. (**c**) After removal of the stomach, the stone canal, radial canals, ring canal, and mouth opening are visible. (**d**) Stone canal and Tiedemann's bodies. (**e**) Stone canal and axial organ at higher magnification

Fig. 9.6 *Asterias rubens*, fixed. Each arm was prepared differently

Fig. 9.7 *Asterias rubens.*
(**a**) Cross-section through the central region of a sea star.
(**b**) Cross-section through a sea star's arm. Illustrations according to wall charts from the Biological Collection of the University of Osnabrück, modified

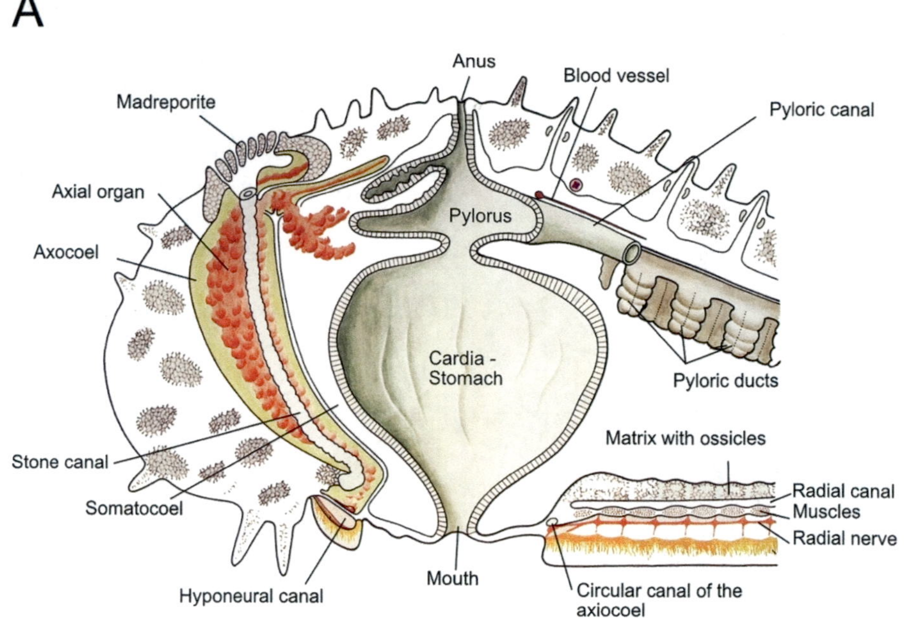

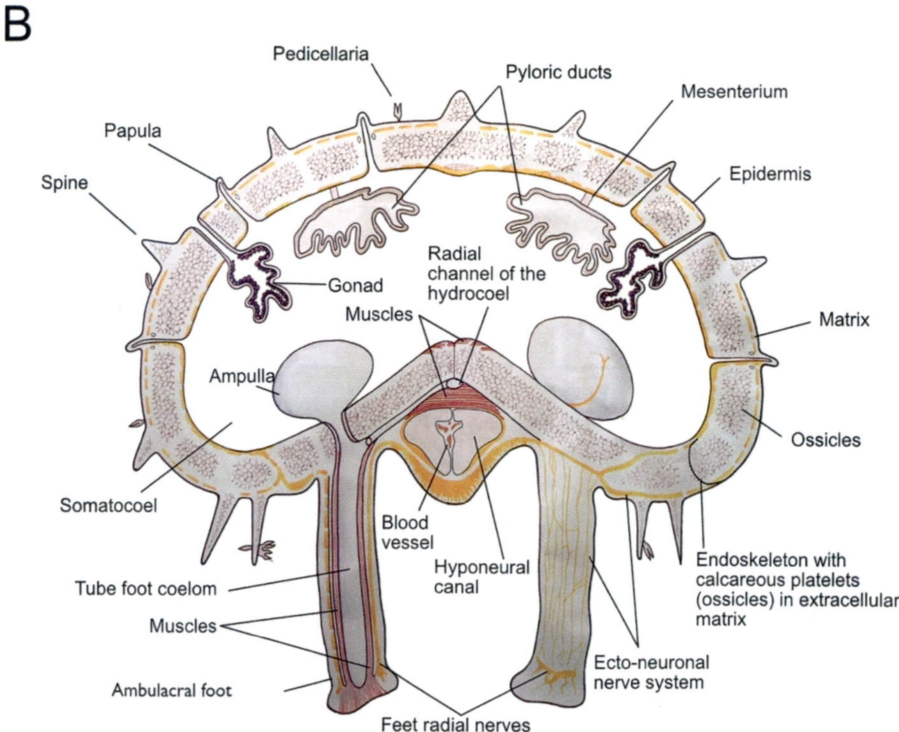

sea stars, a distance of 45 m in 1 h has been recorded. Recently, it was discovered that some sea cucumbers use a special trick to reach new habitats that are further away quickly. They take in large amounts of water through their anus and put themselves into a "floating state" to then drift with the ocean current. This way they may cover distances of up to 90 km per day.

Except for sea urchins, all echinoderms can shed body parts and regenerate them. This ability is called autotomy.

Sea stars can lose and possibly replace whole arms. A movie showing a sea star with regenerating arms can be found at figshare: sn.pub/xr9qhi. Therefore, it sometimes happens that one finds a sea star with less than five arms, or that the arms, due to the still incomplete regeneration, have a different shape and size (Fig. 9.9). When sea cucumbers are seriously threatened, they eject the major part of their intestines, which the predator may then accept as prey,

Fig. 9.8 Generalized scheme of the ambulacral system of a sea star. According to various authors

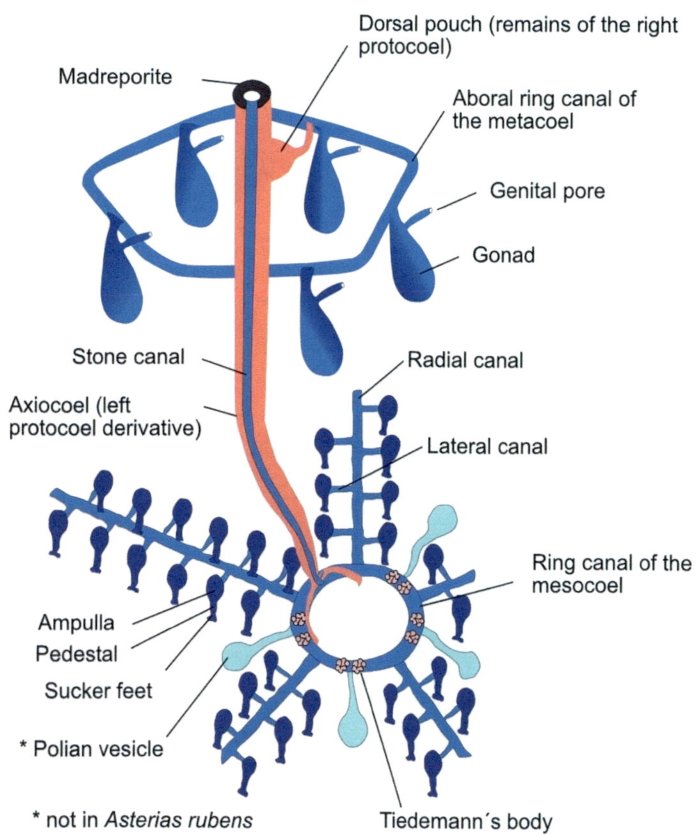

while the sea cucumber itself replaces the missing organs through regeneration, and thus survives the attack. Other sea cucumbers have structures that serve specifically for defense: the Cuvierian tubules. Depending on the species, up to 150 blind-ending protrusions of the rectum near the water lungs (sea cucumbers often have special respiratory organs) are ejected when the animal is threatened. In water, the Cuvierian tubules stretch to extreme lengths, stick together, and are dangerous to many other marine creatures. In some sea cucumber species, the Cuvierian tubules are less adhesive, but very effective due to the venom they contain.

9.1 The Common Sea Star (*Asterias rubens*, Asteroidea)

Sea stars, like all Echinodermata, show an oral and an aboral side of the body (Fig. 9.2a, b). The oral side with the mouth opening is always facing the ground. The mouth marks the anterior end in Bilateria; thus, it is shifted to the underside of the sea star, while the anus has moved from the posterior end to the top. The anus is located in the center of the large aboral central disc (Fig. 9.2a). The five arms of the pentamerically organized sea star extend from this central disc. At the edge of the central disc, between the attachment points of two arms, lies the madreporite. This is a sieve plate made of calcium carbonate, equipped with numerous finest openings that can be easily seen under a stereomicroscope. The madreporite forms the entrance to the sea star's extensive water vascular system. The mesodermal endoskeleton is directly beneath the epidermis. It consists of a large number of small calcareous plates or pieces (ossicles), all embedded in the extracellular matrix (ECM) (Fig. 9.7). This ECM contains—as in all Metazoa—collagen as the main component. Interestingly, in sea stars and other echinoderms, the viscosity of this collagenous matrix can be modified by the inflow and outflow of calcium and sodium ions. By increasing the "stiffness" of the ECM, arms, feet and the posture of spines can gain firmness and, for example, grasp a prey or maintain a body position that supports the filtration of seawater. Thus, the matrix relieves the muscles, which would otherwise have to perform these tasks alone. Such an extracellular matrix is unique to metazoans and is called a mutable connective tissue (MCT).

The central mouth opening is on the oral side of the body (Fig. 9.2b). From there, deep ambulacral grooves run into all arms. Each ambulacral groove carries two or four rows of ambulacral feet. In *Asterias rubens* there are always four such rows. With the help of the ambulacral feet, which

Fig. 9.9 Regeneration of injured or severed arms. *Asterias rubens*, live specimen; Morgat, France. A movie showing this sea star can be found in our movie clip collection at figshare: sn.pub/xr9qhi

belong to the water vascular system, sea stars move on the substrate, hold on to the substrate, grab prey, and transport it to the mouth. The mobility of the ambulacral feet is, unique in metazoans, completely controlled hydraulically by the fluid pressure inside the tube feet. The ambulacral system will be discussed in more detail later in this chapter. The entire body surface of sea stars is densely covered with spines. They protect the sea star from predators. Between the spines are pedicellariae, small, stalked, and muscle-movable pincers or tweezers, which serve to hold onto food particles. In addition, the pedicellariae ensure that parasites, larvae of hard substrate dwellers, or other objects are removed from the body surface and their settlement prevented (Fig. 9.2b). The pedicellariae in *Asterias rubens* are always equipped with two pincers. They are easily recognizable when the sea star is viewed lying in a dissection dish filled with water. Some detached pedicellariae can be microscoped in a drop of water on a slide and covered with a coverslip. The most

beautiful way to observe the diverse movements of ambulacral feet and pedicellariae is, of course, on living sea stars and sea urchins, which, if no live material is available in the course, should be made up for on a marine biology excursion (see Fig. 9.12).

Recommended Material
Sea stars and sea urchins collected on marine biology excursions can be fixed in 4% formaldehyde and collected for zoological courses. Often, dried or preserved animals as total preparations are available for viewing. The sea stars prepared here (*Asterias rubens*) were collected in Brittany (France), fixed in 4% formaldehyde, and stored for about 4 months. Fixed or live sea stars can often be obtained from marine biological stations such as the Biologische Anstalt Heligoland (BAH) /

Fig. 9.10 *Asterias rubens*. (**a**) Arm from the ventral with spines, ambulacral feet and terminal eye spot. (**b**) Eye spot at the tip of the arm at higher magnification, living animal

Alfred-Wegener-Institute on Heligoland or similar institutions. Ask suppliers about the availability of live animals to demonstrate the ambulacral feet and pedicellariae. Sea stars and sea urchins can be kept for several days in well-ventilated, 14–16 °C tempered seawater. All images of sea urchins (*Psammechinus miliaris*) show living specimens. The animals were studied as part of our Developmental Biology and Marine Biology excursions to Roscoff, France, or Heligoland, Germany.

Before fixed animals are dissected in courses, they are soaked for 24 h in slightly flowing tap water to remove formaldehyde.

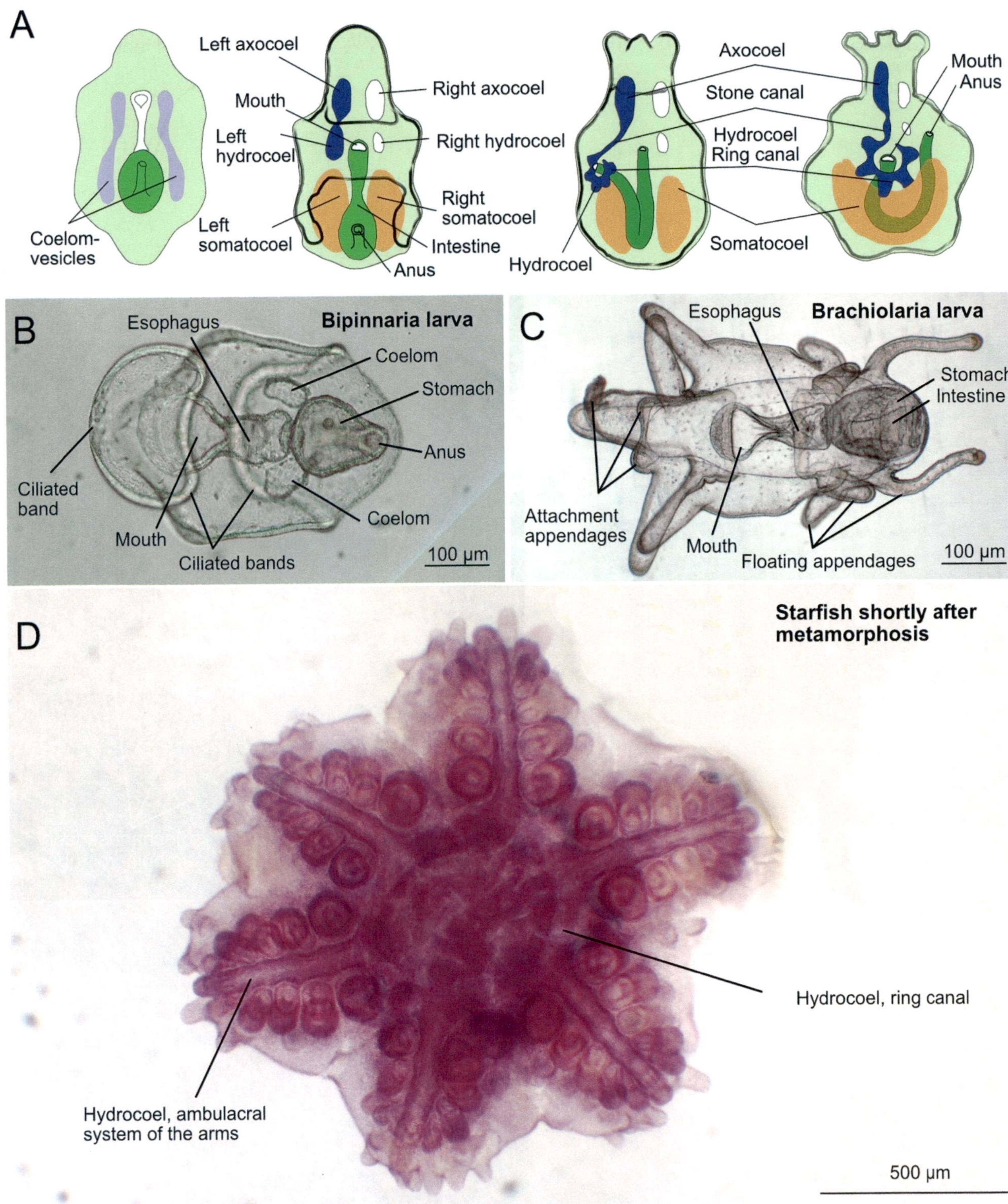

Fig. 9.11 Development of sea stars. (**a**) Schematic representation of the formation of coelomic cavities in Echinodermata. According to various authors. (**b**) Bipinnaria larva of the sea star, live image. (**c**) Brachiolaria larva of the sea star, live image. (**d**) Sea star shortly after metamorphosis, fixed specimen

Food Intake and Intestine

Sea stars are predators that specialize in preying on mussels. They clasp the mussel with their five arms and hold on to it with their numerous ambulacral feet. Although the mussel can initially close the shell halves tightly thanks to its powerful shell adductors, a sea star permanently develops very high tensile forces. In addition, the mussel must open its shell at some point if it does not want to suffocate. If the sea star then manages to open the shell of the mussel a little bit, the sea star will evert its stomach into the mussel and release digestive enzymes. This also attacks the adductor muscle of the mussel. Thus, extraorally predigested food enters the stomach of the sea star via the mouth opening. The stomach consists of the cardia (stomach entrance) and the pylorus (transition to the intestinal diverticula) (Figs. 9.5 and 9.6). The cardia forms a large chamber where the food is further digested and then passed on to the pylorus. From the central pylorus chamber, paired pylorus glands extend into all five arms. They have numerous protrusions, which are referred to as pylorus diverticula (sing. diverticulum). Here, the food is finally digested, and nutrients are absorbed and stored. The pylorus diverticula correspond to the hepatopancreas (often also referred to as the midgut gland for historical reasons) of other invertebrate animals. In some textbooks, the pylorus diverticula are, therefore, collectively referred to as pylorus glands. Since the pylorus diverticula end blindly, food remains must return to the central pylorus chamber and are discharged from there via the anus located on the upper side.

The Water Vascular System, Hydrocoel

The secondary body cavity or coelom in adult echinoderms is a highly complex structure. In addition to the spacious body cavity, which fills the entire body, there are additional coelomic canal systems. In ontogeny, these are formed from paired pouches, which, like in all Deuterostomia, are folded off from the primitive gut roof (Fig. 9.11a). A total of three pairs of such coelomic spaces are formed, from anterior to posterior, referred to as the axocoel (protocoel), the hydrocoel (mesocoel), and the somatocoel (metacoel). Due to the later pentameric body structure, these structures develop differently on the right and left sides of the body of the bilateral larvae. The posteriormost cavities, the right and left somatocoel, line the adult's large body cavity and differentiate oral and aboral canal systems.

The water vascular system of Echinodermata is unique in metazoans. It is called the ambulacral system. In ontogeny, it arises from the left cavity of the second pair of coelomic spaces, which is, therefore, also referred to as a hydrocoel. The water vascular system is responsible for various essential body functions, not only for the gliding movement on the sea floor. The tentacle or tube feet apparatus of the echinoderms is supported by the hydrocoel. This system regulates food acquisition, locomotion, gas exchange, excretion, osmoregulation, and external sensory reception. The high mobility and the thin-walled construction of the feet ensure these functions.

Sea Star Dissection

The sea star is prepared in a dissection dish filled with tap water. The schematic overview illustrates the first steps as cutting lines A and B (Fig. 9.1). However, the animals are examined first from the outside (Fig. 9.2). For the subsequent preparation, coarse scissors, dissecting needles, probes, and tweezers are needed.

(a) The sea star is cut laterally along the arms with scissors (Fig. 9.3a). Start the dissection at the tips of the arms and cut the arms up to the base, but do not cut too deeply.

(b) The aboral body wall can then be folded back by gently loosening the paired pylorus glands of the digestive system with a dissecting needle (Fig. 9.3b, c).

(c) After removing the body wall of the arms, the body wall of the central disc is carefully prepared (Fig. 9.4a). For this, the interradial pillars are cut through. Cut around the madreporite so as to leave it attached to the lower part of the body. This exposes the entrance to the water vascular system. This preparation step allows a view of the internal organs beneath the central disc.

(d) Lift the aboral body wall carefully and detach or cut all material adhering to the surface of the aboral wall. Once the dissection has been done, the digestive system, including the intestine, pylorus diverticula, pylorus, and cardia, can be examined. The gonads, which also extend in pairs into the arms, become visible (Fig. 9.4b, c).

(e) In the final step, the pyloric glands and gonads should be removed and the two-part stomach, consisting of pylorus and cardia, can be studied in detail (Fig. 9.5).

(f) Expose the madreporite with the stone canal and the axial organ. Examine the ambulacral system with ring canal and ampullae. Cut the calcareous stalk near the mouth, transfer it to a microscopic slide with glycerin, and examine it under a microscope.

(g) Prepare some pedicellariae for microscopic inspection.

(h) For demonstration purposes, each arm of the sea star could be prepared differently (Fig. 9.6).

(i) If necessary, histological sections of various organs are used in addition to the in situ examination.

The ambulacral system is connected to the external environment via the stone canal, the axocoel, and its outlet duct, which is covered by the aborally located madreporite (sieve plate) (Figs. 9.2a, 9.4a and 9.5b, c, schematically in Figs. 9.7a and 9.8). The stone canal is located in the axocoel, and forms with it a complex system of parts of the axo- and hydrocoel and intervening blood vessels. The axial organ is considered the most critical vessel in this context. The stone canal, axocoel, and the connection to the madreporite develop in ontogeny from the left anterior coelomic space. Water enters the body through the madreporite and is passed on to the water vessel system via a small cavity (ampulla, dorsal bladder, Fig. 9.8). Even small amounts of water are sufficient to compensate for the water loss caused by pressure changes in the hydrocoel. Although the system is called a water vessel system or hydrocoel, it does not simply contain seawater, but its composition is regulated by the coelothelia.

The water vascular system is located on the oral side of the sea star and consists of a central ring canal that runs around the mouth (Fig. 9.8). This ring canal supplies the arms of the sea star via five large radial canals and thus shows the typical pentameric symmetry. All radial canals in the arms again branch into lateral canals and project through openings in the skeletal plates of the sea star arms into the externally visible, movable ambulacral feet (Fig. 9.8). Each ambulacral foot is connected to a large fluid-filled ampulla, which can contract using fine muscles and pumps water into the feet (Fig. 9.8). A valve in the canal prevents backflow of fluid into the ring canal. In some species, ring-shaped muscles (sphincters) ensure the functionality of the ambulacral system. The ampullae of the ambulacral feet are visible during the preparation after removing the pyloric diverticula in the arms (Figs. 9.4c and 9.6).

In addition to the radial canals, many sea stars also have several larger ampullae, the Polian vesicles, branching directly from the ring canal (Fig. 9.8). They probably also contribute to the regulation of water pressure in the ambulacral system, as well as serving as a water reservoir. However, Polian vesicles are missing in *Asterias rubens*, the species we are discussing here. In contrast, the so-called Tiedemann bodies, small branching bladders, are always connected to the ring canal (Fig. 9.5c, d). They are attributed to the production of coelomocytes, which circulate in the water vascular system and are, among others, phagocytically active. In addition, the Tiedemann bodies probably also serve for a fluid exchange with the adjacent somatocoel, the large body cavity, which surrounds the internal organs and fills most of the sea star body. The stone channel connecting the cavity below the madreporite with the orally located annular channel is visible if carefully prepared (Fig. 9.5d, e). It ensures the connection of the water vascular system with the external environment (Figs. 9.7 and 9.8). The calcareous wall of the stone channel is very firm, which explains its name. Thus, each ambulacral foot is hydraulically supplied by the water vascular system.

The epithelia of the entire coelom system are equipped with monociliary cells, which keep the fluid in motion. Fluid losses, which occur due to the pressure in the water-filled tube feet, can be compensated for via the sieve plate. The permanent activity of thousands of tube feet leads to a uniform sliding movement. If a sea star encounters an obstacle, it immediately "glides" in another direction, with one arm taking the lead. In addition, gas exchange with the surrounding water, excretion, and osmoregulation occurs at the interfaces of the tube feet. Additional structures for gas exchange are the papulae, branched, sac-like protrusions from the epidermis and somatocoel that pass between the arm plates and can be found on the top of the arms. They have significantly thinner-walled epithelia than the feet.

The Hemal Vascular System

Echinoderms have a blood vascular system, the so-called hemal system. It consists of numerous canals and several interconnected narrow-lumen lacunae. The fluid that circulates in the hemal system is also called hemolymph. It transports immune cells (haemocytes) and metabolites from several cellular pathways, among other things. It is also assumed that the hemal system probably plays an important role in the distribution of nutrients in the body. Nutrients are taken up into the hemal system via the epithelium of the intestinal diverticula. Thus, the hemolymph partially takes over the task of vertebrate blood. The hemal vascular system is supplied with fluid by the hydrocoel. Fluid transfer occurs purely osmotically, as the cavity systems have no direct connection. The blood lacunae arise in the extracellular matrix of opposite coelothelia, thus, lacunae and vessels are not lined by an endothelial epithel as typical for invertebrates. The canals and cavities of the hemal vascular system largely run parallel to the water vascular system. The hemal system consists of an oral and an aboral ring canal, from which channels and cavities project into the arms. The vessels emanating from the aboral ring canal supply the gonads, while the oral channels run below the hydrocoel and between the parallel running channels of the hyponeural canals, representing parts of the somatocoel. The two hemal rings are connected by vessels in the axial sinus (axocoel, see below) and via the axial organ. The axial sinus also encloses the stone canal. The axial organ, sometimes called the axial gland, is located in the axial sinus near the stone canal. When dissecting a sea star carefully, the axial organ can be identified by its brown-red color. The vessels of the axial organ continue into the dorsal bladder. The function of the axial organ is not yet fully understood. On the one hand, it might support the fluid flow in the hemal system through contractions. Therefore, the axial organ is occasionally referred to as a "heart," although in echinoderms there is no continuous circulatory system, as the hemal channels end blindly. A real circulation of the fluid in the hemal system is thus not possible, but a rhythmic filling and emptying of individual sections can be observed. The direction of fluid flow in the hemal system therefore changes. Other areas of the vascular system also show contractions, for example, the dorsal bladder or the larger intestinal vessels. On the other hand, recent findings, which we will discuss briefly in the next section, suggest that the axial organ is more likely to be seen as a lymph or excretion organ. In sea stars, chlorocruorin is an iron-containing, respiratory protein, while hemerythrin serves the same purpose in sea urchins.

Excretion and Gas Exchange

Echinoderms do not have dedicated excretory organs. Instead, metabolic byproducts and nitrogen compounds are released directly at the interfaces to the surrounding seawater via the coelomic fluid, for example, at the ambulacral feet or papulae. The same applies, in principle, to gas exchange, which generally occurs at all body surfaces and the interfaces of the coelom. Interestingly, podocytes (coelothelial cells with intertwining foot-like processes) were found in the coelothel of the axial organ, which belongs to the hemal system. In metanephridial systems, podocytes are responsible for the ultrafiltration of the blood and the formation of the coelomic fluid, and thus simultaneously also for the production of primary urine. Therefore, it is speculated whether the axial organ and the hemal system in echinoderms could have a similar function. Then, the madreporite, with its pores, would correspond to an excretory pore. Such an excretory function has already been demonstrated for the larvae of sea cucumbers. In addition, the axial organ produces hemocytes, which circulate in the hemal system and, among other things, have phagocytic tasks. The ciliary wall cells of the hemal system ensure the circulation of the hemal fluid in the hemal channels.

Nervous System and Sensory Organs

Numerous receptor cells are distributed throughout the epidermis, whereas more complex sensory organs are mainly absent in echinoderms. As with many other marine animals, mechano- and chemoreceptors are the most important senses of the sea star for interacting with its environment. In addition, numerous individual photoreceptor cells are present, as most echinoderms react to light and show shadowing reflexes, without complex eyes being present. Sea stars are active predators and use their chemical senses to locate their prey. In an experiment, sea stars move directly toward this source if bait is offered. Chemo- and mechanoreceptor cells were found, for example, in specialized ambulacral feet and tentacles along the arms (Fig. 9.10). The processing of sensory perceptions is taken over by a basiepithelial nervous system, which is arranged in a ring around the mouth and strands in the radii, that is, located below the radial canals. All radial nerves are connected via a nerve ring in the central body. It was previously assumed that this nerve ring represents a superordinate center, the brain, but it only interconnects the radii. A coordinating nerve center, as typical for other bilaterians, that is, a brain, apparently does not exist. Interestingly, during the movement of the sea star toward a target, one arm takes on a guiding function. This is often visible thanks to the curling of the arm tip and increased activity of the terminal tentacles. One hypothesis suggests that the radial nerve in the leading arm takes control of the body by signaling the other radial nerves to coordinate movement toward the leading arm. The description of the sea star's nervous system can only remain rudimentary here; even in advanced zoological courses, the nervous system of the sea stars remains largely hidden from our eyes.

Echinoderms are surprisingly poor in sensory organs. Today, no statocyst has been discovered that coordinates the reversal reflex in sea stars and brittle stars. If one puts a sea star upside down on the aquarium floor, the animal will turn right side up again within a few moments. It is assumed that the reflex center for this reversal reflex is located in the radial nerve and the dorsal skin. A movie showing the turnaround reflex of a sea star can be found at figshare: sn.pub/xr9qhi.

Some sea stars have simple eyes on the tips of their arms. They consist of a few to several hundred pigment cup ocelli. In the living common sea star (*Asterias rubens*), they can be seen as small, bright red spots (Fig. 9.10). Unfortunately, the red color fades away very quickly in fixed animals. So far, it has been assumed that these light sensory organs only allow for distinguishing between light and dark, thus allowing for the orientation toward light and also controlling the animal's day-night rhythm. However, recent studies on crown-of-thorns sea stars (*Acanthaster planci*), which pose a great threat to tropical reefs, suggest that the visual capabilities of these ocelli are also suitable for finding prey.

Reproduction and Development

Asterias rubens, like most echinoderms, is gonochoristic. However, the female and male animals cannot be distinguished externally. The females release their eggs, usually hundreds of thousands (about 2.5 Million in *Asterias rubens*), into the open water at breeding time. The sperm of the males, which are released in clouds comprising millions of sperm (see also Sect. 9.2 on sea urchins), find their way to the eggs chemotactically. Specificity is ensured by species-specific attractants and corresponding membrane receptors. After the successful fertilization of an egg, the first cell divisions, called cleavages, take place. These are holoblastic, the zygote, and the first blastomeres are completely divided and the cytoplasm of the mother cells is distributed to the two daughter cells. This type of cleavage is typical for isolecithal, yolk-poor eggs. The cells (blastomeres) newly formed in each round of cleavage are of different sizes and, therefore, are referred to as macro- and micromeres, both of which later undergo different developmental fates. After a short time, a single-layered hollow sphere, the blastula, is formed. Gastrulation then takes place through an invagination of the epithelium into the interior of the blastula. The resulting first body opening, the blastopore, will differentiate into the prospective anus, as is characteristic for deuterostomes. Following gastrulation, each echinoderm group forms a typical planktonic larva, which can be distinguished by specifically shaped ciliary bands and lobes. In the translucent larvae, the formation of the coelomic cavities is visible (Fig. 9.11). The first coelomic spaces always arise by separating a first coelomic chamber directly on the primitive gut roof (Fig. 9.11a). The first free-swimming planktonic larva of sea stars is a bipinnaria larva (Fig. 9.11b). During further development, arm-like extensions and three temporary attachment stalks are formed. Such larvae are called brachiolaria. They attach themselves to a suitable substrate (Fig. 9.11c). Subsequently, metamorphosis takes place, transforming the larvae into a young sea star (Fig. 9.11d). Most sea stars have a lifespan of several years.

In addition to sexual reproduction, at least 30 of the approx. 1700 known sea star species can regenerate a complete organism from detached body parts (fissiparity). Thus, a completely new sea star can emerge from a detached arm. The common sea star *Asterias rubens* presented here also possesses these amazing regenerative abilities. Our movie clip collection, which includes movies showing sea stars and sea urchins, can be found at figshare: sn.pub/xr9qhi.

9.2 The Shore or Green Sea Urchin (*Psammechinus miliaris*) and the Common or Edible Sea Urchin (*Echinus esculentus*)

Sea urchins are among the most prominent animals in developmental biological research. Groundbreaking insights, for example, regarding how sperm find and fertilize eggs (fertilization) without leading to polyspermy, have been gained from sea urchins. Since sea urchin and sea star larvae are entirely transparent, important observations on the formation and differentiation of the germ layers and the origin of the coelomic cavities were made with them early on. Theodor Boveri already showed, in 1902, through work on sea urchins, that chromosomes are carriers of genetic information. This later led to the "chromosome theory of inheritance." Movies related to this chapter can be found at figshare: sn.pub/xr9qhi.

Therefore, this book also introduces readers to sea urchins, which are easily available during marine biological excursions (Fig. 9.12). The shore or green sea urchin (*Psammechinus miliaris*), measuring approximately 5 cm, possesses a flattened sphere-like shape at its oral pole. Sea urchins continuously search the ocean floor for food. Dead plant remains and dead animals are part of their typical diet, and therefore, they play an essential role in the food web as "recyclers."

Figure 9.12 shows a living sea urchin from the oral (Fig. 9.12a, b) and aboral sides (Fig. 9.12c, d). The centrally located mouth is easily visible thanks to the five strong teeth, representing the only externally visible part of the complex jaw apparatus. The jaw apparatus consists of many individual calcareous pieces, which can be moved against each other with the help of specialized muscles, and is named after its first descriptor as "Aristotle's lantern." With the help of this jaw apparatus, sea urchins crush their food. The mouth is located in a so-called mouth field (peristome), which is equipped with ambulacral feet, gill tufts, and pedicellariae (Fig. 9.12a, b). On the aboral side, we recognize the anus centrally located in an anal field (periproct) (Fig. 9.12c). At higher magnification, the madreporite and the genital openings become visible (Fig. 9.12d). Figure 9.13 shows the exposed Aristotle's lantern using the example of the common or edible sea urchin (*Echinus esculentus*), which, with a diameter of 10–15 cm, is significantly larger than the shore or green sea urchin *Psammechinus miliaris*.

We will omit a detailed description of the anatomy of a sea urchin, crinoid, and sea cucumber here. Usually, there is only time for the preparation of one echinoderm species in our zoological practical courses, most often a sea star. Therefore, we take a closer look at the development of sea urchins since fertilization experiments can easily be done in the context of marine biology field trips.

Fertilization and Embryonic Development of the Sea Urchin

Sea urchins are well suited for carrying out the fertilization of eggs in a glass dish or on a microscope slide in the laboratory. This allows students to observe sperm searching for an egg, and to study the fertilization itself with the formation of the fertilization envelope that inhibits polyspermy. Moreover, it is quite easy to study the further development of sea urchin embryos directly under the microscope. *Psammechinus miliaris*, which can be obtained from several Marine Biology Stations, is sexually mature in late spring and early summer and produces plenty of gametes until about the end of June. To obtain eggs for a fertilization experiment, female animals, which can be recognized by the shape and color of their gonopores (Fig. 9.14a), are placed with the aboral side down on a beaker filled with clean and aerated seawater. The injection of a few milliliters of 0.5 M KCl into the body cavity triggers contractions of the musculature. This causes the gonads to release their eggs (Fig. 9.14b–e). The same procedure is applied to males to obtain sperm. While the eggs are only capable of fertilization for a few hours, the sperm remain vital for 2–3 days when cooled (4–8 °C). They can be used for numerous fertilization and development experi-

Fig. 9.12 Shore or green sea urchin *Psammechinus miliaris*, alive. (**a**, **b**) Oral side with mouth field and teeth. (**c**, **d**) Aboral side with the anal field, madreporite, and genital openings

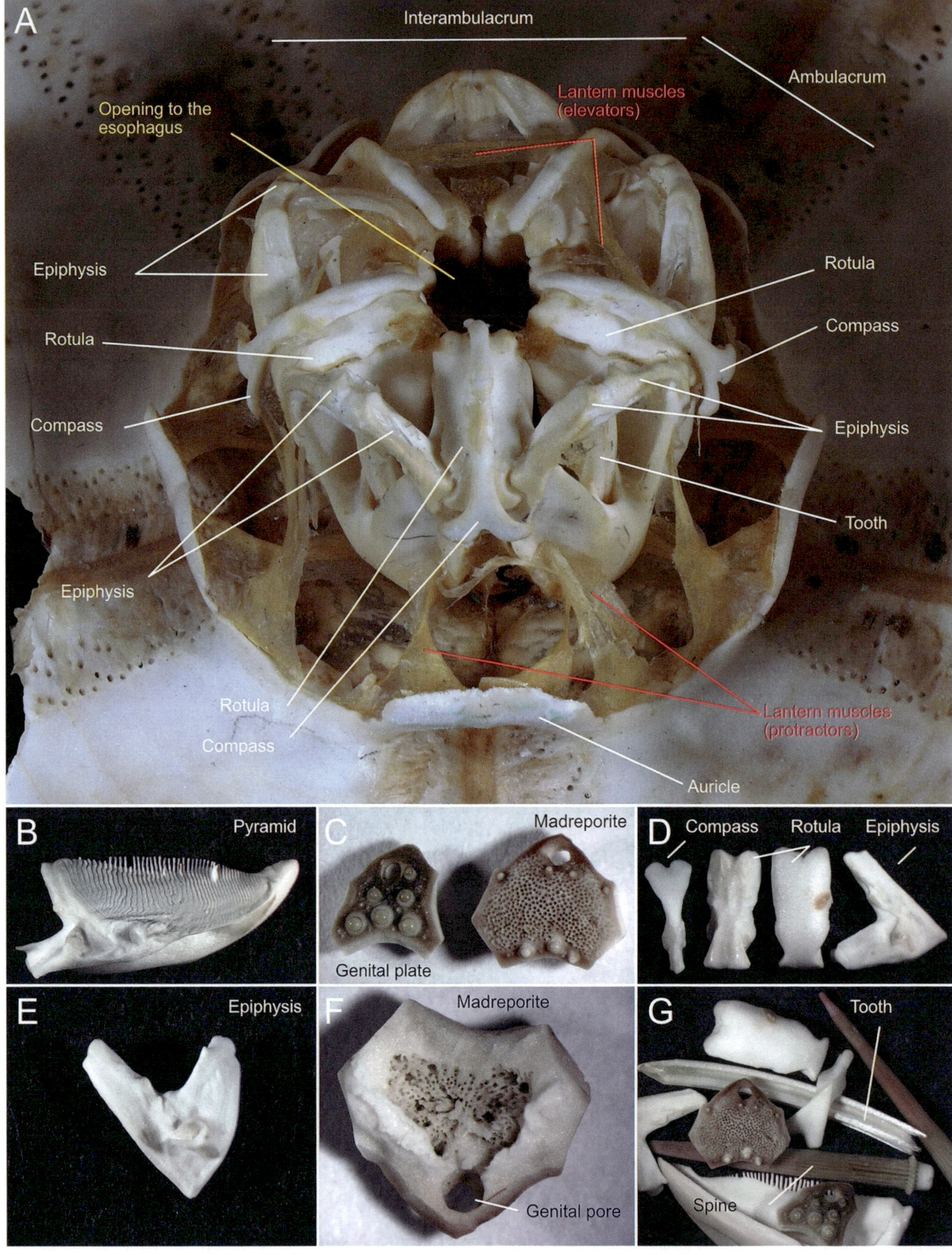

Fig. 9.13 (**a**) Jaw apparatus ("Aristotle's lantern") of *Echinus esculentus*, the edible sea urchin. (**b–g**) Individual calcareous pieces from the jaw apparatus

Fig. 9.14 Fertilization experiments. (**a**) Schematic representation of the externally visible sexual characteristics of *Psammechinus miliaris*. (**b**) To induce the release of gametes, KCl is injected into the coelom. (**c–e**) The released sperm and oocytes are collected in seawater. (**f**) Equipment for fertilization experiments at the Biological Institute Helgoland

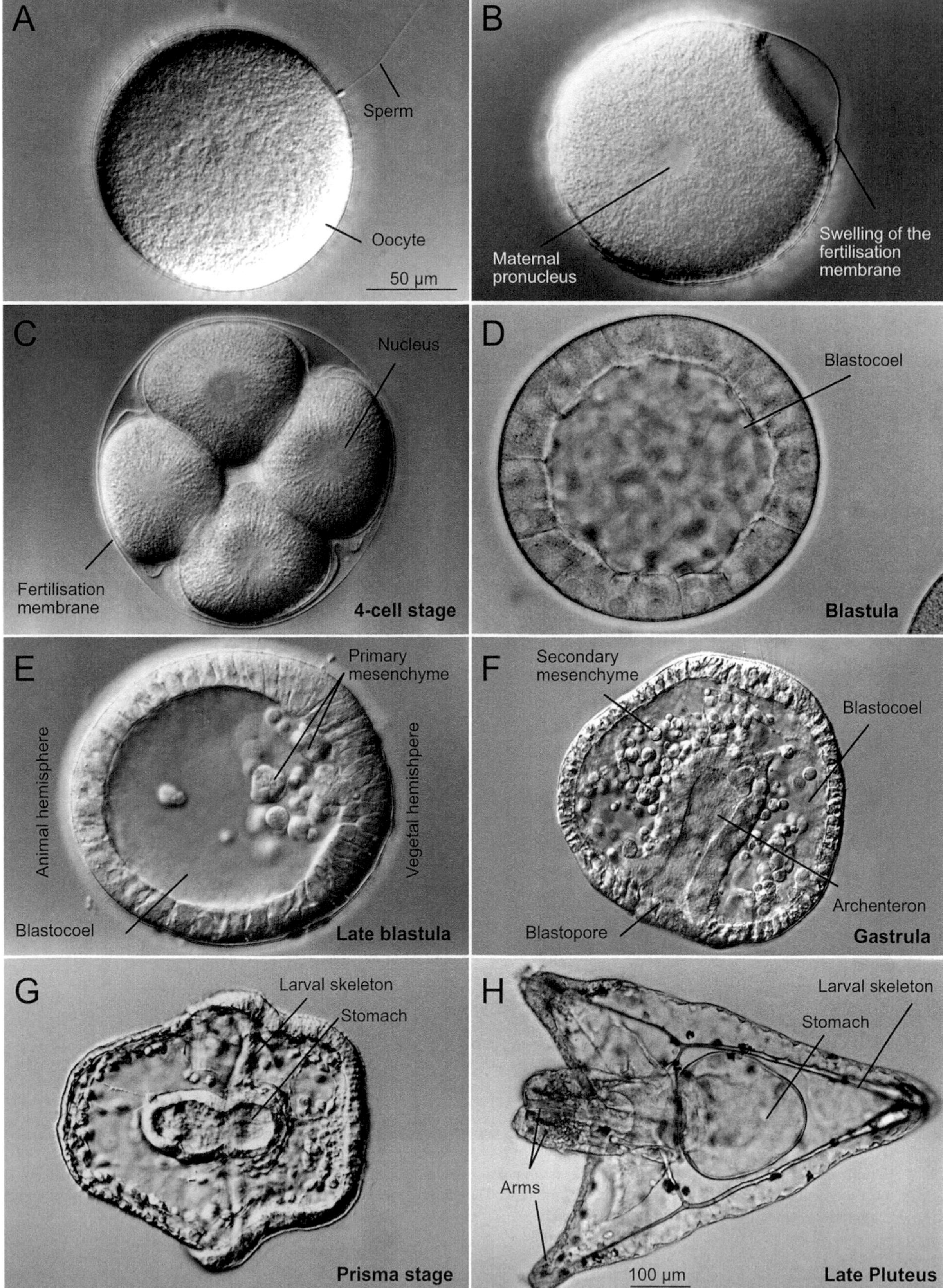

Fig. 9.15 Developmental stages of *Psammechinus miliaris*. (**a**) Entry of a sperm (egg's fertilization envelope removed). (**b**) Lifting of the fertilisation membrane. (**c**) Four-cell stage. (**d**) Blastula. (**e**) Late blastula with the first cells of the primary mesenchyme. (**f**) Gastrula with beginning mesenchyme formation. (**g**) Gastrula in prism stage. (**h**) Early pluteus. All images were kindly provided by Prof. Dr. D. Ribbert, University of Münster

ments. Developing eggs are placed in a seawater basin in such a way that a slight cooling to ambient seawater temperature is achieved (Fig. 9.14f). In recent years, we have used *Paracentrotus lividus* for our fertilization experiments as an alternative to *Psammechnius miliaris*, which is sometimes difficult to obtain. The development times are very similar to those of *Psammechinus* and the sexes can be distinguished externally, albeit not quite as easily. See references for further advise. The same experiment can also be carried out with the larger *Echinus esculentus*; unfortunately, the sexes are externally indistinguishable to us. To obtain both eggs and sperm, it is typically required that one inject more than two animals with a few milliliters of 0.5 M KCl. This procedure can be quite challenging and requires careful attention to ensure the safety and well-being of the animals involved.

To study the development of sea urchin embryos, a few eggs are brought together with a highly diluted sperm solution in an evaporating glass dish (80 × 45 mm) filled with fresh seawater. Be careful, too many sperm cause polyspermia and the death of the embryo. The dish should be covered by half of a petri dish and kept at the appropriate temperature. In some countries, the glass dishes mentioned above are called Boveri dishes, since Theodor Boveri is said to have used such dishes for his historical experiments. To observe the fertilization of sea urchin eggs under the microscope, as well as to study the development of the embryos, place eggs or embryos into a suitable microscopic observation chamber or onto a specialized microscopic slide, e.g., with a hollow loop to avoid squeezing the eggs when placing a coverslip on top. Alternatively, use an inverted compound microscope for observations.

After fertilization and the initial cleavages (Fig. 9.15a–c), a single-layered blastula with a central, fluid-filled blastocoel is formed (Fig. 9.15d). The first developmental steps proceed strictly synchronously over a longer period. Therefore, stages can be taken from a single experimental batch again and again to follow the course of development continuously. Primary mesenchyme cells immigrate at the vegetative pole,

later forming the larval skeleton in the so-called pluteus larvae (Fig. 9.15e). During gastrulation, the primitive gut invaginates and the blastopore becomes visible. As in all deuterostomes, the blastopore later becomes the anus, and the mouth is newly formed (Fig. 9.15f). In the so-called prism stage, enterocoely occurs at the tip of the primitive gut (Fig. 9.15g), where the primitive gut bulges out and forms three coelomic sacs on each side: the axocoel, the hydrocoel, and the somatocoel, also called the protocoel, the mesocoel, and the metacoel (we have already discussed this with the sea star). Now, the pluteus larvae is formed, characterized by thin floating appendages (Fig. 9.15h). The spinning or floating planktonic pluteus larvae still have a strongly pronounced bilateral symmetry. On the appendages, which are internally supported by a larval skeleton, ciliary bands serve for locomotion and the capture of smaller food particles, which are then transported to the mouth on the ciliary bands. The pluteus shows a typical trimeric coelomic structure. The left hydrocoel cavity becomes the ring canal, which is directly connected to the axocoel and the madreporite. It establishes the connection to the external environment. The axocoel forms, among other structures, the axial organ, which is now seen as an excretory and lymphatic organ. The right hydrocoel degenerates in some species or becomes the madreporite vesicle (dorsal vesicle). The two somatocoel spaces lie against each other, enclose the intestine, and thereby form mesenteries, which suspend the intestine and enclose the secondary body cavity. In Eleutherozoa (sea stars, brittle stars, sea urchins, and sea cucumbers), radial canals branch off from the orally located somatocoel, located between the hydrocoel and the epidermis and between which the hemal system of the radii is formed.

Table 9.1 summarizes the duration of the first developmental steps in the two species discussed. During marine and developmental biology field trips, the entire development of the sea urchin, from fertilization to the formation of the pluteus, can be traced (Fig. 9.15). Our movie clip collection, which includes movies showing sea stars and sea urchins, can be found at figshare: sn.pub/xr9qhi.

Table 9.1   Timeline of normal development of sea urchins at 10 °C water temperature (from Fischer, 2013)

	Psammechinus miliaris	*Echinus esculentus*
Fertilization	0 h	0 h
Fusion of the pronuclei	30–40 min	–
1st cleavage	50 min	2 h
2nd cleavage	1 h 20 min	3 h 30 min
3rd cleavage	1 h 50 min	5 h
4th cleavage	2 h 20 min	7 h 30 min
Blastula	7 h	20 h
Early gastrula	10 h	30 h
Prism stage	24 h	40–50 h
Late gastrula	–	68 h
Pluteus	48 h	93 h

▶ **Summary** **Despite lacking a bony skeleton, vertebral column or spine, a skull, and a distinct brain, Acrania are close relatives of vertebrates (Vertebrata) and share several characteristics with them as a basal group of Chordata. Thus, Acrania have a notochord, a dorsal nerve cord, and a muscular postanal tail. Only a few species exist, and all of them are marine chordates of slender, lanceolate shape. The lancelets are the only living representatives of Acrania.**

Acrania (Cephalochordata, lancelets) is a small group of exclusively marine chordates that comprises only about 30 species. Since their description in the middle of the eighteenth century, the European species *Branchiostoma lanceolatum*, initially published under the name *Amphioxus lanceolatum*, has become particularly well-known in evolutionary biology and zoology. It has aroused great interest among biologists, as Acrania, although they do not have a spine, already show numerous characteristics of vertebrates. If evolutionary biologists were forced to sketch or describe a possible ancestor of vertebrates without knowledge of the lancelets, this fictitious animal would probably look very similar to *Branchiostoma*. The animals can grow to about 6 cm long. They live as microphagous filter feeders in occasionally high population densities (up to 6000 individuals per square meter) in well water-permeated coarse sands. However, these habitats are relatively rare. So, lancelets are not among the more common marine animals and are even considered endangered in some places. The animals live in the sublittoral down to about 80–100 m depth, where they constantly filter small food particles from the breathing water. They are slightly buried in the sediment, with the mouth opening upward and reaching the surface. Corresponding habitats are so characteristic for the occurrence of the lancelets that these regions were often named after them as "Amphioxus sands," such as the Amphioxus ground (Loreley Bank) a few kilometers northeast of Heligoland in the German Bight.

Cephalochordata have a series of characteristics in common with Craniota (Vertebrata, vertebrates). These include an axial elastic support rod, the dorsal notochord or chorda, the dorsal hollow nerve cord, including its formation process, the neurulation, the neural canal (canalis neurentericus), the pharynx with the pharyngeal gill slits with an endostyle, the segmentation of the musculature, a postanal tail, the basic structure of the blood vascular system, and a hepatic cecum. The somatic musculature, which consists of mononucleated longitudinal muscle fibers, arises from segmental, dorsolaterally arranged mesoderm pockets, somites, of the embryonic primitive gut. In contrast, the somatic muscle cells of Craniota are multinucleated and syncytial, but also segmentally arranged and longitudinally oriented in basic craniotes. The multitude of morphological similarities has led to Acrania being seen as the closest relatives of Craniota since their discovery. However, we now know, especially from phylogenomic analyses, that the third chordate taxon, Tunicata (sea squirts and planktonic relatives, see next chapter), is the sister group of Craniota in all probability, despite the less obvious morphological similarities. It is mainly the larvae of tunicates that show a similar multitude of chordate features. Sea squirts are considered an essential evolutionary biological model as well. In contrast to their other relatives, the adult animals no longer resemble craniotes externally after metamorphosis, leading to a completely sessile lifestyle. This implies that the high degree of morphological similarities between Acrania and Craniota must represent ground pattern features of all Chordata, which are secondarily missing in the adult Tunicata. Nevertheless, Acrania are still well suited to introduce the basic organization of Craniota and therefore continue to be important course objects.

© The Author(s), under exclusive license to Springer-Verlag GmbH, DE, part of Springer Nature 2025
A. Paululat, G. Purschke, *Metazoa – Morphology and Evolution of Animals*, https://doi.org/10.1007/978-3-662-69904-1_10

Observations

The lanceolate body of *Branchiostoma* is divided into a trunk and a tail region (Figs. 10.1 and 10.2). Some characteristic features, such as mouth cirri, pharynx, gonads, and muscle segments, can already be recognized in living animals under the stereomicroscope (Fig. 10.1e, 10.2a, b). When you take a living animal in your hand, you can easily feel the strength of its longitudinal muscles. The tail is exclusively used for locomotion. A movie showing *Branchiostoma* swimming can be found at figshare: sn.pub/xr9qhi. Accordingly, the digestive tract ends with the anus in front of the tails. The ventrally located mouth opening is surrounded by numerous curved tentacles (oral cirri) in a basket-like manner. They prevent the unwanted intake of larger particles (Figs. 10.1e, 10.2a, b, c, 10.3a, b, and 10.4a). Next to the mouth, densely ciliated cells form the loop-shaped wheel organ, which drives the food particles to the closable velum. This marks the entrance to the pharynx proper and is also equipped with sensory tentacles (Figs. 10.2d, 10.3a–c, and 10.4a). The ciliary pit located at the roof of the mouth cavity, Hatschek's pit, is not always clearly visible. It is considered a precursor of the adenohypophysis and was named after the Austrian zoologist Berthold Hatschek (1854–1941), who described the organ first. The pharynx with the pharyngeal gill slits, which extends to about one-third of the body length, follows the velum. The gill slits do not open directly to the outside, but rather into a peribranchial space or atrium, which only opens to the outside through the atrioporus, located somewhat

before the last third of the body. The atrium thus significantly exceeds the length of the pharynx. The anus is located relatively far back and opens to the outside on the left side of the body. It marks the beginning of the postanal tail (Figs. 10.2a, b and 10.3a). A continuous small fin runs over the entire dorsal side of the animal, transitioning into a rostral fin at the front and into the tail fin at the back. The tail fin continues ventrally between the anus and atrioporus (Figs. 10.2a, b and 10.3a), while the fin is further replaced at the front by the paired metapleural folds. Acrania are gonochoristic, but females and males are externally indistinguishable.

The Pharynx, Respiration, and Food Uptake

The pharynx in mature animals has up to 180 gill bars and as many gill slits. The gill bars run obliquely from the ventral to the dorsal, inclining from posterior to anterior. Primary and secondary gill bars can be distinguished, formed one after the other in ontogenesis. The primary gill bars arise from the body wall in the region between the gill slits. Therefore, they contain a coelomic space and two skeletal rods, whose ventral ends are bent anteriorly and posteriorly. The primary gill slits formed in this way are divided into two gill slits during further development by a secondary gill bar growing out dorsally, so that primary and secondary gill bars alternate (Figs 10.2d, 10.4b, c, and 10.6c, d). The secondary gill bars contain only a simple skeletal rod and no coelom (Fig. 10.6c, d). Thus, one can distinguish the two types of gill bars quite well in the differentiated animal. The gill bars are firmly connected by cross-connections, the synapticles. They also contain skeletal elements and thus form a solid gill basket (Figs. 10.2e and 10.4c). The epithelium of the gill bars comprises ciliated cells directed toward the lumen of the pharynx (Fig. 10.6c, d). Skeletal elements and the ciliated epithelium can also be well observed in whole mounts. The lateral cells of the gill bars have longer cilia, which are responsible for driving the water through the slits. The shorter cilia inside the bars transport the food particles trapped in mucus dorsally into the epibranchial groove (Fig. 10.6a). From there, food is forwarded to the intestine by cilia, while the water leaves the pharynx via the atrium and atrioporus. The essential mucus net for food acquisition is produced by glandular cells in the hypobranchial groove (endostyle) and transported dorsally by the cilia via the gill bars (Figs. 10.4 and 10.6e). The endostyle consists of various ciliated cells with gland cells in between. These cells are arranged in a longitudinal band-like manner, and the unpaired median cell row stands out due to its particularly long cilia (Fig. 10.6e). The endostyle is also supported by skeletal elements. Below it is the endostyle coelom with the ventral aorta, from which the blood vessels extend laterally into the gill bars (Fig. 10.6e). The main function of the pharynx is thus feeding rather than respiration, the latter of which takes place over the entire body surface.

Fig. 10.1 Collecting lancelets on a marine biology excursion; here, with the research vessel *Neomysis* of the Station Biologique de Roscoff in Brittany, France. (**a**) Emptying a dredge filled with sediment. (**b, c**) Hand selection of animals from the sediment and placing them in bottles filled with seawater. (**d**) Some freshly collected, not yet buried specimens on the sediment from the habitat, which mainly consists here of broken mollusk shells (shell debris). The serial bright dots are the gonads. (**e**) A single animal with gonads

Epidermis and Musculature

As with almost all invertebrate animals (exception: arrow worms, Chaetognatha), the epidermis is monolayered and covered with a glycocalyx made up of acidic mucopolysaccharides. Below the epidermal cell layer follows an extracellular matrix (ECM) with several layers of collagen fibers. Underneath the prominent sheets of ECM follows mesodermal connective tissue in which epithelial coelomic spaces, nerves, and blood vessels are embedded. The connective tissue continues into the myosepta and the axial connective tissue around the notochord (Figs. 10.4 and 10.7a, b). Dorsally, above the hollow nerve cord, there is a central connective tissue body that extends into the fin rays. These are located in the fin boxes, derivatives of the myomeres, which are lined with mesothelium. There are between three and five fin rays per muscle segment. The fin rays may also store nutrients, so they change during life and do not look the same in all examined individuals (Fig. 10.5).

The musculature of the trunk is divided into about 60 muscle segments or myomeres in mature animals. The muscle segments are V-shaped, with the tip facing anteriorly, and separated from each other by myosepta (Figs. 10.2a, b, f and 10.4d). In addition, the left and right myomeres are out of register. Thus, in cross-sections, several myomeres are always visible (Fig. 10.5). The myofibrils are oriented longitudinally in the individual muscle plates. Extensions of the

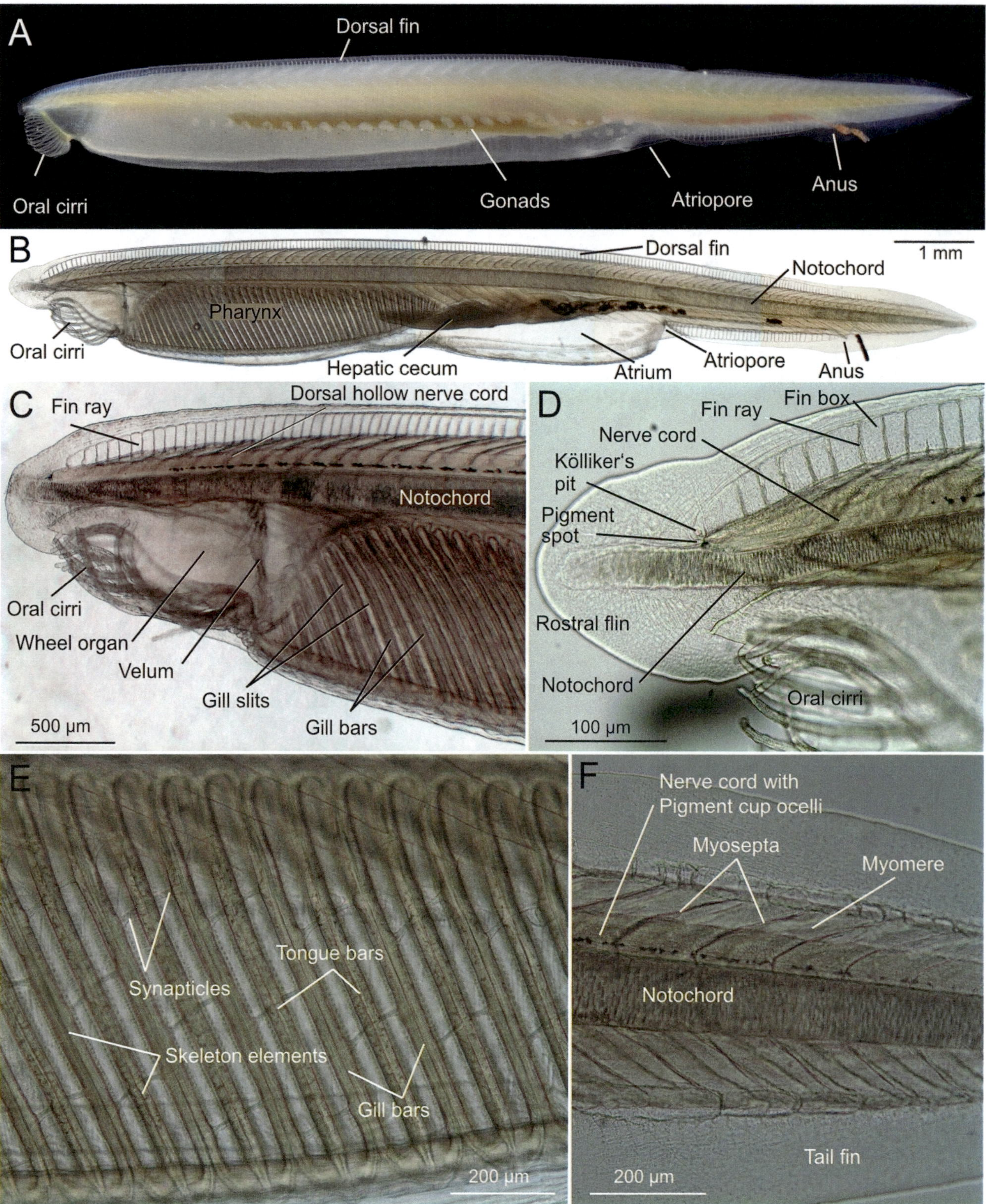

Fig. 10.2 *Branchiostoma lanceolatum*. (**a**) Living animal at low magnification, total length 6 cm, anesthetized. (**b**) Juvenile animal, anesthetized. (**c**) Juvenile animal. Anterior end with rostral fin, mouth cirri, velum, and anterior part of the pharynx. (**d**) Anterior, notochord, nerve cord with pigment cup cells, rostral and dorsal fin; dorsal fin with fin rays and fin boxes. (**e**) Pharynx with main and secondary bars. (**f**) Tail with nerve cord, notochord, myosepta, and myomeres

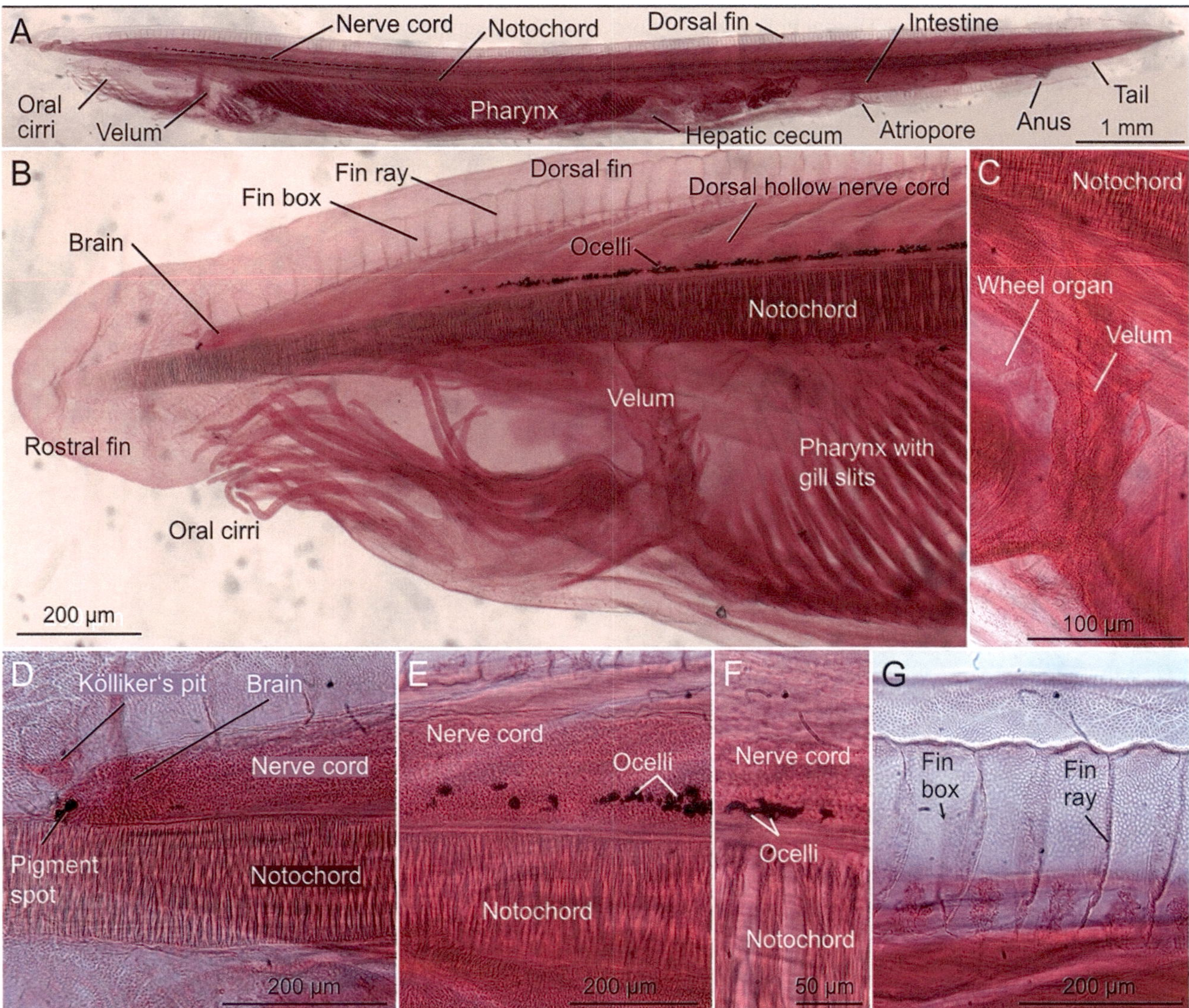

Fig. 10.3 *Branchiostoma lanceolatum.* Juvenile, total preparation, borax carmine stained. (**a**) Whole animal at low magnification. (**b**) Anterior end with rostral fin, mouth cirri, velum and anterior pharynx, notochord, and nerve cord with pigment cup cells. Dorsal fin with fin boxes. (**c**) Wheel organ, velum with velar tentacles. (**d**) Brain vesicle with pigment spot and Kölliker's pit, nerve cord, notochord. (**e**) Nerve cord with pigment cup cells and notochord with the clear disc-shaped arrangement of the chorda cells. (**f**) Higher magnification; the two lower ocelli oriented ventrally. (**g**) Skeleton of the fin; fin boxes and small fin rays. Specimens from the Biological Collection of the University of Osnabrück

muscle fibers project directly to the nerve cord (Fig. 10.7a, b). In addition to the myomeres, the myocoel and sclerocoel are usually recognizable (Figs. 10.5 and 10.7a, b). As with the primarily aquatic vertebrates, the so-called fish, the trunk musculature is essentially longitudinal and is responsible for the undulating swimming movements and the wriggling movements during burrowing into the substrate.

Chorda and Nerve Cord

All organisms that establish a notochord (synonym: chorda dorsalis) during their early embryonic development are referred to as chordates (Chordata). The notochord is a flexible tissue rod of mesodermal origin that runs longitudinally through the animal's body. In addition to its function as a supportive structure, the notochord plays a central role in early development, representing a signaling center in the differentiation of all surrounding tissues. The notochord itself consists of consecutive, disc-shaped cells, which are surrounded by a cell-free, collagen-containing notochordal sheath (Figs. 10.2a–e, 10.3a, b, d–f, 10.4d, 10.5, and 10.7a, b). The notochord cells in Acrania are highly specialized, transversely striated muscle cells, whose myofilaments run

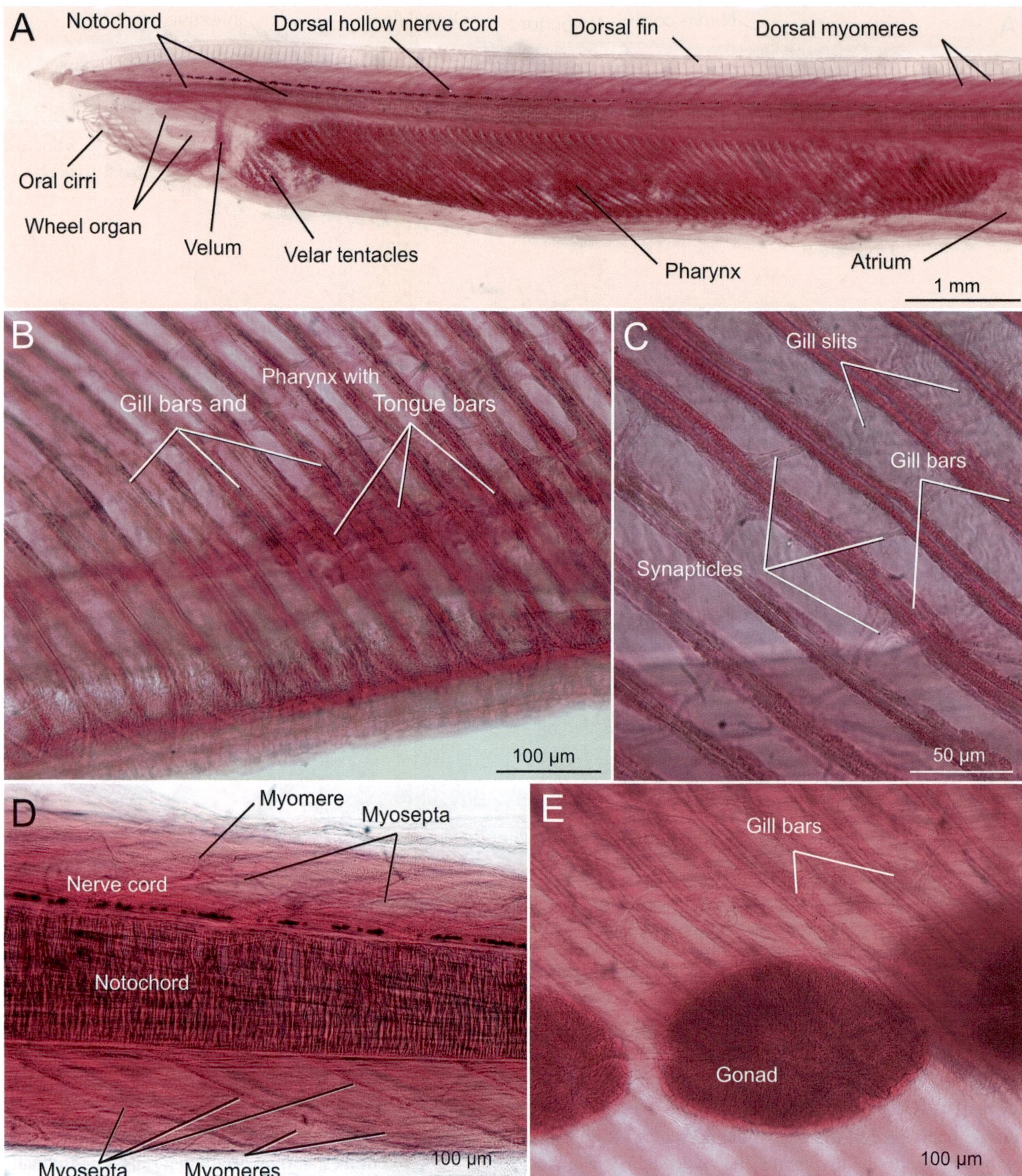

Fig. 10.4 *Branchiostoma lanceolatum*. Juvenile, total preparation, borax carmine stained. Pharynx with mouth cirri and velar apparatus. (**a**) Overview, Pharynx with approximately 65 pharyngeal gill slits. (**b**) Gill bars and gill slits; main bars with inverted Y-shaped splitting of the skeletal elements. (**c**) Higher magnification; epithelium, synapticles, and skeletal elements. (**d**) Notochord and nerve cord in the tail. (**e**) Gonads at the beginning of sexual maturity as small egg-shaped structures. Specimens from the Biological Collection of the University of Osnabrück

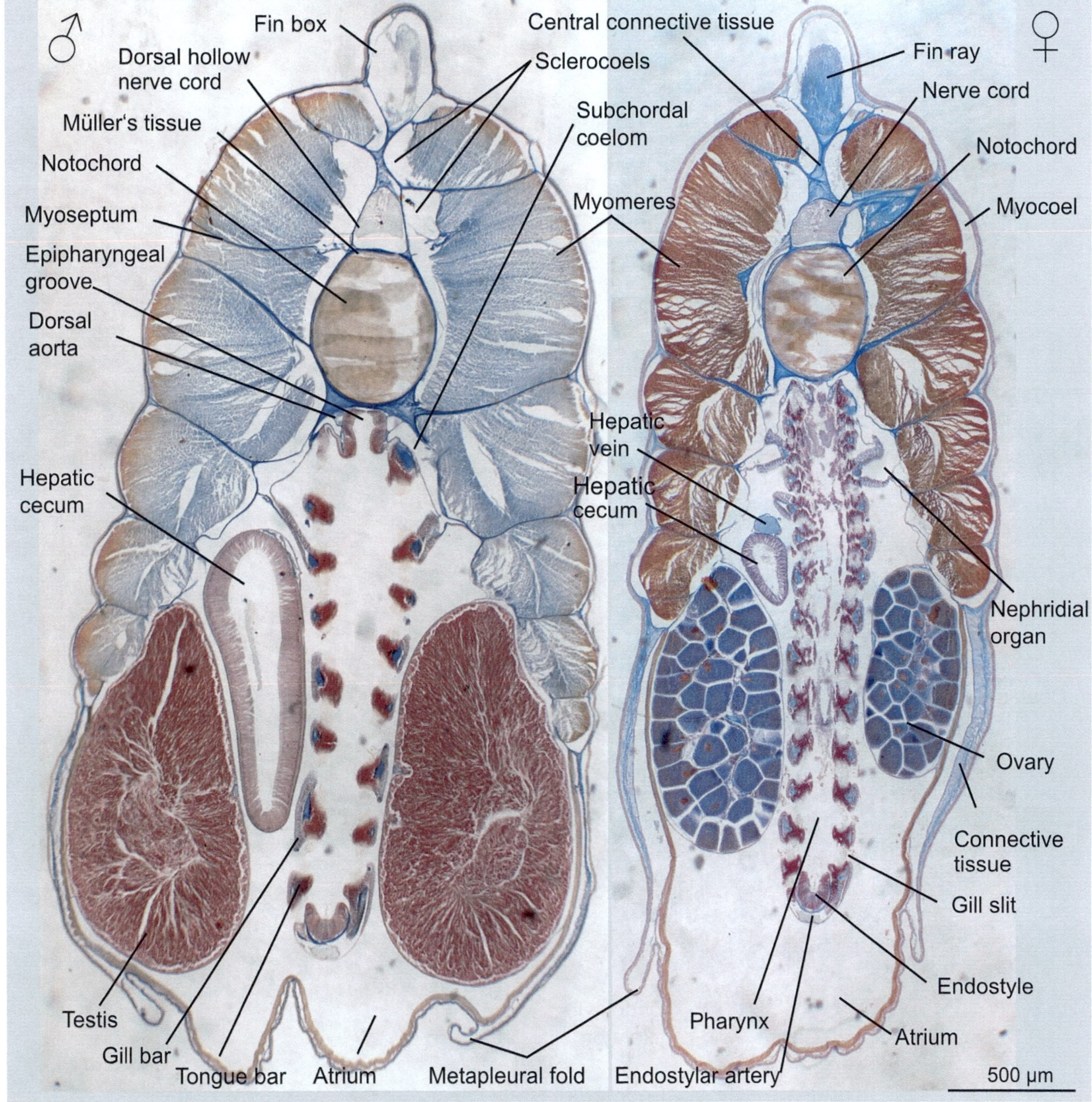

Fig. 10.5 *Branchiostoma lanceolatum*. Azan staining; male on the left, female on the right. Cross-sections through the pharynx, the liver blind sac, and the gonads. Specimens from the Biological Collection of the University of Osnabrück

transverse to the body's longitudinal axis and are connected to the notochordal sheath via hemidesmosomes. Through the so-called notochordal horns, they are connected to the nervous system. In this way, the animals can modify the elasticity and strength of the notochord while moving. The structure of the notochord in *Branchiostoma* differs at the cellular level from that of Craniota and Tunicata, which consists of highly vacuolated cells. Net-shaped cells, the so-called Müller cells, lie dorsally and ventrally on the notochord and form flat bands. These cells form extensions, which project into the neural tube and establish the neuromuscular connections. In Acrania, the notochord extends beyond the front end of the neural tube and continues into the rostral fin (Figs. 10.2a–d and 10.3a, b). This tapering part forms a rostrum, a snout-like structure used to push sand grains aside in initial burying.

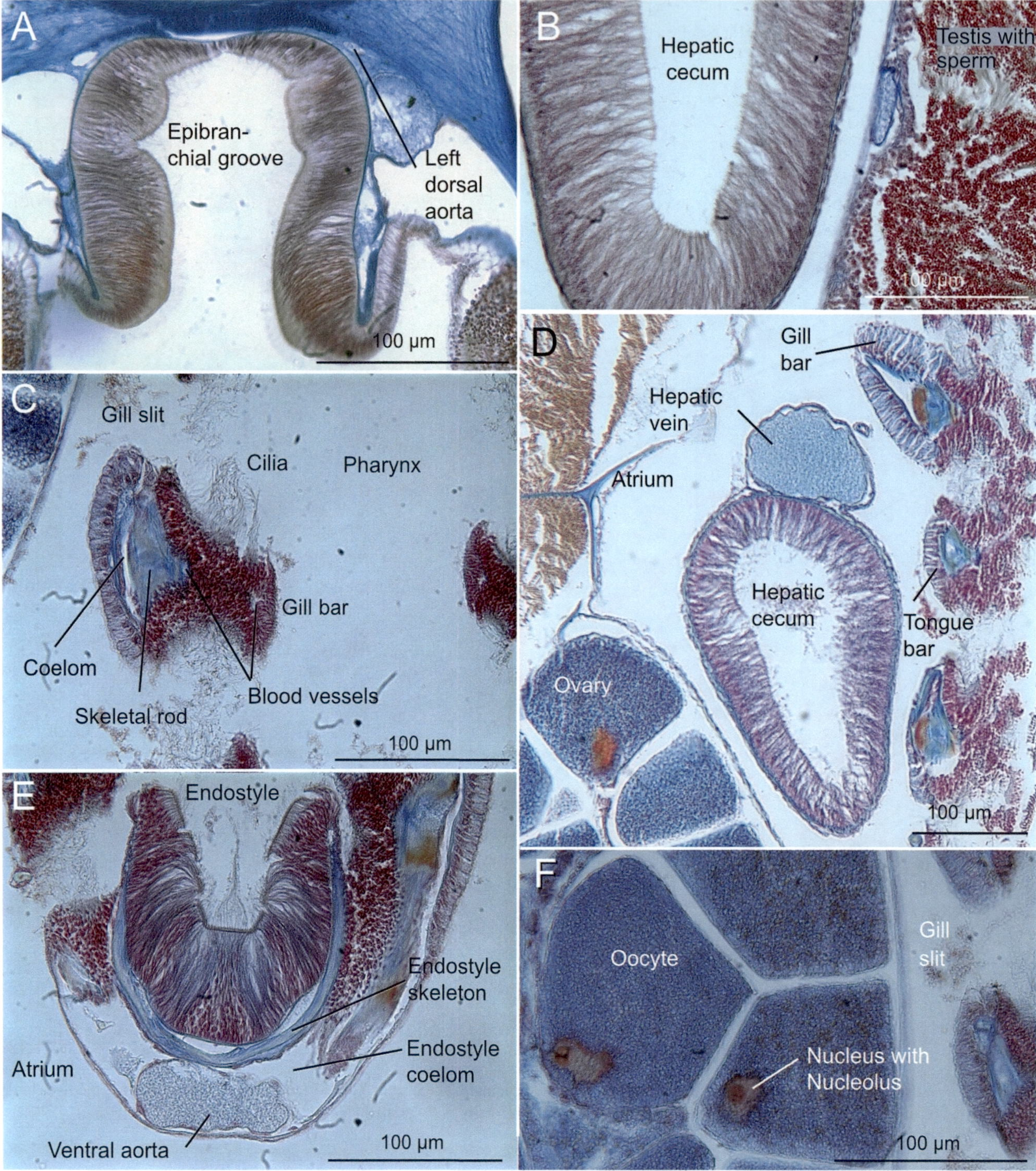

Fig. 10.6 *Branchiostoma lanceolatum.* Azan staining. Pharynx, digestive cecum, and gonads. (**a**) Epibranchial groove with aortic roots and subchordal coelom. (**b**) Digestive cecum consists of high prismatic ciliated epithelium; on the left testis with sperm. (**c**) Main bar with differently ciliated epithelia, left peribranchial space, right pharynx. (**d**) Pharynx with main and secondary bars and hepatic cecum with the hepatic vein (vena hepatica) shortly before the sinus venosus. (**e**) Hypobranchial groove, made of rows of ciliated epithelial cells and unciliated gland cells. Median ciliated cells with particularly long cilia. Endostyle coelom with ventral aorta. (**f**) Ovary with oocytes. Specimen from the Biological Collection of the University of Osnabrück

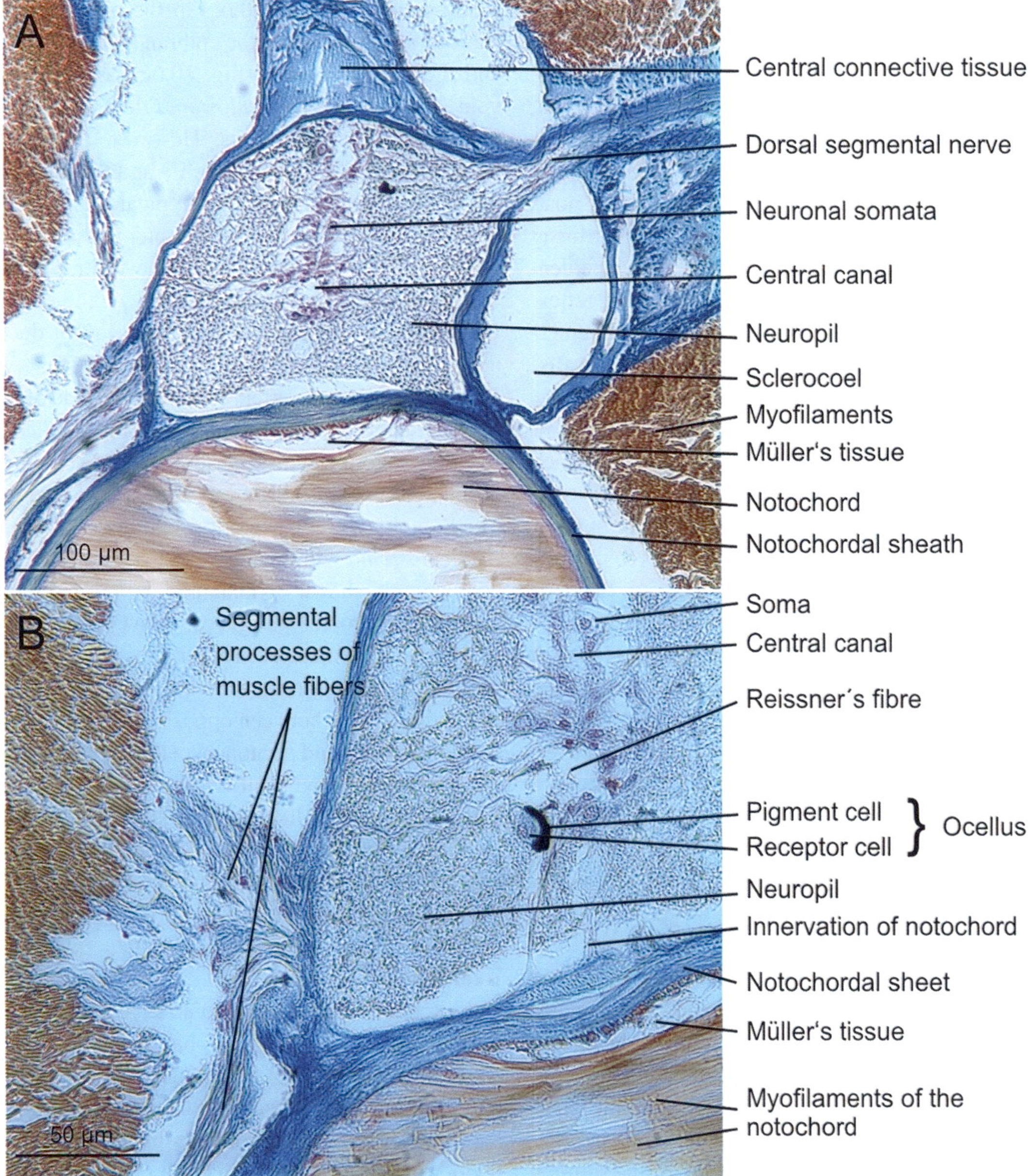

Fig. 10.7 *Branchiostoma lanceolatum*. Azan staining; nerve cord and notochord. (**a**) Nerve cord and notochord. Nerve cord with dorsal segmental nerve and ventral bundle of muscle cell extensions. In the nerve cord, the arrangement of the neuronal somata around the central canal and the neuropil is visible. In the neuropil, there are some particularly strong neurites. (**b**) Higher magnification, muscle cells with extensions that run to the spinal cord and fibers from the notochord plates. Notochord, muscle plates, and dorsally located Müller cells. In the nerve cord, a medianly sectioned ocellus with shading pigment cell and receptor cell. Specimen from the Biological Collection of the University of Osnabrück

The nerve cord is located dorsally to the notochord (Figs. 10.2a–d and 10.3a, b). It terminates somewhat behind the notochord. Anteriorly, the central canal forms a small hollow expansion, the brain (Fig. 10.3b). The central canal of the nerve cord and brain is connected to the exterior via the neuropore and the ciliated Kölliker's pit at the front end of the rostral fin (Figs. 10.2d and 10.3d). It probably serves as a chemosensory organ. Along the neural tube, dorsal nerve roots originate between the myomeres (Fig. 10.7a). Like the muscle cell extensions, they are segmentally arranged.

At the anterior end of the brain is a frontal ocellus consisting of several pigment cells and monociliary receptor cells, recognizable in whole mounts as a pigment spot (Figs. 10.2c, d and 10.3d). About 1500 pigment cup ocelli are notable, distributed all along the entire nerve cord. Each consists of only two cells, a pigment cup cell with shading pigment and a photoreceptor cell (Figs. 10.2c, f, 10.3b, d–f, 10.4d, and 10.7a, b). The number of these ocelli varies from front to back, as well as in the direction in which they are oriented; the smallest number is found in the middle of the trunk. These arrangement of simple eyes likely plays a crucial role in orientation in the substrate. Light-sensing organs on the back possibly help in adopting an optimal swimming position, with the ventral side facing the bottom and the dorsal side facing the water surface. Numerous fish use incident light for body orientation, known as the dorsal light reflex. Other sensory structures include the mouth cirri and numerous, mostly single-cell receptor cells in the epidermis, distributed throughout the body.

The transition from the pharynx to the digesting part of the alimentary canal takes place via a short esophagus, which continues into the absorbing midgut. In the epibranchial groove, the food is transported to the esophagus. The straight midgut then transitions into the rectum and opens to the outside through the left-sided anus. At the beginning of the midgut, the digestive cecum (also called hepatic cecum) branches off. It has been shown that the cecum functions in glycogen and lipid storage. Thus, it is considered a precursor of the vertebrate liver. Digestion in *Branchiostoma* is primarily extracellular and takes place in the gut lumen. However, digestion is completed in cells of the intestine and cecum wall (Figs. 10.2a, b, 10.3a, 10.5, and 10.6b, d). The entire intestinal canal is formed of ciliated epithelial cells (Fig. 10.6c). The transport of the intestinal contents is exclusively carried out by these cilia.

Mesoderm, Somites, and Coelom

The mesoderm of Acrania originates from segmentally arranged, paired coelomic cavities, the somites. From the epithelial parts of the somites, the musculature (from the myotome), all skeletal elements (from the sclerotome), and the connective tissue of the skin (from the dermatome) arise during further embryonic development. Dorsally, the somites remain segmental and form the musculature of the trunk, while ventrolaterally, they merge into the uniform perivisceral coelom. From the coelomic spaces of the dorsal somites, the sclerocoel lying in the median, the usually strongly compressed lateral myocoel, as well as the fin boxes, remain (Fig. 10.5). The perivisceral coelom forms the subchordal coelom between the myotome, as well as the coelomic spaces of the main gill bars, the endostylar coelom, and the pterygocoel in the metapleural folds. It surrounds the intestine, including the hepatic cecum (Fig. 10.6b, d). Also, the gonads, which originate from the ventral area of the myotomes, are each located in a coelomic space (Fig. 10.6b, f).

The blood vascular system of Acrania is closed and already comprises the same vessels and general structure as in the primarily aquatic vertebrates. On sections, various blood vessels can be assigned, such as the ventral aorta (endostyle artery, Fig. 10.6e) in the pharynx, or the paired roots of the aorta next to the epibranchial groove (Fig. 10.6a), and the gill bar vessels (Fig. 10.6c), as well as the hepatic vein (Fig. 10.6d). *Branchiostoma* does not possess a complex heart like vertebrates. However, the ventral vessels below the intestine are contractile and thus form a very simple, tubular heart. It lacks the typical features of vertebrate hearts, such as valve structures. Later in the evolution of vertebrates, at the posterior end of the ventral aorta, the heart with which we are familiar evolved. At the end of the pharynx, the hepatic vein and the left and right ductus Cuvieri unite here to form the so-called sinus venosus in *Branchiostoma* and vertebrates. On each side of the body, the ductus Cuveri, or common cardinal vein, collects the blood from the anterior and posterior cardinal veins, which transport the blood returning from the body toward the pharynx. The homology of the posterior section of the ventral aorta of *Branchiostoma* with the heart of the Craniota could be very convincingly shown by matching expression patterns of the involved developmental genes. However, the blood vessels of Acrania, like those of invertebrates in general, do not possess an endothelium; they are only limited by an extracellular matrix (ECM) between opposing cell layers, for example, cardiomyocytes and peritoneal cells (or, in general, coelothelial cells). The blood itself is colorless and contains very few cells. Thus, it is unlikely that it significantly contributes to respiration. Instead, water flow through the atrium and diffusion through its entire epithelial lining is mostly responsible for ventilating the tissues.

Excretion

The excretory organs are also arranged segmentally and are located dorsally next to the pharynx below the subchordal coelom. They represent a biological peculiarity. The ultrafiltrate is produced by special cells near the aortic roots, called cyrtopodocytes. These cells combine the characteristics of podocytes and cyrtocytes that are characteristic of protonephridia. Cyrtopodocytes possess interdigitating finger-like (or foot-like) processes lying on the ECM at the dorsal aorta. Through the clefts between the cell processes, the ultrafiltrate passes into the subchordal coelom. From there, it strains through the weir portion of the cyrtopodocytes into the nephridial canal, driven by the beating flagella inside the weir. Thus, these cyrtopodocytes represent a particular type of weir/flagellar/flagellate cells that form the wall of the blood vessels and have basal foot-like extensions. The urine is transformed in the subsequent short excretory canal (nephridioduct) and ultimately discharged into the atrium. Parts of the nephridial organs can be detected on most cross-sections (Fig. 10.5). The excretory organs of Acrania are therefore not the expected typical metanephridia, since they combine characteristics of two elements typically located in different cell types. Podocytes are typical of metanephrial systems and are found at certain blood vessels in any organism with metanephridia. Their interpretation as protonephridia, as assumed in older literature, is also incorrect, as the primary urine is formed into a coelomic cavity. Thus, consistent with metanephridial systems, ultrafiltration occurs on the walls of blood vessels. Whether a second filtration step also occurs on the weir parts of the excretory organs is being discussed.

Gonads

The gonads of the gonochoristic animals are quite simple and sack-like (Figs. 10.4e and 10.5). They are located in a small coelomic space and border directly on the atrial space (Figs. 10.5 and 10.6b, d). In the testes of sexually mature animals, the flagellated sperm can easily be recognized (Fig. 10.6b), while the ovaries are filled with large oocytes, of which the large, bright cell nuclei with prominent nucleoli are characteristic (Fig. 10.6f). There are no special genital ducts; the mature gametes are released through the bursting of the gonad walls and are discharged via the atrioporus. Lancelets live for several years, but reproduce only once in their annual reproductive phase in Central Europe, depending on environmental conditions between May and July. The water temperature probably determines the synchronous release of the gametes in a population. Our movie clip collection, which includes movies showing lancelets, can be found at figshare: sn.pub/xr9qhi.

▶ **Summary** As adults, the anatomy of the sessile sea squirts and their planktonic relatives does not resemble that of chordates. Nevertheless, they, and especially their larvae, possess several characteristic features that morphologically identify them as members of the taxon Chordata. Recent molecular data have shown that the urochordates are the closest relatives (sister group) of vertebrates. Their free-swimming tadpole larvae develop a notochord, a hollow dorsal nerve cord, and a swimming tail, which are critical characteristics of chordates. However, sea squirts lose all these structures during metamorphosis and transform into sessile animals. All known 3000 species of tunicates are marine and are present in both the plankton and the benthos.

Tunicata (tunicates, sea squirts, and relatives, animals with a mantle) played a crucial role in elucidating the evolution of chordates and humans. Even though these sessile, tubular animals do not possess familiar body features, such as a head or limbs (Fig. 11.1), they represent the animal group of the closest living relatives of vertebrates. Together with cephalo-chordates (Acrania) and vertebrates (Craniota), of which we present the lancelet, the trout, and the rat as examples, tunicates form the taxon Chordata (chordates). They all possess a notochord (in some textbooks called chorda or chorda dorsalis), the primary internal axial skeleton of all chordates. The notochord consists of mesodermal cells and extends on the dorsal side of the body from anterior to posterior through the entire body. In *Ciona intestinalis*, the species we use here for dissection, the notochord is formed during embryonic development, as is typical of chordate embryos. However, it is only present in the free-swimming tadpole larva and is limited to the tail. During metamorphosis and transition to a purely sessile lifestyle, the notochord, including the musculature and the corresponding parts of the nervous system, is reduced, as it serves exclusively and essentially for locomo-tion of the tadpole larvae and becomes functionless after metamorphosis.

Tunicata, Acrania, and Vertebrata share several common characteristics. These include the dorsal central nervous system located above the notochord. In Acrania, the tadpole larvae of Tunicata and Craniota (craniotes, vertebrates), it is formed as a dorsal neural tube, and in vertebrates, it further develops into the brain and spinal cord. Common structures also include the segmented trunk and the tail musculature, which is exclusively formed as longitudinal musculature. Muscles are innervated by segmented spinal nerves. Another common feature is the pharynx with pharyngeal gill slits. Through the gill slits, small gaps in the pharyngeal wall, food particles are filtered from the water and transported to the stomach. This occurs with the help of a mucus film, which is secreted by the cells of the pharyngeal epithelium, especially by the gland cells of the hypobranchial groove (endostyle) at the bottom of the pharynx. This mucus film is continuously transported dorsally by the cilia of the pharynx epithelium, rolled up in the epibranchial groove, and moved posteriorly to the digesting part of the digestive tract. The pharynx, with its pharyngeal slits, is, therefore, primarily used for food acquisition. The water passing through the numerous gill slits is first collected in a space surrounding the pharynx, the atrium or the peribranchial space, before it is released to the outside through a cloacal pore. Since this mode of nutrition is typical for both, Acrania and Tunicata, chordates are very likely primarily filtering suspension feeders. From the structural elements of the pharynx and the pharyngeal gill slits, the purely respiratory gills have developed in vertebrates. Thus, a structure serving nutrition became a respiratory organ later in evolution.

However, it should be mentioned that the jaw apparatus of vertebrates also originated from the pharynx with its gill bars. Are there craniotes in which the pharynx still has a filtering function? Among the recent forms, the ammocoetes larvae of the lampreys, which belong to the jawless Cyclostomata, basal craniotes on their own, feed in a very

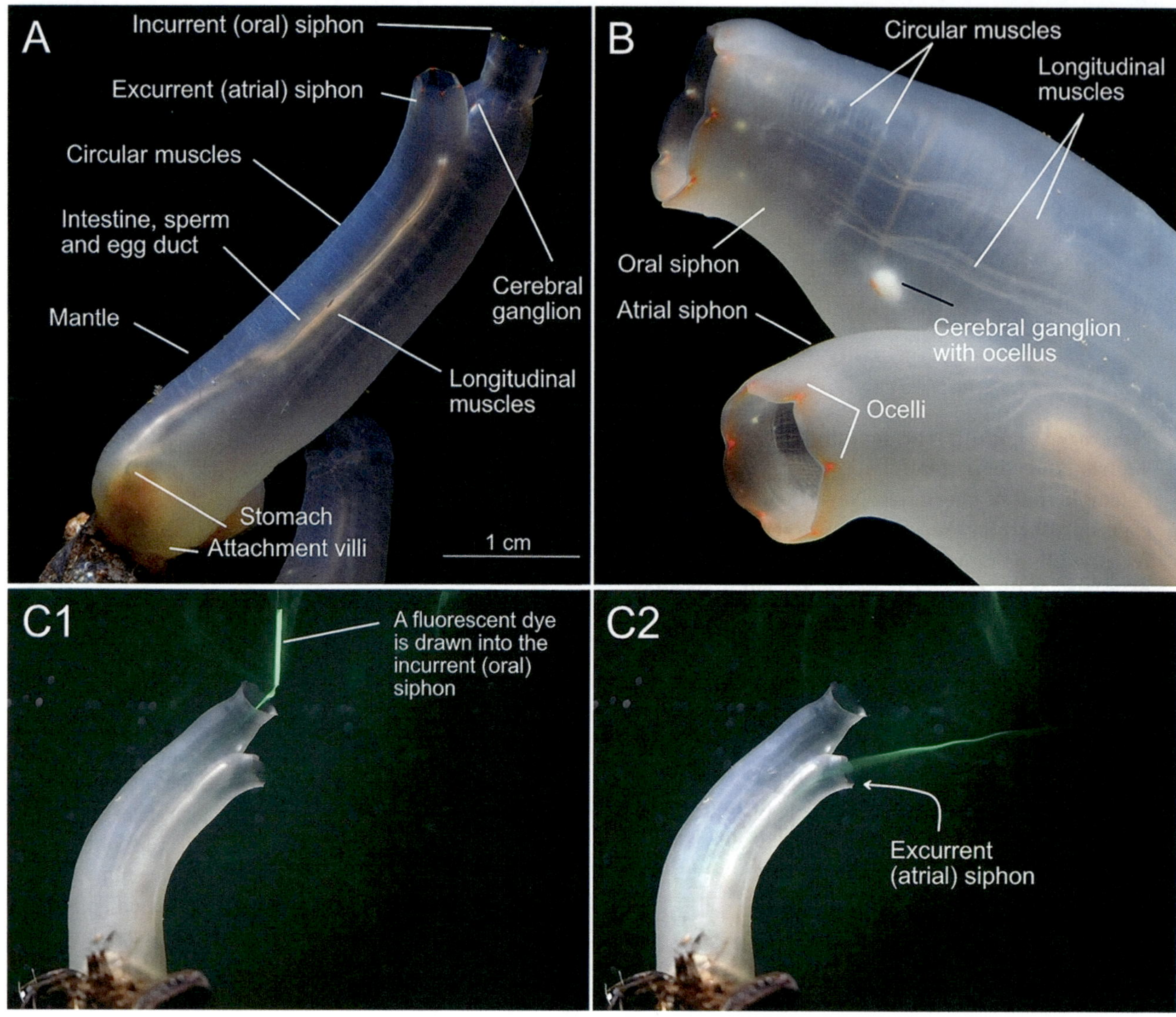

Fig. 11.1 The common solitary ascidian *Ciona intestinalis*. (**a**) The transparent tunic and mantle allows observation of internal organs such as muscles, the cerebral ganglion, and the intestine in the living animal. (**b**) Incurrent and excurrent siphon with ocelli. (**c**) Fluorescein is drawn into the incurrent siphon with the water flow (**C1**) and, after passing through the gill gut slits, exits the body seconds later through the excurrent siphon (**C2**). A movie showing the fluorescein experiment can be found at figshare: sn.pub/xr9qhi

similar way to the marine filtering chordates. This mode of nutrition is also assumed for many fossil paleozoic Craniota. Only with the evolution of the jaw apparatus did the pharynx and the structural elements of the gill bars and slits become purely respiratory organs.

Another common characteristic of chordates is the endostyle and its derivatives. The thyroid gland of vertebrates developed from the endostyle. Interestingly, even in Acrania and Tunicata, certain cells of the endostyle produce the hormone thyroxine, and others bind the element iodine. This indicates that the functions of the vertebrate thyroid gland have existed for a long time and were initially located in cells of the endostyle.

Almost all tunicates live attached to the substrate and feed as filter feeders. However, around 130 pelagic tunicate spe-

cies also swim in the open sea. These include the salps (Thaliacea), which occur in long chains of numerous single individuals at depths of up to 800 m and float through the open water. Among these pelagic forms, only Larvacea (Appendicularia) also have a swimming tail with a notochord as adults. These animals are interpreted as so-called progenetic forms, which remain in a tadpole larval state but nevertheless become sexually mature—hence the scientific name Larvacea.

Many of the 3000 known tunicate species are capable of asexual reproduction and form large animal colonies by budding. Not so *Ciona intestinalis*, which probably reproduces exclusively sexually. Due to their swimming larvae with a well-developed sensory system many ascidians can quickly colonize suitable habitats and outcompete competitors through faster growth. *Ciona intestinalis* is a globally distributed species found on all Atlantic and Pacific coasts and in the Mediterranean Sea, the North Sea, and the Baltic Sea. Like about 100 other tunicate species, it also occurs in the fouling of ship hulls, which has certainly contributed to today's global occurrence. Therefore, the worldwide distribution of many species is most likely of anthropogenic origin rather than true cosmopolitism. The animals adhere with their tunica via adhesive zots to rigid substrates, such as rocks, stone fragments, shells, pontoons, quay walls, or ship hulls, and often form surprisingly large mass populations. For example, *Ciona intestinalis* can reach densities of 1500–5000 individuals per square meter in protected habitats. The two large body openings in living animals, with their primarily red or yellow-colored ocelli, are immediately noticeable (Figs. 12.1 and 12.2). These are the incurrent, ingestion, or branchial siphon and the excurrent, egestion, or atrial siphon pointing to the side. The incurrent siphon leads into the mouth and pharynx, and the excurrent siphon represents the opening of the atrium, the cloacal chamber, into which the anus and the genital openings also open. The two siphons mark the anterior side of the body, with the incurrent siphon extending directly anteriorly. The opposite side of the animal with the adhesive disc or base marks the posterior side of the body. *Ciona intestinalis* is always slightly curved. The concave side below the excurrent siphon marks the dorsal side of the body, while the opposite is the convex ventral side of the body. This itself results in the third body axis, the left-right axis.

Recommended Material
Ciona can easily be collected on marine biology excursions. Numerous animals can always be found on the floaters of jetties, on anchor buoys, or on quay walls in marinas. The animals are first collected alive. To fix them in the extended, relaxed state, they should be narcotized before fixation to prevent any sudden contraction. Allow *Ciona* to become fully expanded in a dish of seawater. Then, slowly add menthol or magnesium chloride crystals or a small amount of 95% alcohol up to a maximum of 1 part alcohol to 9 parts seawater. When *Ciona* reacts only slightly to a touch, after a couple of hours, the seawater is replaced slowly by 70% alcohol. Change the alcohol two or three times during the next few days. Delicate species should be fixed after narcotization in a mixture of 1 part commercial formalin and 10 parts seawater, and then stored in the same solution. Usually, the animals are stored in 4% formaldehyde. The narcotization and fixation methods suggested here refer to Bullough, W.S., 1958, Practical Invertebrate Anatomy. Before the dissection course starts, fixed animals are soaked in seawater or tap water for at least 2 h to remove formaldehyde.

The Anatomy of *Ciona intestinalis*

The tunicates owe their name to the tunica, which due to its chemical composition is unique in the animal kingdom and is a "coat" that completely encloses the body. The tunica can vary in thickness and consistency. The foot or base area forms a villous, solid structure with which the animal adheres to the substrate (Fig. 11.1a). The matrix of the tunica contains fibers and is mainly composed of the cellulose-like polysaccharide tunicin and other polysaccharides. It is secreted by the epidermal cells of the body wall. Thus, Tunicata are the only animals that can synthesize cellulose. In addition, there are numerous proteins, glycoproteins, and glycosaminoglycans, as well as migrating mesodermal cells, which, in sum, give the tunica its connective tissue-like properties and peculiarities. The tunica thus differs fundamentally from the otherwise common cuticle of other invertebrates. Since the tunica is transparent in *Ciona*, internal organs such as the intestine can already be observed in the living animal. The function of the tunica is diverse. It offers the animal safe protection against predators due to its often tough, leathery structure. In addition, as an exoskeleton, it ensures a stable, yet flexible body shape. The transition to a sessile lifestyle and the procurement of food by filtering the surrounding water made it possible to dispense with energy-consuming active food searches, but required increased protection against predators, as escape was no longer possible. Therefore, forming a solid tunica can be considered an essential evolutionary achievement that is advantageous for a sessile lifestyle. Tunicin is produced by the enzyme cellulose synthase. Recent studies suggest that this unique ability was probably achieved through horizontal gene transfer of the cellulose synthase gene *CesA* from bacteria.

The body of *Ciona* is closely attached to the tunica in the area of the incurrent and excurrent siphons, and the same applies to the lower part of the animal near the base, which forms the adhesive cilia. Below the epidermis, there are numerous muscle bands in *Ciona* that can be observed in the living animal due to the transparency of the tunica (Fig. 11.1). A complete muscular body wall, as seen, for example, in the earthworm, is missing in *Ciona*. The longitudinal muscles extend from the siphons to the adhesive base of the body. In case of danger, the muscles contract and quickly pull the siphons back to the body. Circular muscles can open and close the siphons. If living animals are available, the activity of the muscles can be easily observed. Nerves of the cerebral ganglion and the neural plexus innervate the muscles.

Ciona: Preparation
The animals must be fixed in a fully stretched state for proper dissection.

(a) The tunica is opened from the base to the incurrent siphon with scissors.

(b) Now, carefully push the animal out of the coat with your fingers. The animal is now pinned down in a dissecting dish on its right side (Fig. 11.3). Cover the specimen with water.

(c) Starting from the incurrent siphon, the pharynx with the outer membrane of the peribranchial space is opened with scissors by a longitudinal cut all the way down (Fig. 11.3b).

(d) The pharynx and epidermis, with their body wall muscles, can now be unfolded and pinned down in the dissecting dish. The exposed organs are then examined (Figs. 11.4, 11.5, and 11.6). By cutting the excurrent siphon along the right side, the cloacal chamber and the gonads with their ducts are exposed.

(e) The structure of the pharynx should be studied with a microscope. Take the pharynx and cut out a cent-sized piece (Fig. 11.9a). The cells should be stained with methyl green or neutral red.

(f) Set aside the intestine and gonads (Fig. 11.5a, b). The transparent, 2–3 cm long, and relatively thin heart is often hard to recognize. (Fig. 11.7).

(g) Histological sections of various organs are used as a supplement.

The Importance of the Pharynx, the Pharyngeal Basket, and Pharyngeal Gill Slits

Among chordates, only tunicates (Tunicata), lancelets (Acrania), and the ammocoetes larvae of lampreys (Craniota, Cyclostomata) harbor a pharyngeal chamber with slits used for food filtration (also known as a branchial chamber, branchial basket, or pharyngeal basket). Through the gill slits in the pharyngeal basket, which are visible as narrow windows in the wall of the pharynx, tiny food particles are filtered from the water in *Ciona* and all other tunicates. For this purpose, ambient water is continuously swirled into the pharynx via the incurrent siphon (Figs. 11.1, 11.2, 11.5, and 11.8) and pushed through the gill slits into the atrium. From there, the swirled water leaves the body via the excurrent siphon. The continuous water flow is generated by the densely ciliated epithelial cells of the gill bars (Fig. 11.9), whose cilia are in constant motion. A movie showing the cilia of the epithelial cells at the gill bars of Ciona can be found at figshare: sn.pub/xr9qhi. As with the lancelets, the pharynx is relatively large in relation to the body. Furthermore, in tunicates more than in lancelets, the bars are firmly connected by cross-bridges to keep the size of the slits constant. Likewise, the pharynx is connected via tissue strands, which run through the narrow atrium in tunicates, to its outer epidermis with the body wall and the tunica. This ensures the shape constancy of the entire phar-

ynx, prevents a collapse of the barrel-shaped inflated pharynx, and, at the same time, increases the flow speed in the atrium.

A longitudinal groove runs along the pharynx, easily visible in the living and the dissected animal (Figs. 11.5a and 11.6a). This is the hypobranchial groove or the endostyle. The cells of the endostyle secrete an iodine-containing mucus net, which is distributed by the cilia over the entire branchial basket and retains the food particles when water flows through the slits of the basket. In *Ciona intestinalis*, the meshes are about 400 × 600 nm wide, with a "thread thickness" of about 25 nm. On this net, the so-called nano- and ultraplankton as well as other particulated matter with a diameter under 1 μm are filtered from the water. The filtration performance of ascidians is enormous; in a single individual of a larger species, more than 100 liters of water can pass through the pharynx per day, with a filtration efficiency between 70% and 100%.

Opposite the endostyle is the epibranchial groove with lobe-shaped "tongues" (Fig. 11.6b). With the help of the tongues, the mucus-trapped food particles are formed into a "sausage" and transported by cilia to the also ciliated esophagus and further into the stomach. The digestive enzymes are produced by glandular cells of the stomach wall. Undigested food leaves the stomach and enters the rectum, which runs anteriorly and ends about halfway up the body, with the anus near the excurrent siphon in the peribranchial space. With the water stream, the indigestible food residues now leave the animal very quickly via the excurrent siphon.

The pharynx is of great evolutionary importance. It serves Tunicata and Acrania mainly for food acquisition. From components of the pharynx, with its gill slits, the gill arches later developed in aquatic vertebrates, becoming the well-known gills, however, with a considerable reduction in the number of slits. Thus, a structure primarily used for feeding became a purely respiratory organ. During embryogenesis of all vertebrates, a pharynx with gill slits is initially formed, with gill pouches that become gill slits when they break open. In humans, four pouches are already created in the 20-somite stage, corresponding to the third to fourth week of development; a fifth is added a little later. However, in further development, the components of the pharynx are remodeled and acquire various new functions. For example, parts of the face, the upper and lower jaw, and the middle ear are formed from the gill arches; among other structures, the auditory ossicles, such as hammer (malleus), anvil (incus), and stirrup (stapes) or columella, are based on embryonic gill arches. The evolution of the pharynx is presented in more detail in chapters on Craniota.

Circulatory System, Heart, Excretion

The circulatory system consists of a heart and numerous tiny vessels. The approximately 2–3 cm long, delicate, and transparent U-shaped heart is located posteriorly in the body near

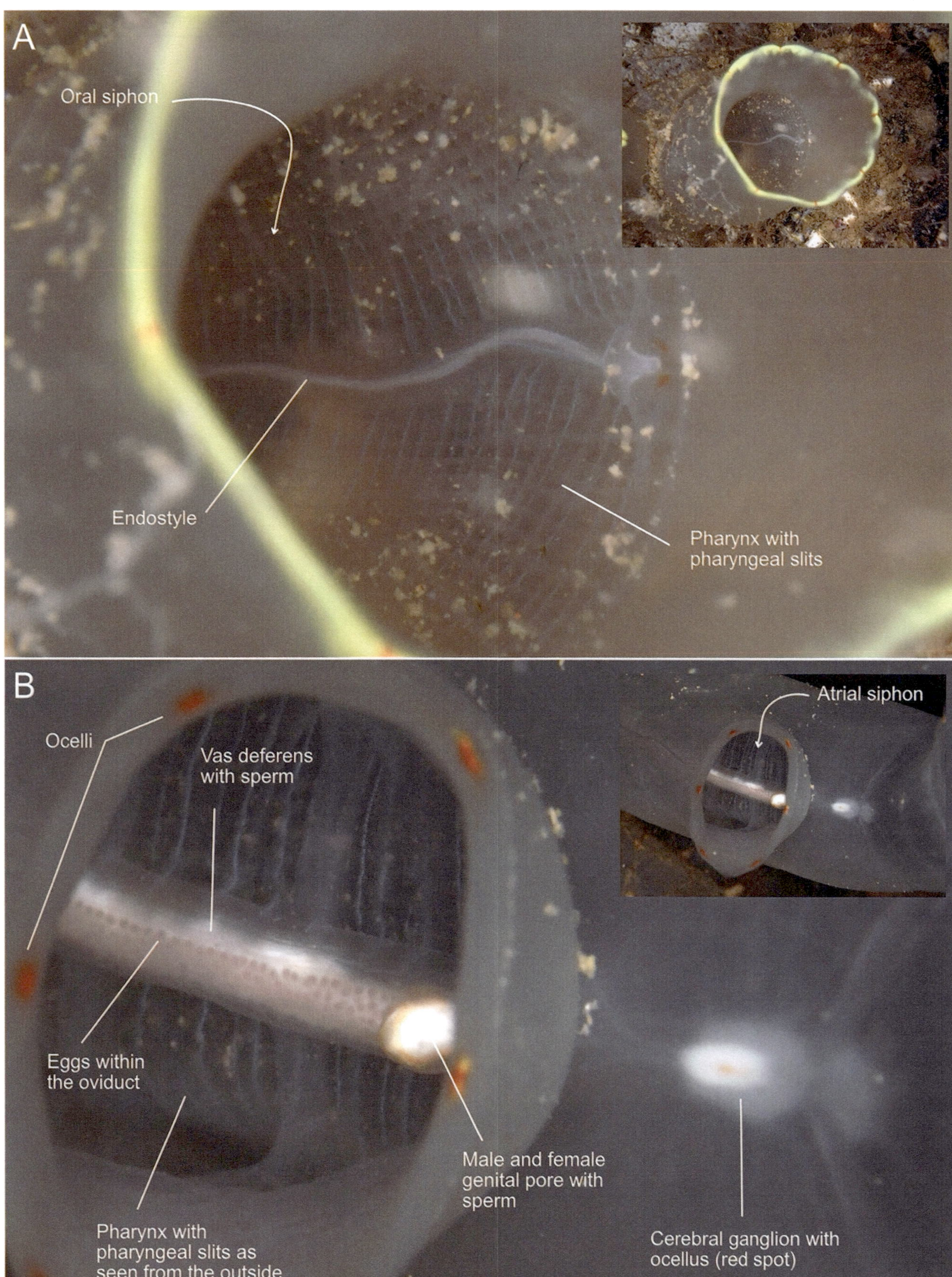

Fig. 11.2 *Ciona intestinalis*, sea squirt. (**a**) View from outside through the incurrent opening into the pharynx with gill slits and endostyle (**b**) View from the outside through the excurrent opening into the cloacal chamber with intestine, oviduct, and vas deferens. The whitish mass is exiting sperm fluid

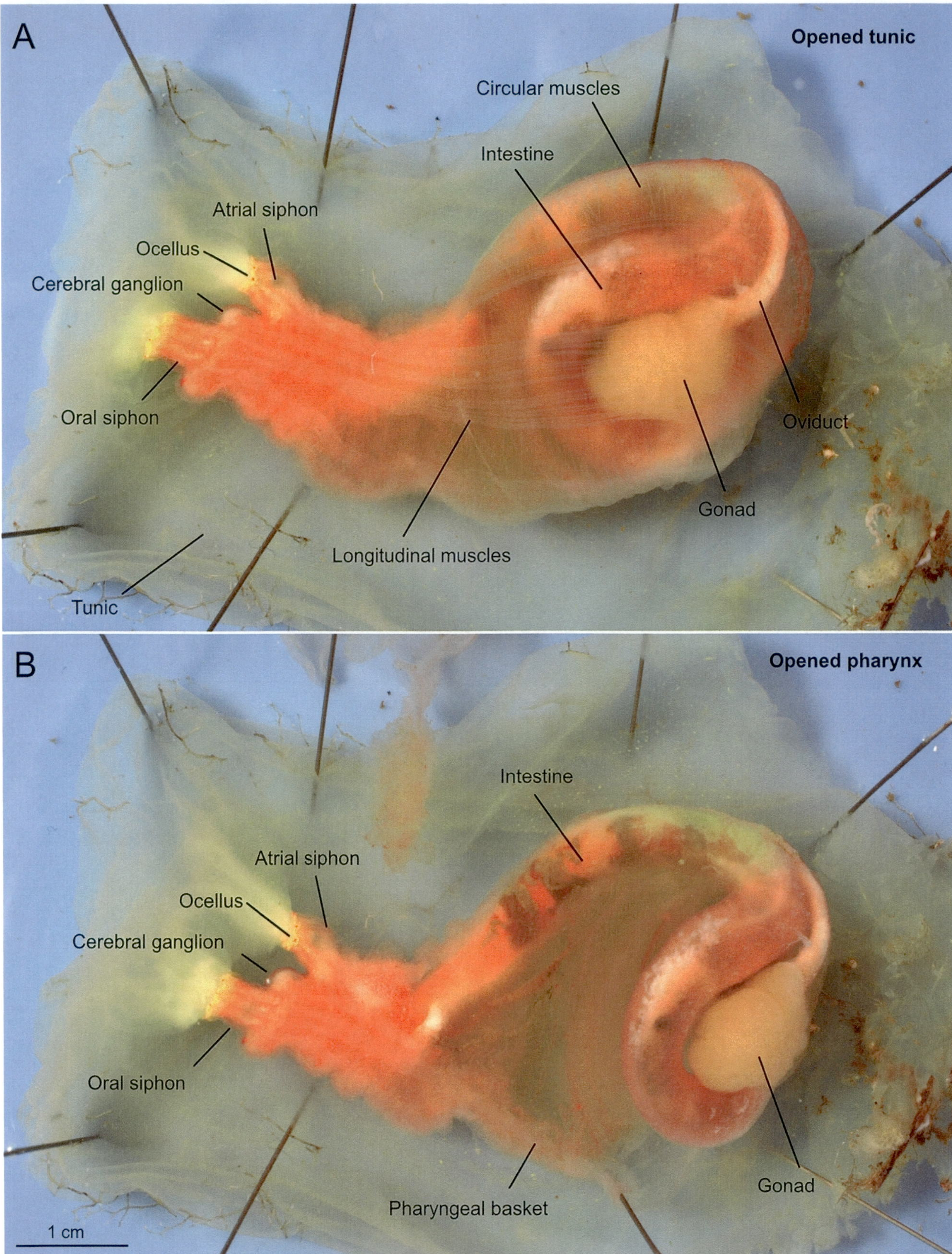

Fig. 11.3 *Ciona intestinalis*. (**a**) Body exposed after cutting the tunic and pinning it down. (**b**) Cutting open the pharyngeal basket

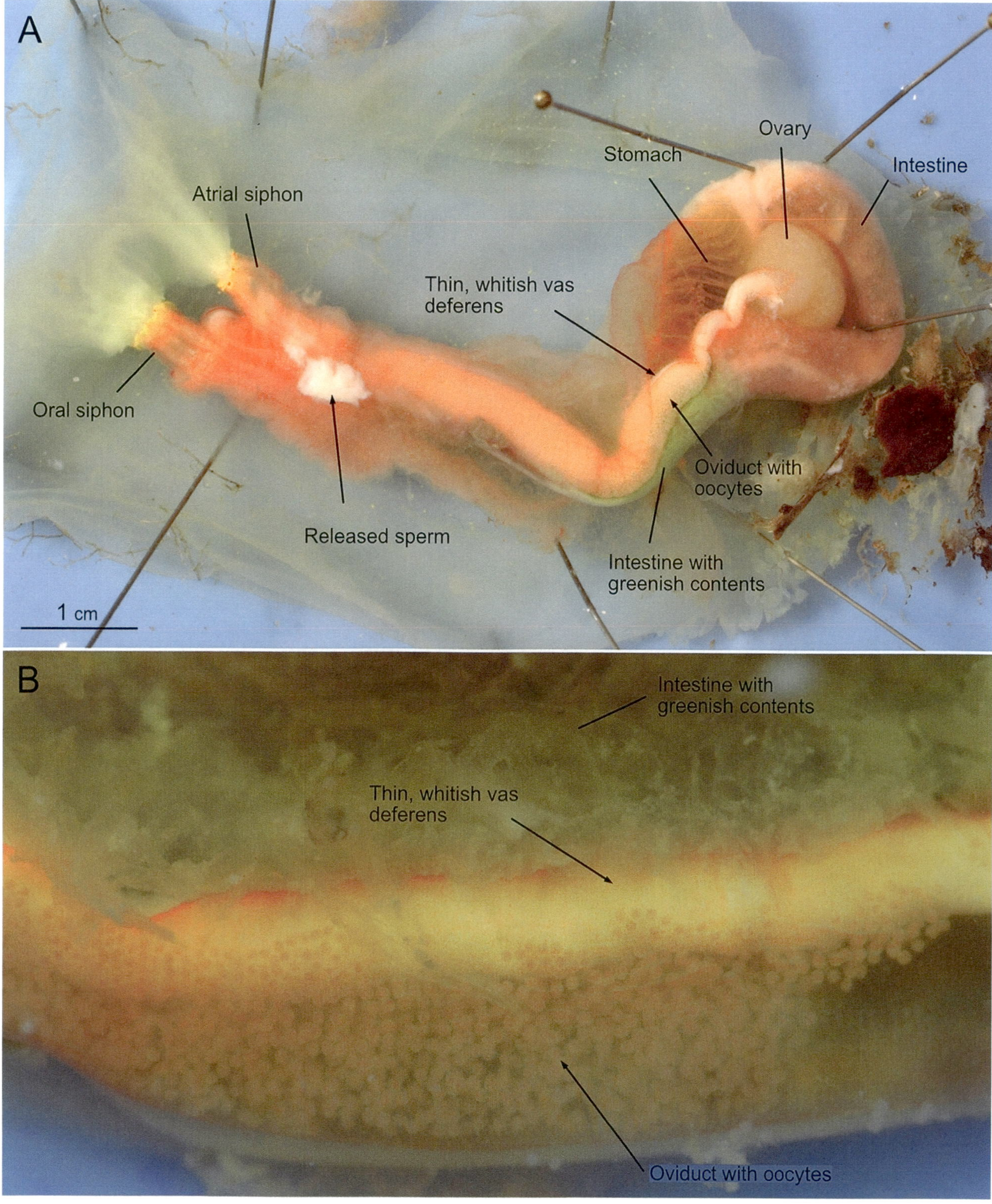

Fig. 11.4 *Ciona intestinalis*. (**a**) Internal organs, stomach, ovary, intestine, egg and sperm ducts are visible. (**b**) Egg and sperm ducts at higher magnification. The slightly orange-colored egg cells can be seen in the oviduct

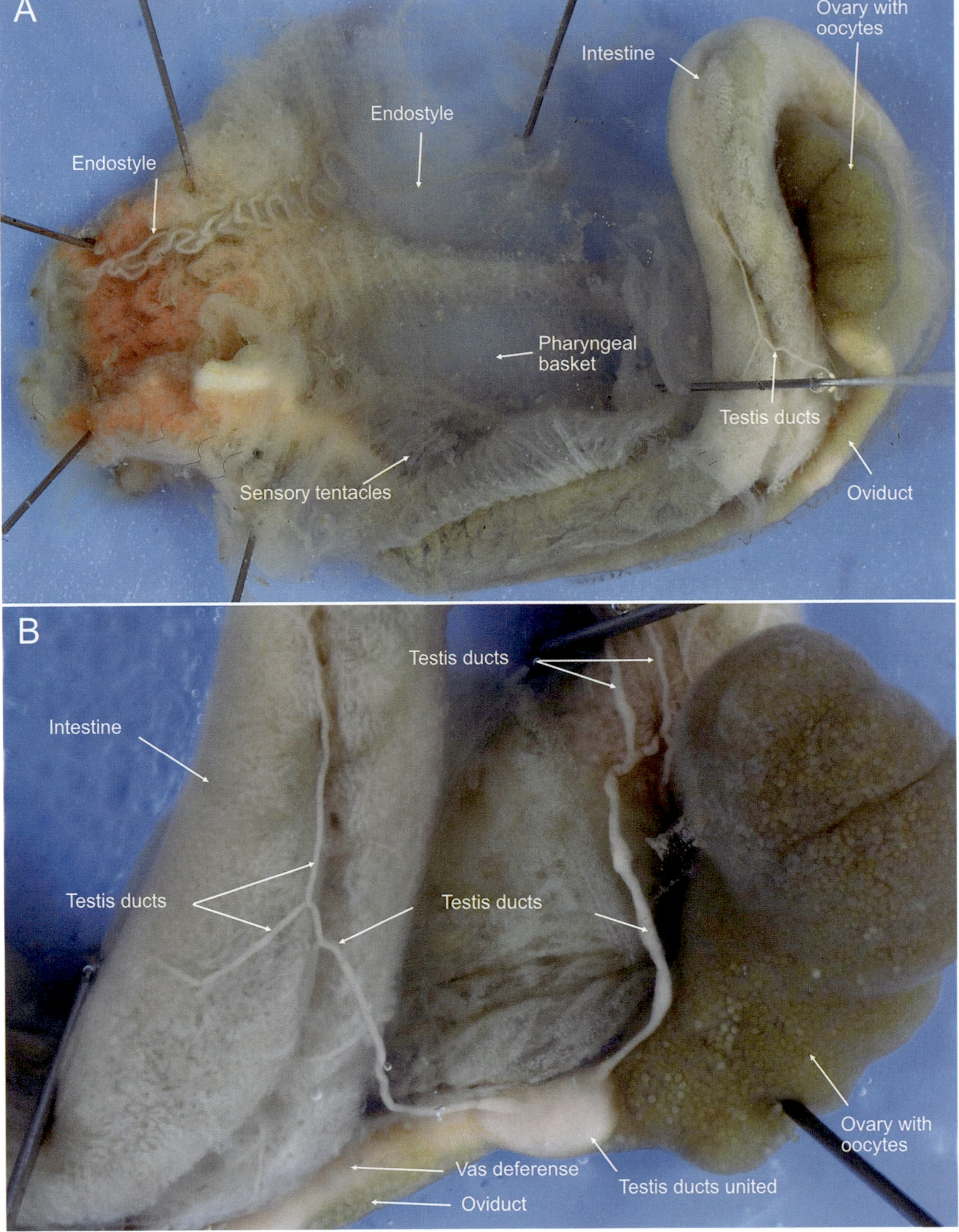

Fig. 11.5 *Ciona intestinalis*. (**a**) Preparation of the pharynx with endostyle. (**b**) Intestine, ovary, and testes at higher magnification

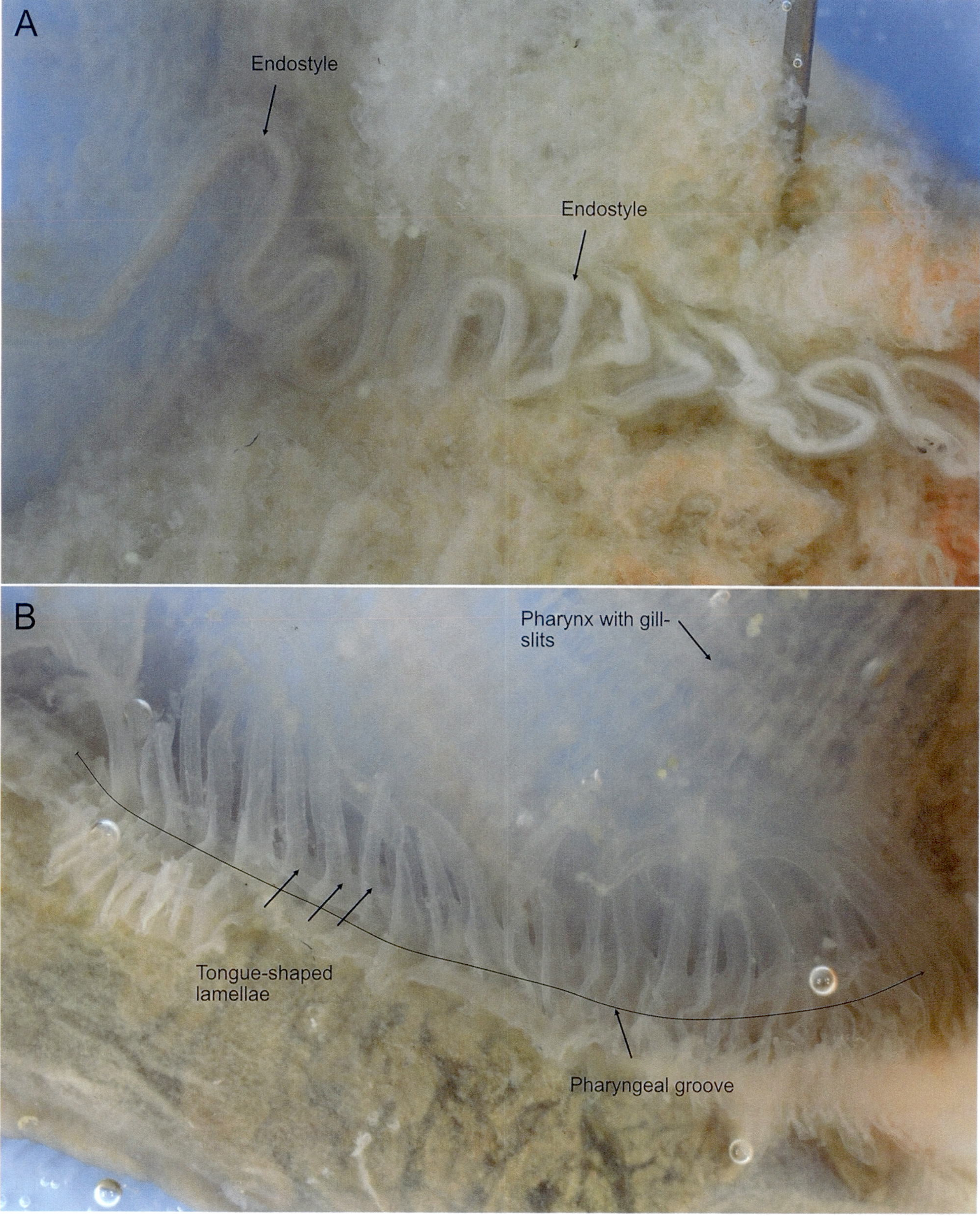

Fig. 11.6 *Ciona intestinalis*. (**a**) Endostyle. (**b**) Epibranchial groove with tongue-shaped lamellae

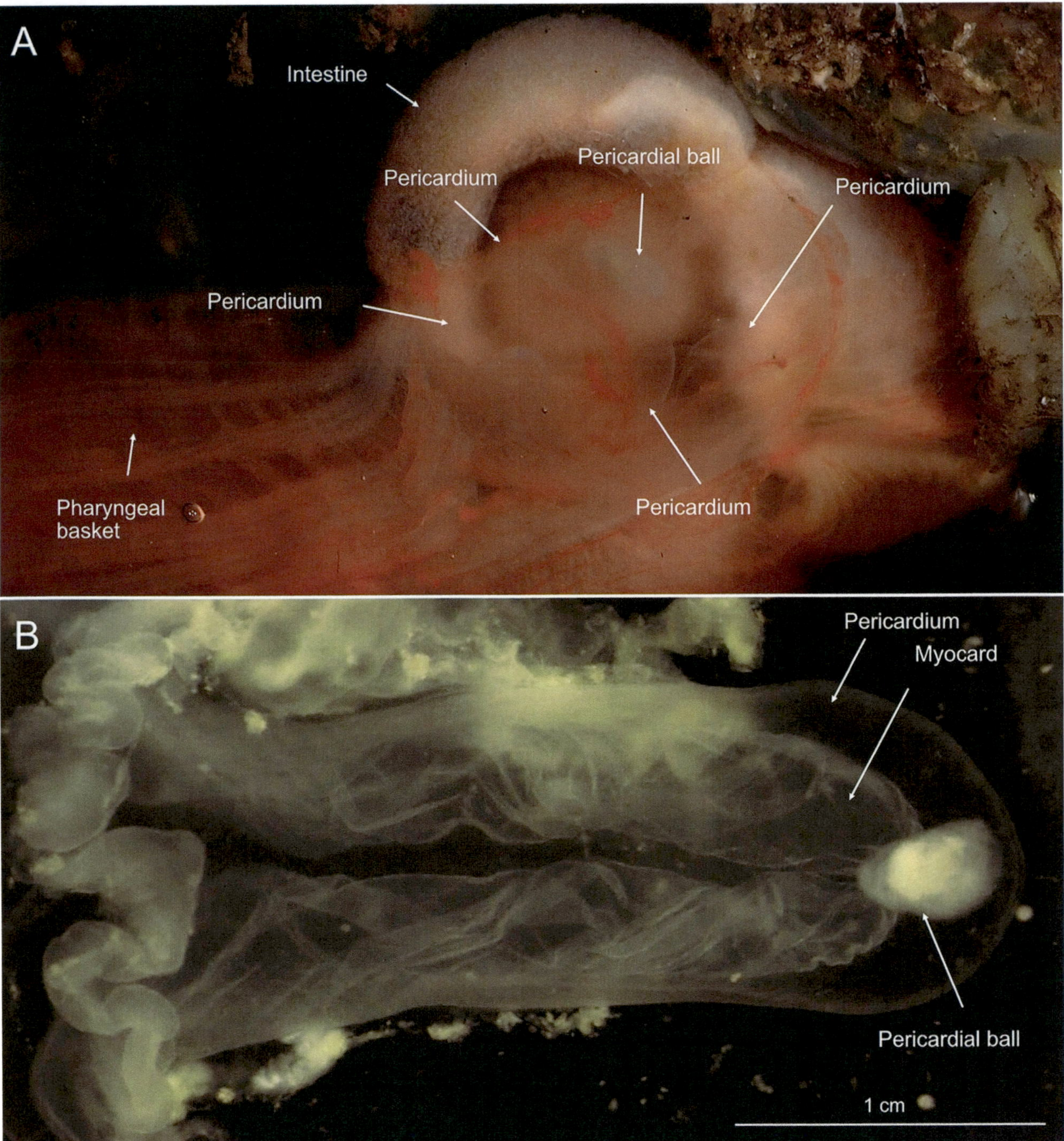

Fig. 11.7 Ciona intestinalis. (**a**) The heart is highly delicate and translucent and can hardly be depicted in a printed images. (**b**) The dissected heart with pericardium is easier to recognize against a black background

the stomach and under the pharynx (Fig. 11.7a). It is surrounded by a heart sac (pericardium), which is a coelomic space (Fig. 11.7b). A myogenic pacemaker at each end of the tubular heart triggers peristaltic contraction waves in alternating directions. Various cell types are found in the blood, including amoebocytes, which play a role in the transport of nutrients or in the accumulation of waste products. A movie showing the heart of *Ciona intestinalis* can be found at figshare: sn.pub/xr9qhi.

The vascular system of *Ciona* is a closed circulatory system with vessels that run from the heart to all organs. However, the blood vessels are not lined with endothelial cells, a typical and exclusive feature of vertebrate blood vessels. A true endothelium consists of a single layer of flat or

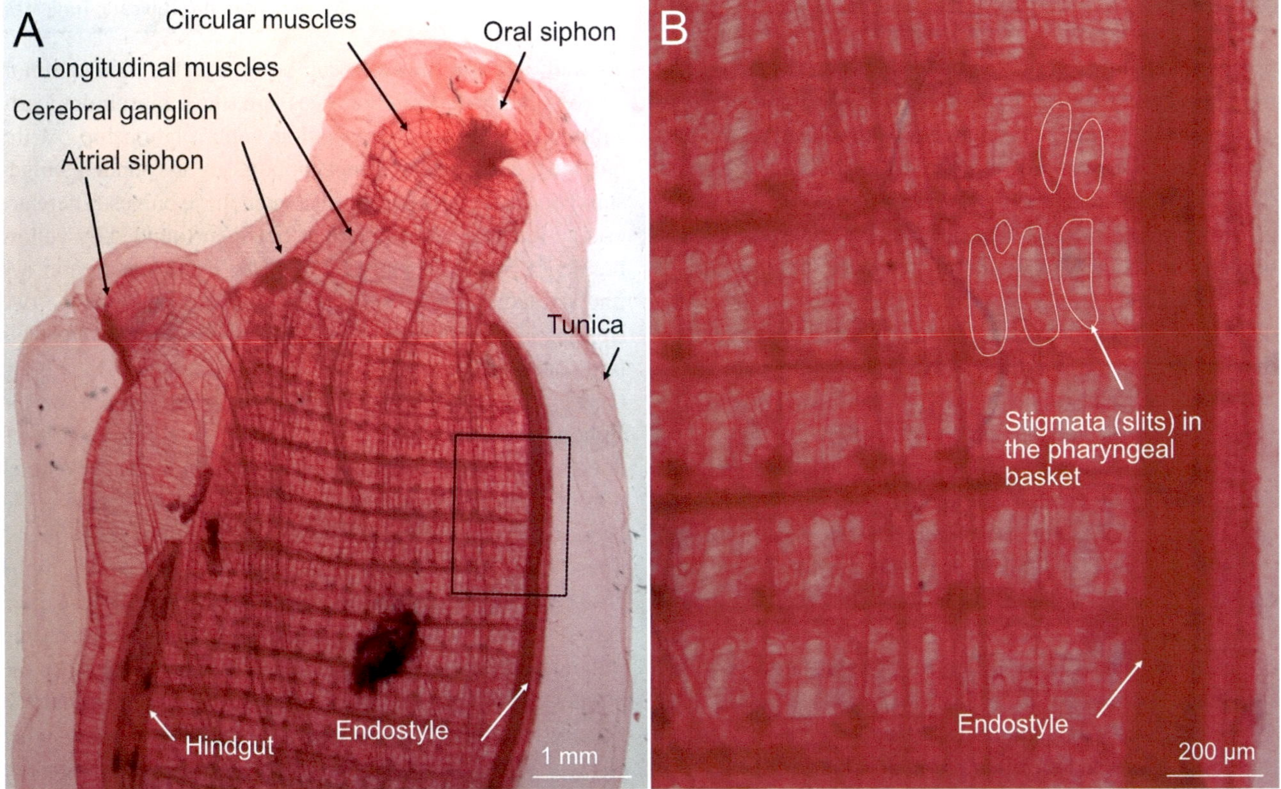

Fig. 11.8 *Ciona intestinalis*. (**a**) Young animal, borax carmine stained. (**b**) Detail. Specimen of the Zoological Teaching Collection, Philipps University Marburg

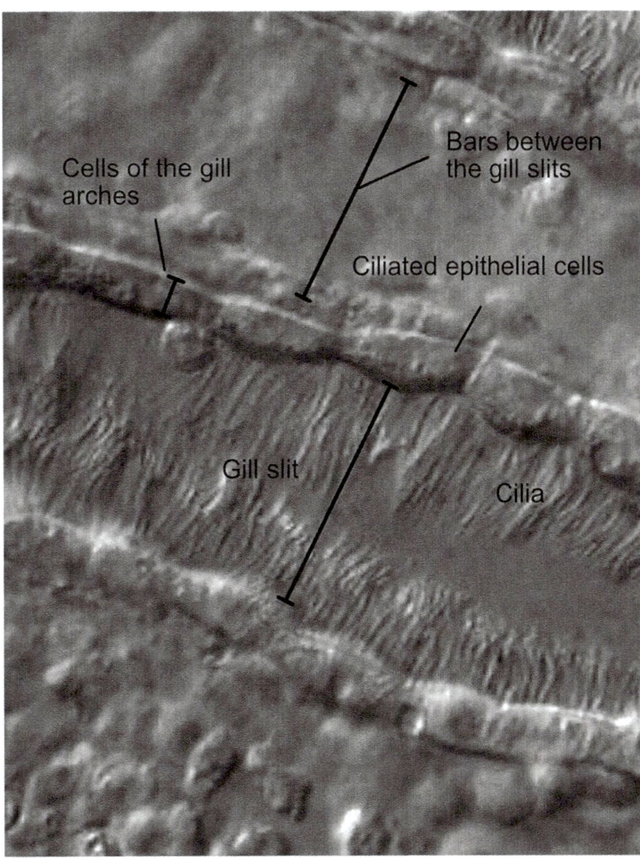

Fig. 11.9 *Ciona intestinalis*. Microscopic preparation of the pharynx, freshly prepared, unstained. The slit-forming cells are ciliated. They generate the water flow in the pharyngeal slits

prismatic cells that forms a barrier between the blood and the surrounding tissue. Endothelial cells are connected to each other by so-called tight junctions that regulate the permeability and function of the barrier. In addition, the endothelium always rests on the ECM that supports the cells, provides structural stability and, in invertebrates, solely borders the blood. Some authors postulate that the lack of endothelial cells in tunicates makes the vessels of these animals more permeable, but this is very likely compensated by the lining of the cardiac and vascular lumen with a proteoglycan-rich extracellular matrix and the opposite coelomic lining which also forms an epithelium.

According to current knowledge, the gas exchange itself takes place at the entire body surface, and especially at the epithelial cells of the pharynx. This is supported by the fact that the pharynx is supplied by many blood vessels and, unlike in the body wall, diffusion is not hindered by a tunica. The diffusion paths in the pharynx are also very short. An originally discussed hypothesis that oxygen transport and storage are carried out by hemocyanin as a respiratory protein has been called into question by recent works. *Ciona* has three genes that code for hemoglobins. One of these hemoglobins in particular, *CinHb2*, is strongly expressed in the heart tissue and lymphoid-like blood cells of *Ciona*.

Very little is known about the disposal of no longer needed metabolites or toxic degradation products. Since the animals have no dedicated excretory organs, they must dispose of all excretions directly via cells on the body's surface. Due to its large surface and direct contact with the surrounding water, the pharynx could be a possible organ for such a function. Recently, cells were found in the bloodstream of tunicates described as typical nephrocytes (kidney-like cells) due to their histological structure. Possibly, these cells break down toxic nitrogen compounds. Similar cells were also found in the endostyle and the epibranchial groove.

How Do Tunicates Perceive Their Environment?
Tunicates possess various receptor cells, which are attributed with mechano-, vibration-, gravity-, chemo-, photo-, and proprioceptive functions. These are ciliary primary sensory cells. In simple terms, primary sensory cells are just neurons with a dendrite, soma, and axon that can generate an action potential. The dendrite is usually unbranched, embedded in the epithelium, and carries the stimulus-receiving projections, typically a cilium, a few cilia, or microvilli. An exception is the so-called tunicate coronal organ, in which—unique among invertebrates—secondary receptor cells, i.e., sensory cells that only function as sensors but require a downstream neuron to generate an action potential. The coronal organ is a mechanosensory organ at the base of the incurrent siphon. Such receptor cells are otherwise only found in vertebrates in the vestibular system, the lateral line organ of fish, and the electroreceptors. The occurrence of a coronal organ in tunicates can be interpreted as a common derived character with craniotes and, thus, morphologically support their sister group phylogenetic relationship. At the outer edge of the incurrent siphon, eight rounded bulges stand out in the indentations between these bulges. There are orange-red spots, interpreted as ocelli surrounded by yellow pigment (Figs. 11.1 and 11.2). On the excurrent siphon, we find six such pigment spots. The brain lies between the two siphons and marks the dorsal side of the animal. Here we find another ocellus lying directly next to the brain (Figs. 11.1, 11.2, and 11.3). However, the ocelli of the siphons probably play no role in light perception, as, even after removing all ocelli, exposure to light leads to contraction of the siphons without delay. In addition, opsins could be detected in the brain, but not in the ocelli of the siphons. These are the same opsins as in vertebrates and belong to the group of ciliary opsins. To trigger this reaction in *Ciona*, exposure to the cerebral ganglion area is sufficient. According to some authors, it is, therefore, questionable whether the pigment spots on the siphons are in fact eye structures. Possibly, the ocelli on the edges of the siphons have other tasks. *Ciona* also has mechanosensory cells, which are arranged in a ring around the siphons and some authors compare them with the mechanoreceptive hair cells in the ear and in the lateral line organ of vertebrates.

The cerebral ganglion (brain) forms the central nervous system of an adult *Ciona*. It can already be seen from the outside in living animals through the transparent tunica (Figs. 11.1a, b, 11.2b, and 11.3a). From the cerebral ganglion, paired main nerves run to the front and back body walls and further to the siphons. As recent studies with fluorescence markers have shown, nerve pathways run to eight or six centers of the in- and excurrent siphons. This corresponds to the number of ocelli in the siphons (Fig. 11.1a, b). Additional nerves project to the gonads and the gonoducts with oviduct and vas deferens and to the rectum. The nervous system of *Ciona* also plays a role as a neurosecretory system, among others, while the neuropeptide tachykinin is released at the ovaries. It regulates the growth of the yolk-containing follicles. The brain is connected to the neural gland immediately below it. Little is known about its function. The neural gland opens into the pharynx. Many textbooks often summarize the cerebral ganglion and neural gland as the neural complex.

Reproduction
Ciona intestinalis is a simultaneous hermaphrodite, which develops testes and ovaries at the same time. The fertilization of the eggs takes place in open water. The follicle cells, which always accompany the eggs of *Ciona*, secrete attractants to lure sperm (Fig. 11.10). In the North Sea, the reproduction of *Ciona intestinalis* takes place from May to September. The mature egg cells and differentiated sperm

are found in the gonads of the animals. The ovary is visible as a larger ellipsoid organ on the stomach's left side (Figs. 11.3 and 11.5b). In sexually mature animals, the large oocytes can already be seen from the outside (Fig. 11.5b). The testis consists of a system of branched tubes that extend far beyond the stomach and then unite. The vas deferens is embedded by the larger oviduct, and both gonoducts ascend together along the rectum to the cloacal chamber and to the excurrent siphon. (Figs. 11.4 and 11.5). Compared to the oviduct, which is abundantly filled with egg cells during the reproductive period, the vas deferens appears significantly thinner and of a whitish color (Fig. 11.4a, b). The openings of the gonoducts extend beyond the anus and release their eggs and sperm into the water stream coming from the pharyngeal basket. The eggs released during spawning often adhere to mucus in the mother animal's vicinity, where fertilization and larval development occur.

Although self-fertilization occurs in tunicates, *Ciona* is largely self-sterile, meaning that eggs fertilized by sperm from the same animal do not develop. The chorion, the extracellular vitelline layer deposited on the egg (extracellular matrix, ECM), plays a decisive role. Only foreign sperm can bind firmly to the ECM and initiate fertilization. This requires specific recognition and receptor proteins on the part of the sperm and eggs. Therefore, the underlying molecular mechanism has been the subject of intensive research in recent years.

Ciona reproduces exclusively sexually. However, many other tunicates can reproduce asexually. Thus, bud formation (blastogenesis) is widespread and allows for the rapid spread of the species on the available substrates. The sexually produced oozoids produce the first asexual generation of animals, the so-called blastozooids, by budding. These multiply rapidly through further budding. Budding is advantageous for many colony-forming species, for example, to be able to recover quickly after losing parts of the colony due to predators. *Ciona* and other tunicates also have exceptional regenerative abilities, which decrease with age. Thus, young animals can regenerate the entire incurrent siphon, the neural complex, or just the ocelli of the incurrent siphon in case of loss. This ability is also subject to loss with increasing age. The close relationship between tunicates and vertebrates makes *Ciona* an attractive model system for regeneration and aging research, as an age-related loss of regenerative ability is also observed in higher vertebrates.

Development

As already mentioned, most of the typical chordate features are only present in the free-swimming tadpole larva of *Ciona* (Fig. 11.10). During marine biology excursions, fertilization experiments with the sperm and eggs of *Ciona* can be easily carried out. For this purpose, gametes are isolated from several sexually mature animals. Mature eggs (oocytes) of *Ciona* are about 130 μm in size. When released into the open water, they are accompanied by protective testa cells, an ECM-based chorion, and elongated follicle cells. The latter allows the oocyte to float in seawater. At 13 °C water temperature, the first cleavage divisions take about 3–4 h. After about 24 h, the tadpole larva hatches, and after only 2 days, the animals already begin with the metamorphosis and the transformation into a sessile lifestyle. The tadpole larvae possess a notochord (chorda dorsalis) that runs throughout the swimming tail. The notochord is a supportive structural element and an important signaling center for early developmental processes. It is formed by a constriction from the roof of the early primitive gut and is located between the gut and the neural tube. The foremost section of the neural tube is expanded into the brain. In the dorsal brain wall, there is a light-sensing organ (ocellus), and in the ventral brain wall, a static organ (otolith), which enables the larvae to perceive light and gravity and control the swimming behavior before the larva attaches to the substrate. The larval neural tube and the notochord are regressed during metamorphosis. Essentially, only the brain remains from the central nervous system of the larva after metamorphosis, and all other structures must be newly formed during or after metamorphosis, as the larva and the adult form differ so fundamentally (Table 11.1).

***Ciona*: Collection of Gametes**

(a) The animal is placed in a dissecting dish filled with some seawater. The right side of the body is at the bottom. In this position, the incurrent siphon points forward, and the excurrent siphon points to the left. Now, the tunica is carefully cut, starting at the attachment disc or base of the animal. Work toward the excurrent siphon using scissors. The tissue is folded to the side and pinned in place. From now on, it is best to work with three to four diopter reading glasses or under the stereo microscope.

(b) The sperm duct and the oviduct run parallel to each other. The sperm duct can be recognized as a thin white tube and the oviduct as a yellow-green, thicker tube (Fig. 11.4). First, mature eggs are isolated from the *distal* section of the oviduct. To do this, the oviduct is opened with a watchmaker's tweezers or a fine pair of scissors. Now, stroke along the oviduct with a probe and squeeze the eggs out of the oviduct. About 50–100 eggs are taken with a clean Pasteur pipette and stored in seawater.

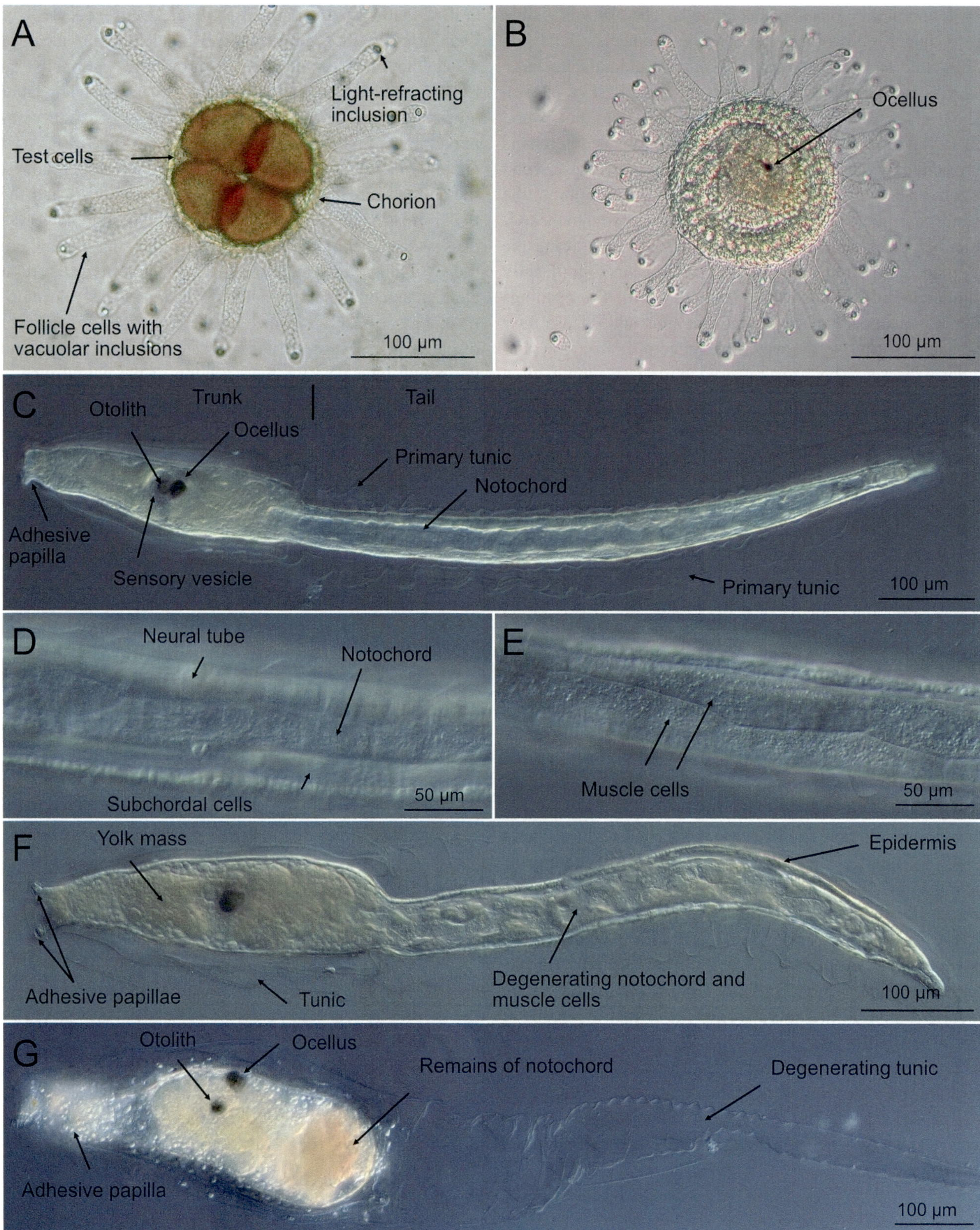

Fig. 11.10 *Ciona intestinalis*. (**a**) Four-cell stage. (**b**) Tadpole larva still in the eggshell, alive. (**c**) Swimming larva with tadpole tail, ocellus, and otolith, alive. (**d**) Tadpole tail, high magnification with chord cells. (**e**) Tadpole tail, high magnification, muscle cells. (**f**) Larva, shortly after settling on the substrate. (**g**) Degradation of the larval tail. Prof. Ernst-Martin Füchtbauer, University of Aarhus, Denmark provided the images A and B

(c) Once all the eggs have been removed, the sperm duct can be opened with fine tweezers, and sperm can be removed using a new pipette. The sperm fluid is transferred to a 1.5 ml Eppendorf vessel with a bit of seawater. Before the fertilization experiment, a small drop of sperm fluid is diluted in about 50 ml of seawater. For fertilization experiments, the sperm solution must be highly diluted, otherwise there is a risk of polyspermia and the termination of further development. Label the vessels containing oocytes and sperm individually due to self-infertility.

(d) Sperm and eggs from several animals are needed.

Table 11.1 Normal development of *Ciona intestinalis* at 13 °C water temperature, according to Fischer (ed.) *The Helgoland Manual of Animal Development*, own observations, and other sources

	Ciona intestinalis
Fertilization, fusion of the pronuclei	0 h
First cleavage	1–2 h
Second cleavage	3 h
Third cleavage	4 h
Morula	6 h
Gastrulation	14 h
Hatching of the larva	24 h
Metamorphosis	48–72 h
Juvenile development	10 d
Lifespan	12–18 m

Ciona intestinalis and *Ciona robusta*

Ciona intestinalis was described in older studies as a single, globally distributed species. Molecular studies have now shown that the European and North American populations probably comprise at least two species, subsequently referred to as *Ciona intestinalis* clade A and *Ciona intestinalis* clade B. Morphologically, the two species cannot be distinguished. Hence, the current literature refers to them as so-called cryptic species. Further investigations have now shown that clade A is *Ciona robusta* Hoshino & Tokioka, 1967 (see Brunetti et al., 2015), a species originating from Japan, while clade B corresponds to *Ciona intestinalis* (Linnaeus, 1767). The latter was incidentally described by Linnaeus in 1767, under the name *Ascidia intestinalis*. Both species are now widespread, most likely due to human activity, so they occur in some regions in sympatry (together in the same habitat). Our movie clip collection, which includes movies showing tunicates, can be found at figshare: sn.pub/xr9qhi.

▶ **Summary Within Chordata, Craniota or Vertebrata are the largest and most ecologically successful taxon, with about 60,000 recent species. They have colonized all habitats on earth, from the sea to freshwater, from the land to the air. In addition to primarily aquatic forms, mostly cartilaginous and bony fish, a whole range of secondary aquatic vertebrates have evolved from terrestrial forms and have again adopted a fish-like body. For these reasons, vertebrates show extraordinary morphological ecological diversity**.

Compared to other taxa, such as Arthropoda or Gastropoda, the over 60,000 species of Craniota, Vertebrata, are relatively modest in number. Nevertheless, these animals always have a particularly prominent place in our life. When we talk about animals, many of us often only think of vertebrates, mammals, or birds. One reason is that we humans also belong to this group of animals. From a biological point of view, another outstanding characteristic of vertebrates is probably more important: Their relatively large body size and, thus, their general conspicuousness, which far exceeds that of invertebrate animals. While body sizes in the centimeter range are already in the upper size range of the respective invertebrate groups, with only a few exceptions, such as cephalopods, vertebrates with these body dimensions always belong to the smallest forms of their taxon. Due to their often ossified internal skeleton, which is almost predestined for preservation as a fossil, an extraordinarily extensive fossil record of vertebrates exists. Fossils trace back more than 500 million years of evolutionary history, and make possible the reconstruction of their Cambrian marine habitats. As a result, the evolutionary history of vertebrates has been reconstructed much more accurately than that of all other Metazoa. Fossils have also shown that only 1% of all vertebrate species that have appeared in the history of the Earth have survived up to today.

The size and weight ranges of the vertebrates span several orders of magnitude. Probably the smallest mammal species, the Etruscan pygmy shrew (*Suncus etruscus*) measures 20 mm at 1 g weight. In contrast, the largest species, the blue whale (*Balaenoptera musculus*), measures 30 m at 190 t weight (or 30,000 mm at 190,000,000 g!). The males of the largest recent land animal, the African elephant (*Loxodonta africana*), reach up to 7.5 m body length and 7 t weight (or 7500 mm and 7,000,000 g!). In the case of fossil land animals, these dimensions were far exceeded by extinct mammals and sauropods.

Large animals have several advantages over smaller ones, such as intraspecific competition, fewer direct predators, better chances as predators, and favorable metabolic conditions due to an optimal volume/surface ratio. Large body size is achieved in craniates by increasing the number of body cells, rather than through their enlargement. More nerve cells, therefore, also allow for more complex cognitive performance, often a longer lifespan, and thus the possibility to shape behaviors through flexible learning. The phenomenon of extraordinary body sizes is linked with several biological questions and problems rooted in craniates' body plan. First and foremost is their cartilaginous and bony endoskeleton, which does not set narrow limits to size increase. It can grow with the animal, and can carry the mass of the soft tissues. In addition, there are several other characteristics, such as the high degree of cell differentiation, with over 200 different cell types distinguished, the segmental chorda-myomere system, the cross-striated skeletal musculature consisting of up to 10 cm long syncytia, a saltatory, and thus very fast, conduction of excitation in the nervous system, a highly developed central nervous system with corresponding sensory organs, efficient kidneys, and respiratory organs, as well as a powerful vascular system, with its endothelium, red blood cells (erythrocytes), and a heart consisting of several chambers. Thus, the blood vessels can withstand relatively high pressure, the blood can transport more oxygen per unit volume, and the heart works largely autonomously. It adjusts its pumping performance to the requirements and supplies all body regions with oxygenated blood. However, large body dimensions are not purely advantageous. For example, with their large biomass, vertebrates are particularly attractive to various parasites, disease-causing microorganisms, and predators. No other metazoan group has to defend against a comparable diversity of parasites. This fact has undoubtedly led to the evolution of Metazoa's most highly developed immune system.

General Characteristics of Craniota

The vertebrate body is subdivided into a head, trunk, and tail (Fig. 12.3a). Each body part has specific functions, even if these body parts are hardly distinguishable externally in primarily aquatic vertebrates due to their streamlined shape. The head carries the most important sensory organs, the brain, and the structures for food intake. These are effectively protected by a bony or cartilaginous skull, which can be functionally subdivided into three sections: the neurocranium, the viscerocranium, and the dermatocranium (Fig. 12.4a, b). In fish, the gill region (viscerocranium) represents the posterior region of the head. The trunk and tail form the central part of the locomotion system, containing the primarily segmentally arranged longitudinal muscles with the connective tissue and the axial skeleton, the notochord. The presence, composition, and structure of the notochord and the musculature correspond to the basic pattern of all Chordata.

A spine does not yet exist in the most primitive vertebrates, only some precursor structures. The name of this taxon, Vertebrata, does not refer to a commonly derived character of all members. Therefore, many zoologists today prefer the name Craniota (Craniata) for this taxon. This view considers that some vertebrates, such as the cyclostomes with hagfish and lampreys (Fig. 12.1a), do not have a vertebral column or a spine as their axial skeleton, but rather the persistent notochord. However, all vertebrates have an ossified or cartilaginous skull in common; its presence is, therefore, one of the common derived characters (synapomorphies) of vertebrates. The spine evolved from the fusion of arcuate cartilaginous and bony elements that formed around the notochord. These arches surround and protect the dorsal nerve cord and the most important ventrally located longitudinal blood vessels. From this so-called arch stage of the spine, compact vertebrae have repeatedly emerged convergently in the evolution of the vertebrates, eventually replacing the notochord in several lineages. This evolutionary transition is still visible in those vertebrates that still possess a notochord as adults, such as certain Actinopterygii (basal ray-finned fish, e.g., bichirs, sturgeons; Fig. 12.1d–f), Actinistia (Coelacanthimorpha, coelacanths), and Dipnoi (lungfish) (Fig. 12.1k, l).

Below the axial skeleton, the trunk contains the large body cavity (coelom) with the internal organs. Trunk and coelom end at the ventrally located anal and genital openings (Fig. 12.3a, e). Within the coelom, one compartment is almost always completely separated from the large trunk coelom, the pericardium. Thus, the heart's activity is not affected by other organs, such as the intestine, which is extremely important for the steady supply of blood to the body. The tail is the original propulsion organ of the primarily aquatic vertebrates, commonly called fish; it consists of the axial skeleton and musculature. It carries unpaired fins to increase the effective surfaces and to stabilize the body position (Figs. 12.3d and 12.4a, c). In terrestrial vertebrates, the tail often undergoes functional and structural remodeling and can sometimes be completely absent (apes). Interestingly, this section, indeed the entire body, has reverted to fish-like shape in secondarily aquatic vertebrates (†Ichthyosauria, Sirenia, manatees, whales).

As in Acrania, the locomotor system consists of the segmental longitudinal musculature, which transmits its energy to the connective tissue-like myosepta, the outer muscle fascia, and, last but not least, the flexible axial skeleton. Each paired muscle segment is innervated by the segmentally arranged spinal nerves. The muscles of the left and right body sides work antagonistically, and thus put the trunk and tail into an oscillating wriggling movement in the sagittal plane (dorsoventral plane) the forward-directed force component of which is converted into propulsion. Basal branching vertebrates such as lampreys, hagfish, and their extinct ancestors only possess unpaired fins, the anterior and posterior dorsal fins, and the caudal fin (Fig. 12.1a). In Gnathostomata (jaw-bearing craniates), two additional pairs of fins are present ventrolaterally that can be actively moved, the paired pectoral and pelvic fins (extremities or limbs; Figs. 12.1b, l, 12.3a, b, d–f, and 12.4a–c). These paired fins primarily serve as control and stabilization elements, best seen in sharks and sturgeon (Fig. 12.1b, f); only in derived aquatic forms do they contribute to propulsion. Ultimately, the paired extremities of terrestrial vertebrates evolved from these paired pectoral and pelvic fins. During the transition to terrestrial life, these extremities and their anchoring in the body underwent extensive evolutionary adaptations, so that they could eventually carry the entire body and lift it off the ground. Similar adaptations apply to the dorsal and anal fins: Only representatives with long dorsal fins can generate enough propulsion through wave-like movements alone, like the bowfin or the European eel, for example (Fig. 12.1h).

Vertebrates show a high degree of differentiation in the anterior-posterior axis; this is achieved by, among other things, a duplication of the *Hox* gene cluster. However, other essential gene families or large genome regions have also been duplicated during the evolution of vertebrates. Through complex series of duplications or multiplications, but also deletions of large genome regions or even just individual genes, the newly arising gene copies could now take on new or more differentiated functions without giving up their original functions. This additional complexity in the genome is closely related to the evolution of the derived vertebrate characteristics. These include, among others:

- The neural crest cells arise during embryonic development at the transition area between the neural tube and the trunk ectoderm. Numerous structures of the vertebrate body are formed from them, including parts of the skull and components of many sensory organs.
- An integument consisting of a multilayered epidermis and a mesodermal subcutis (cutis). The former can keratinize, and skeletal elements can form in the cutis (the so-called dermal skeleton, dermatocranium, scales, etc.). An epidermis consisting of several cell layers is only found in vertebrates and arrow worms (Chaetognatha).
- Skeletal structures in the mesodermal connective tissue, including cartilage and bone tissue. Both structures consist of living cells and can grow as the body grows. Cartilage is highly elastic, impact- and tear-resistant, and can proliferate quickly. Many bony elements are, thus, initially formed as cartilage.
- A complex, multipart brain, which is morphologically subdivided into five sections from anterior to posterior: telencephalon (cerebrum, endbrain, olfactory brain), diencephalon (interbrain), mesencephalon with tectum opticum (midbrain), metencephalon with cerebellum (hindbrain with cerebellum), and myelencephalon (posterior brain, afterbrain).
- Highly developed paired sensory organs, including the nose, the lens eyes, vestibular sensory organs, and the lateral line system (Figs. 12.1a–l and 12.3a–d). In addition, there are two unpaired median eyes, the parietal and the pineal organ. Both undergo a functional change during the evolution of Craniota; the pineal organ, in particular, becomes a pure hormone gland (epiphysis, pineal gland) and controls the diurnal rhythm in vertebrates through melatonin secretion.
- The neuromast organs are a crucial element of the lateral line system (flow and pressure perception) and the vestibular system which are unique to the Metazoa. They are also found in the electro-sensory system of fish. The vestibular system, or inner ear, is more complex than its name suggests; in addition to hearing, it serves for perceiving accelerations, gravity, and static orientation in a three-dimensional space. In other words, the vestibular system constitutes the sense of balance and spatial orientation to coordinate movement with balance. With the cochlea (mammals) or lagena (other vertebrates) as the auditory system, the vestibular system constitutes the labyrinth (or labyrinth organ) of the inner ear. This name

was coined referring to its physiological and morphological complexity. The lateral line system detects both reflections of pressure waves that the animal emits and those of other organisms nearby. This sense, therefore, allows for three-dimensional orientation in space and supports and supplements the visual sense in dim water. Moreover, it is extremely sensitive to the smallest changes in pressure. The electric sense detects electric fields generated by other organisms' nerve and muscle activity. However, many ray-finned fish (Actinopterygii) have lost their electric sense. Since the functions of the lateral line and electric systems are entirely linked to an aquatic environment, these organs became functionless during the conquest of land by vertebrates (air does not conduct electricity) and, therefore, have been completely reduced in tetrapods. Of course, their sensory perceptions are beyond our imagination. They represent critical sensory structures for these organisms, without which the evolutionary success of the vertebrates would not have been possible.

- External gill slits (openings) in the pharynx and the restriction to respiration. The number of gill arches and their supporting skeletal elements became significantly reduced compared to Acrania and Tunicata. For example, most Gnathostomata have five gill arches.
- Remodeling of the circulatory system: vessels with endothelium, multichambered heart, erythrocytes with hemoglobin as respiratory protein.
- An efficient excretion system, consisting of primarily segmentally arranged nephrons (renal corpuscles), can be derived from metanephridia.
- Functionally linked reproductive and excretory systems. Nearly all vertebrates are gonochoristic, with or without sexual dimorphism; hermaphrodites or dwarf males are rare.

Starting from this basic pattern, extensive evolutionary differentiations occurred within vertebrates, ultimately giving rise to the land vertebrates. Many new characteristics were acquired in the process that were decisive for this evolutionary success. To illustrate this differentiation, we will discuss here one representative of the primarily aquatic vertebrates, a bow-fin fish, and one of the terrestrial vertebrates, a mammal.

In most dissection courses, only 1–2 days are available for studying vertebrates. Therefore, we present only two animals in detail. The rainbow trout (*Oncorhynchus mykiss*), belongs to the most species-rich and youngest vertebrate group, Teleostei. As an example from the group of terrestrial vertebrates, Tetrapoda, we chose the laboratory rat (*Rattus norvegicus*). The rat belongs to the placental mammals, one of the two groups of tetrapods showing the highest degree of adaption to the terrestrial environment. A detailed discussion of vertebrate histology would far exceed the scope of this book, so we limit ourselves to the dissection of these two selected vertebrate species.

12.1 The Rainbow Trout (*Oncorhynchus mykiss*, Teleostei, Bony Fish)

The vertebrates comprise two sister groups, the jawless fishes, Agnatha, Cyclostomata, Cyclostomi, cyclostomes and the jaw-bearing Gnathostomata (Fig. 12.2). However, only two small, species-poor groups have survived off the Agnatha, the lampreys and hagfish. These still show many of the ground pattern characters of vertebrates. While hagfish (Myxinoidea) are pure marine dwellers, the eel-like lampreys (Petromyzontidae) inhabit sea- and freshwater. Native species include, for example, brook and river lampreys (Fig. 12.1a). However, the vast majority of recent vertebrates belong to Gnathostomata. This group is divided into cartilaginous fish (Chondrichthyes, Figs. 12.1b, c and 12.2), ray-finned fish (Actinopterygii, Figs. 12.1d–i and 12.2), and lobe-finned fish (Sarcopterygii, Figs. 12.1k, l and 12.2). The first two groups consist exclusively of aquatic taxa. The latter taxon, the ray-finned fish, is the most speciose in Craniota and comprises more than half of all vertebrates. Sarcopterygii are characterized by prominent muscular fins and a specific skeleton within the fins and also include the terrestrial vertebrates (Tetrapoda, Fig. 12.2). However, most of the aquatic representatives among the lobe-finned fish have become extinct, except for just two marine species of coelacanthids (*Latimeria chalumnae*, *Latimeria menadoensis*) and six species of lungfish (Fig. 12.1k, l).

Gnathostomes are characterized by the remodeling of the branchial skeleton, in which two anterior gill arches (mandibular and hyoid arches) were transformed into tooth-bearing jaws (Figs. 12.3g and 12.4a, b). This allowed for acquiring and exploiting numerous new food sources, as mobile and tooth-equipped jaws are more versatile. This also includes a predatory lifestyle, as the movable jaws are suitable for tearing pieces out of the prey. At the same time, the to-date dominant jawless vertebrates probably became attractive prey organisms for these jaw-bearing organisms, which could also have meant the "end" for most agnathan representatives. These two gill arches transformed into jaws are then followed by five more, gill-bearing arches retaining the respiratory function (Figs. 12.3h and 12.4a, b).

Similarly, the locomotor system of these organisms has improved; the paired extremities (pectoral and pelvic fins, including the pectoral and pelvic girdles, Fig. 12.2a, d, e) have evolved. The unpaired fin seams were dissolved into a few individual fins. The fins contain a skeleton of bony fin rays, so they are still deformable, but better supported (Figs. 12.3f and 12.4a–c).

The ground pattern of Actinopterygii (Fig. 12.2), the ray-finned fish, includes, in addition to the paired limbs, a dorsal fin, the caudal fin, and the anal fin (Fig. 12.3a), just as in the rainbow trout. The fins have no external muscular part in actinopterygians (Figs. 12.3d, f and 12.4a–c). In salmonids (Salmonidae), there is also an adipose fin dorsally on the tail (Fig. 12.3a, d). As can easily be seen from the skeleton preparation, this fin does not possess bones (Fig. 12.4a, c). In other teleosts, such as the codfish (Gadidae), the dorsal fin can be divided into two or three fins.

After developing a jaw apparatus with teeth, the next crucial step in evolution was the emergence of lungs, which allow for survival in oxygen-poor waters. In parallel, such an air-filled chamber always causes buoyancy and thus also functions as a swim bladder. Oxygen can be absorbed directly from the air. One hypothesis assumes that this evolutionary novelty arose in oxygen-poor freshwater about 400 million years ago during the Devonian period. Both, the ray-finned fish (Actinopterygii) and the lobe-finned fish (Sarcopterygii), but not the cartilaginous fish (Chondrichthyes), possess such a lung-swim bladder organ, and it is considered a decisive preadaptation for the conquest of the land. In both groups, recent representatives still use the lung-swim bladder organ for air breathing. For example, bichirs (Fig. 12.1d, e) and lungfish (Fig. 12.1l) must regularly come to the surface to breathe air. They would otherwise drown, as their gills are not efficient enough for a complete oxygen supply. Also, the more advanced gars and bowfins possess a highly vascularized lung-swim bladder, which supplements gill respiration (Fig. 12.1g, h). The fact that such an organ also generates buoyancy was initially an evolutionary side effect. The exclusive use of the lung-swim bladder as a purely hydrostatic organ is only realized in teleosts, like the rainbow trout (Figs. 12.5a, b, e and 12.6a, b). However, the ability to absorb gases into the blood and to release them into the lung-swim bladder always remains. Such a buoyancy organ allows for an energy-saving stay at a certain water depth and its precise maintenance. The organisms do not have to work against sinking, due to the slightly greater density of their bodies compared to the surrounding medium water. However, buoyancy can also have disadvantages. In some bottom fish, deep-sea fish, and fast enduring swimmers like mackerels, the gas-filled swim bladder has probably been lost.

With about 34,000 species, Teleostei, or bony fish, represent by far the most species-rich and evolutionarily youngest group within the vertebrates. The earliest forms of Teleostei are known from the upper Jurassic. The phylogenetically older, remaining ray-finned fish are represented by only 44 extant species in our fauna and belong to four superordinate taxa: Cladistia (Polypteriformes, bichirs), Chondrostei (Acipenseriformes, sturgeon-like), Ginglymodi (Lepisosteiformes, gars), and Halecomorphi (Amiiformes, bowfins) (Figs. 12.1d–i and 12.2). From an evolutionary point of view, they are particularly interesting, as they allow for a fairly accurate reconstruction of the early evolutionary

Fig. 12.1 Diversity of fish. (**a**) Agnatha, jawless fish: *Lampetra fluviatilis,* European river lamprey, 0.25–0.4 m, anadromous species, migrates from the sea into fresh water for spawning. (**b–l**) Gnathostomata, gnathostomes, jaw-bearing craniates. (**b, c**) Chondrichthyes, cartilaginous fish, Neoselachii, sharks, and rays. (**b**) *Triakis semifasciata,* leopard shark, 1.2–1.5 m, marine. (**c**) *Raja montagui,* young spotted ray, 0.8 m, marine. (**d–i**) Actinopterygii, ray-finned gnathostomes. (**d**) Polypteriformes, bichirs: *Polypterus senegalus,* Senegal bichir, 0.5 m, freshwater, respiration by lungs and gills, drown if prevented from breathing. (**e**) Polypteriformes: *Erpetoichthys calabaricus,* reedfish, ropefish, or snakefish, 0.2–0.4 m, freshwater, respiration by lungs and gills, drown if prevented from breathing. (**f**) Chondrostei = Acipenseriformes, sturgeon, and paddlefish: *Acipenser transmontanus,* up to 2.1 m, white sturgeon, anadromous species, migrates from the sea into fresh water for spawning, respiration through gills. (**g**) Ginglymodi, gars, and fossil relatives: *Lepisosteus oculatus,* spotted gar, 0.6–0.9 m; freshwater, highly vascularized swim bladder lung supplements gill respiration. (**h**) Halecomorphi, bowfins, and fossil relatives: *Amia calva,* 0.5–0.7 m, bowfin, mudfish, mud pike, dogfish, grindle, grinnel, swamp trout, and choupique; freshwater, gas exchange by gills and pulmonoid swim bladder. (**i**) Teleostei, teleosts: *Squalius cephalus,* 0.3–0.6 m, common chub; freshwater, respiration exclusively by gills, swim bladder purely hydrostatic. (**k, l**) Sarcopterygii, lobe-finned gnathostomes. (**k**) Actinistia, Coelacanthimorpha, coelacanthids: *Latimeria chalumnae,* up to 1.8 m, West Indian Ocean coelacanth, gombessa, marine; lung highly modified, filled with fat, note typical lobe-fins, archipterygia, comprising articulated skeleton and muscular limb lobes within the fins. (**l**) Dipnoi, lungfish: *Protopterus annectens,* up to 1 m, West African lungfish or Tana lungfish; lungs and gills, air breather, freshwater. Images of animals were taken at the river Ems, Germany (**a**), in the Aquarium de Lyon, France (**b**), the Muzee Aquarium Delfzijl, Netherlands, (**c**), Aquarium NaturaGart Ibbenbüren, Germany (**d, e, f, g, h, l**), and the Aquarium Berlin, Germany (**i**). The fixed *Latimeria chalumnae* specimen is on display in the Musée de Toulouse, France (**k**)

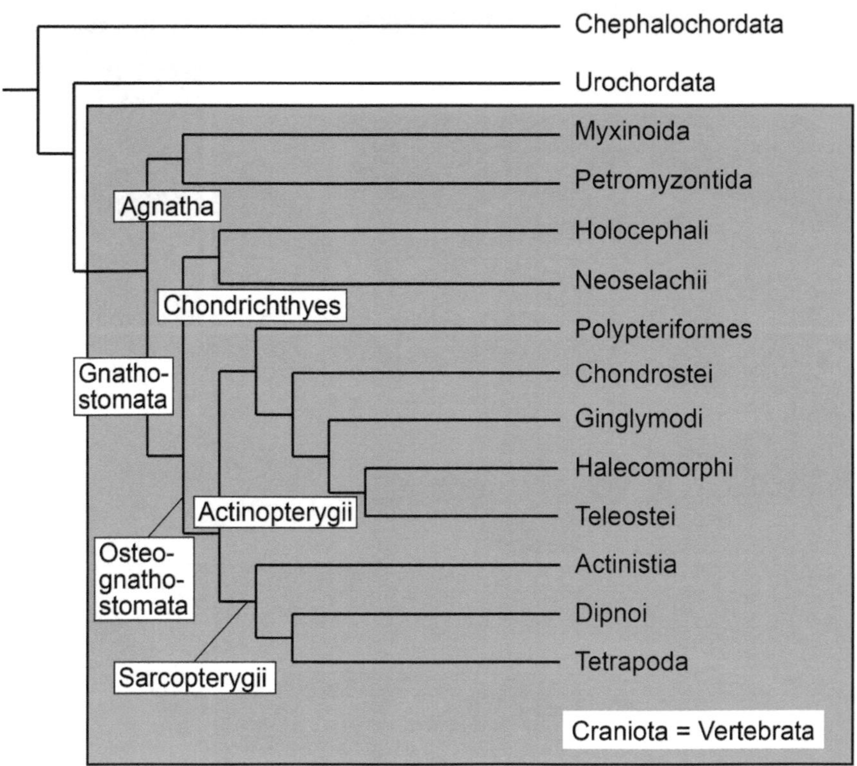

Fig. 12.2 Phylogenetic relationships of Craniota. Only primarily aquatic groups are included in detail. Extinct taxa omitted. After various sources

history of the ray-finned fish, which has been reconstructed congruently both, molecularly and morphologically. All groups are primarily freshwater inhabitants; some members of Chondrostei are anadromous forms, returning to the river systems for reproduction after they have spent some time in the open sea. Such forms also exist within Teleostei. The most famous are the salmons, which return to their former birth waters for spawning. Also, the representatives of the basal branches in the teleost phylogenetic tree are freshwater inhabitants, so this fact, like their physiology, also speaks to an origin in freshwater. Among the basal actinopterygians, sturgeons are the only representatives of economic importance; the eggs of some species are used as caviar. Due to economic use and increasing water pollution and construction, these species are highly endangered or almost extinct. This also applies to the formerly very common European species, the Baltic sturgeon, *Acipenser sturio.*

Today, teleosts inhabit all aquatic habitats, from the deep sea to high mountain regions. Among the teleosts are the smallest known vertebrates, such as *Paedocypris progenetica* (Cyprinidae, carp fish), which reside in the peat swamp forests of Sumatra, with a body length of 10 mm and a weight of about 0.2 g. The largest representatives reach a body length of up to 17 m (*Regalecus glesne,* giant oarfish), while the sunfish (*Mola mola*) and the blue marlin (*Makaira nigricans*) each weigh about 900 kg, probably the heaviest tele-

osts. Accordingly, teleosts show the greatest diversity within the aquatic vertebrates. Almost all economically useful fish species belong to this group; they are of outstanding importance for human nutrition. Only around 50 marine species account for more than half of the fish caught. Among them, the herring-like and cod-like species are particularly affected. However, they are so intensively hunted that many species are extremely endangered due to more advanced fishing methods. Other species are equally endangered or extinct due to marine pollution, the damming of river systems, or other environmental changes caused by humans. Since many edible fish only reach the food trade in processed form (for example, fish sticks), the appearance of these animals is largely unknown to many people.

Preparation of a Rainbow Trout (*Oncorhynchus mykiss*)
The frequently farmed rainbow trout (*Oncorhynchus mykiss*) originally comes from the North American Pacific region (Alaska to Northwest Mexico). It belongs to the salmons (Salmonidae) and, like many species from this group, is originally an anadromous migratory fish that ascends rivers from coastal marine waters for spawning. In addition, there are, naturally, also pure freshwater forms of the rainbow trout; in culture, both have been mixed. The rainbow trout shows a very high temperature, oxygen, and pH tolerance. This high tolerance to fluctuations in the environmental parameters

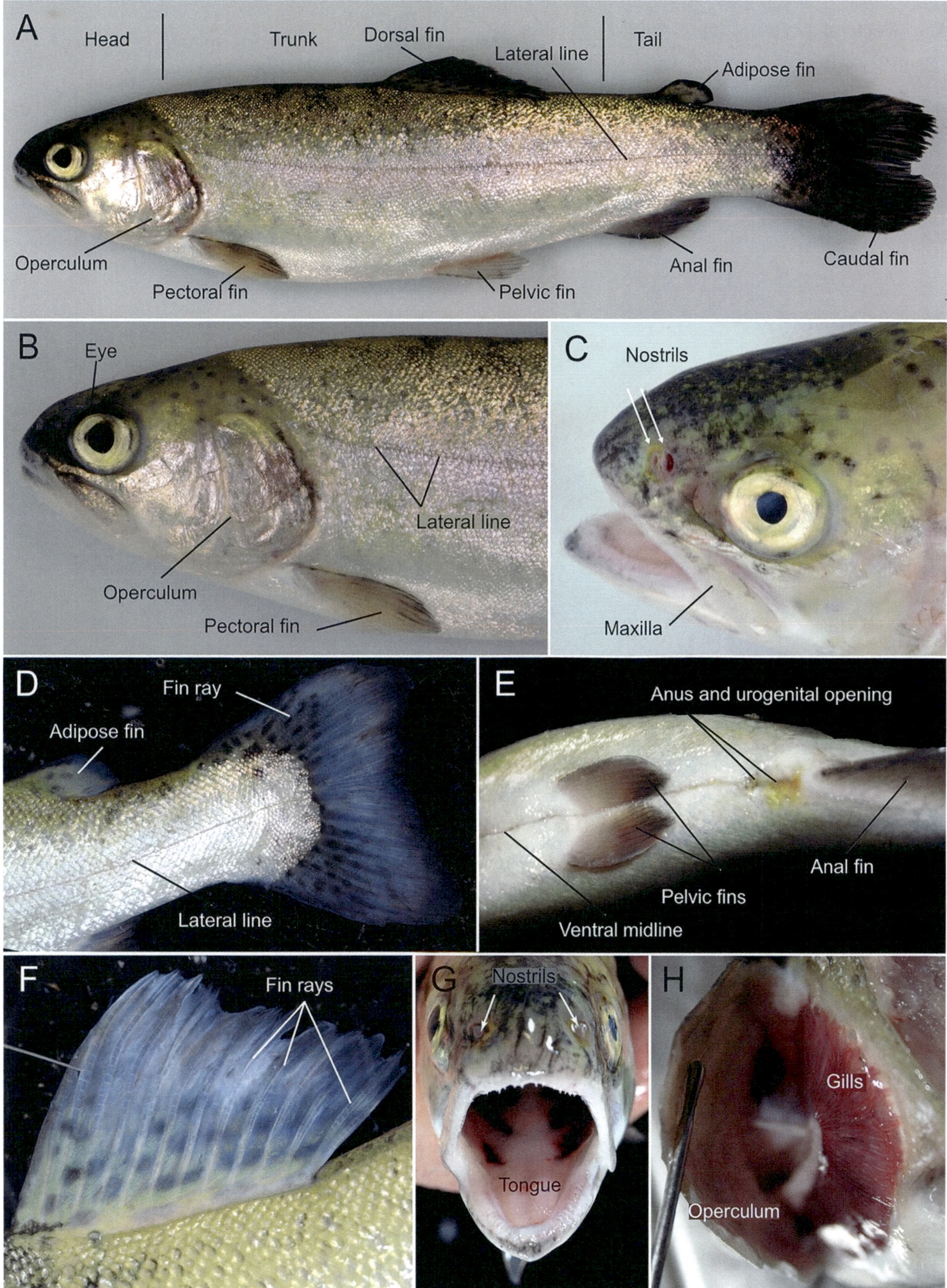

Fig. 12.3 *Oncorhynchus mykiss,* Rainbow trout. Habitus and external morphology. (**a**) Whole animal; length of the individual fish is approximately 25 cm; typical coloring. (**b**) The head region with gill cover, pectoral fin, and lateral line. (**c**) Head with anterior and posterior nostrils (arrows), eye; the movable maxillary is clearly visible. (**d**) Tail region with adipose and caudal fin; scale coat. (**e**) The abdominal region, ventral fins, anus, genital opening, ventral midline. (**f**) Dorsal fin with spread fin rays. (**g**) View into the opened mouth; arrows point to forward-facing anterior nostrils. (**h**) Spread gill cover reveals view of gills

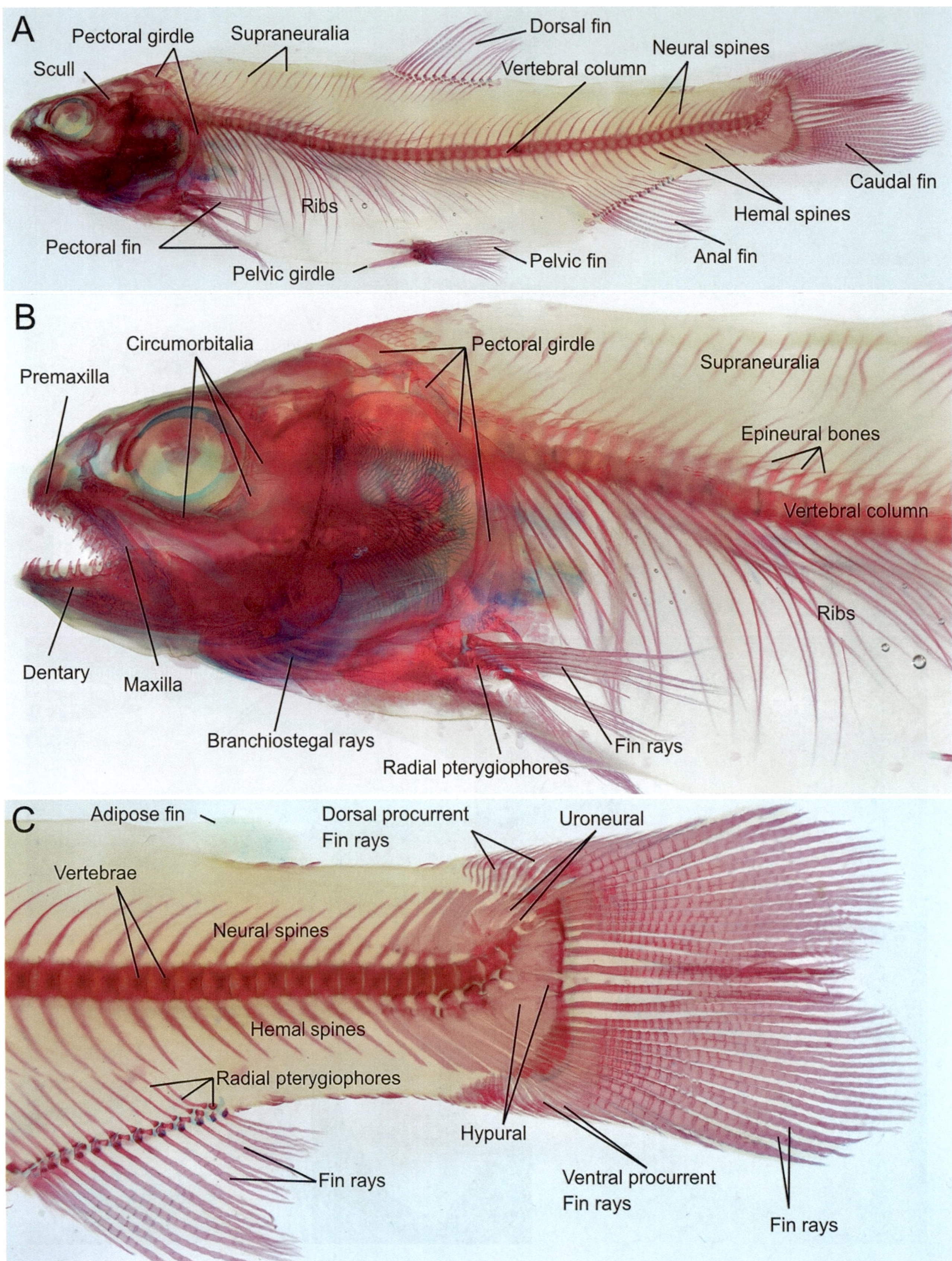

Fig. 12.4 *Oncorhynchus mykiss*. Skeleton. Alcian blue-Alizarin preparation. (**a**) Whole skeleton with skull, spine, ribs, and all fins. (**b**) Front region enlarged, skull, pectoral girdle, pectoral fins, and anterior region of the axial skeleton. Beneath the largely transparent gill cover, the gill skeleton, including the cartilage elements supporting the gill lamellae, becomes visible. (**c**) Tail vertebrae, tail, and anal fin; the adipose fin contains no skeletal elements

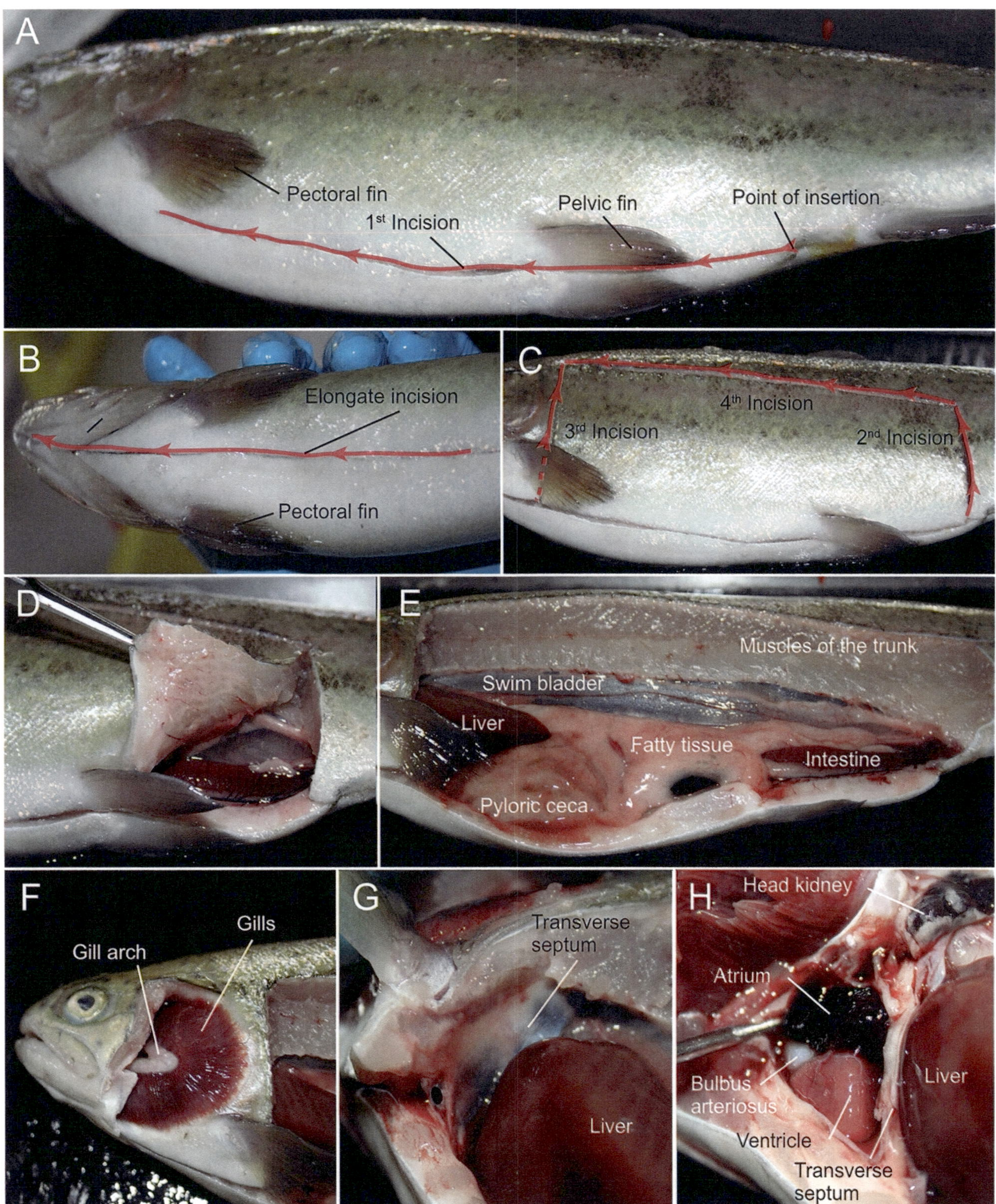

Fig. 12.5 *Oncorhynchus mykiss*. Dissection steps. (**a**) The first cut is made ventrally in the median from back to front up to the pectoral fin. (**b**) Extend the cut to the mouth opening. (**c**) Incisions 2–4 to open the body cavity. (**d**) Removal of the trunk muscles from back to front. (**e**) Opened body cavity. (**f**) The gills are exposed by cutting off the operculum. (**g**) View from the back of the transverse septum. (**h**) Opening of the pericardium by removing the anterior body wall with pectoral girdle

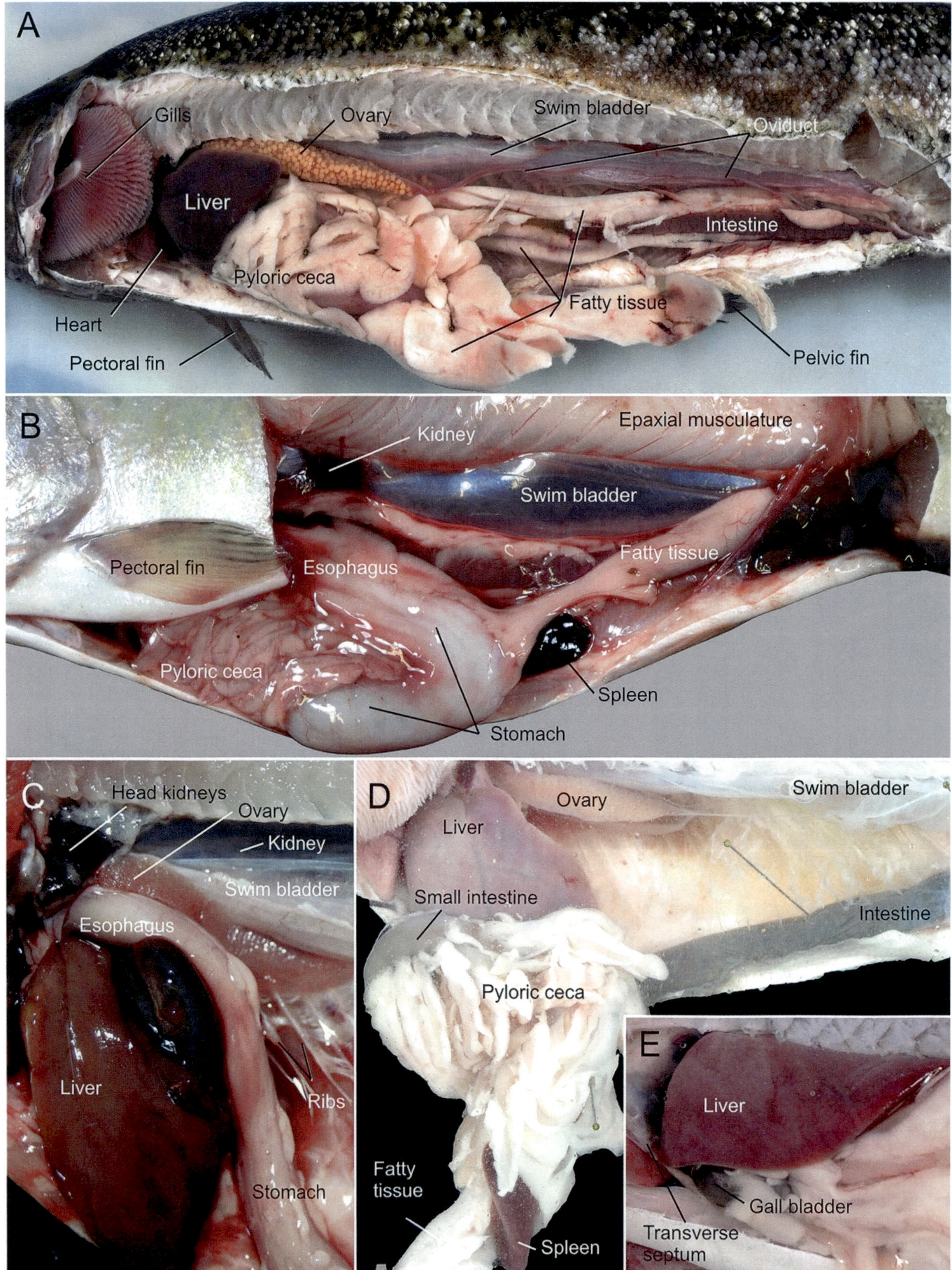

Fig. 12.6 *Oncorhynchus mykiss*. A female trout. (**a**) Body cavity after removal of the lateral body wall, internal organs still almost in their natural position, liver, pyloric ceca, and adipose tissue. (**b**) After careful dissection, individual parts of the digestive tract, such as the stomach and the spleen, become distinguishable. (**c**) Esophagus, stomach, and liver. (**d**) After removing the connective tissue, the intestine is laid ventrally: midgut with pyloric ceca. The ovary is relatively small, below the gas-filled swim bladder. (**e**) Liver with gallbladder

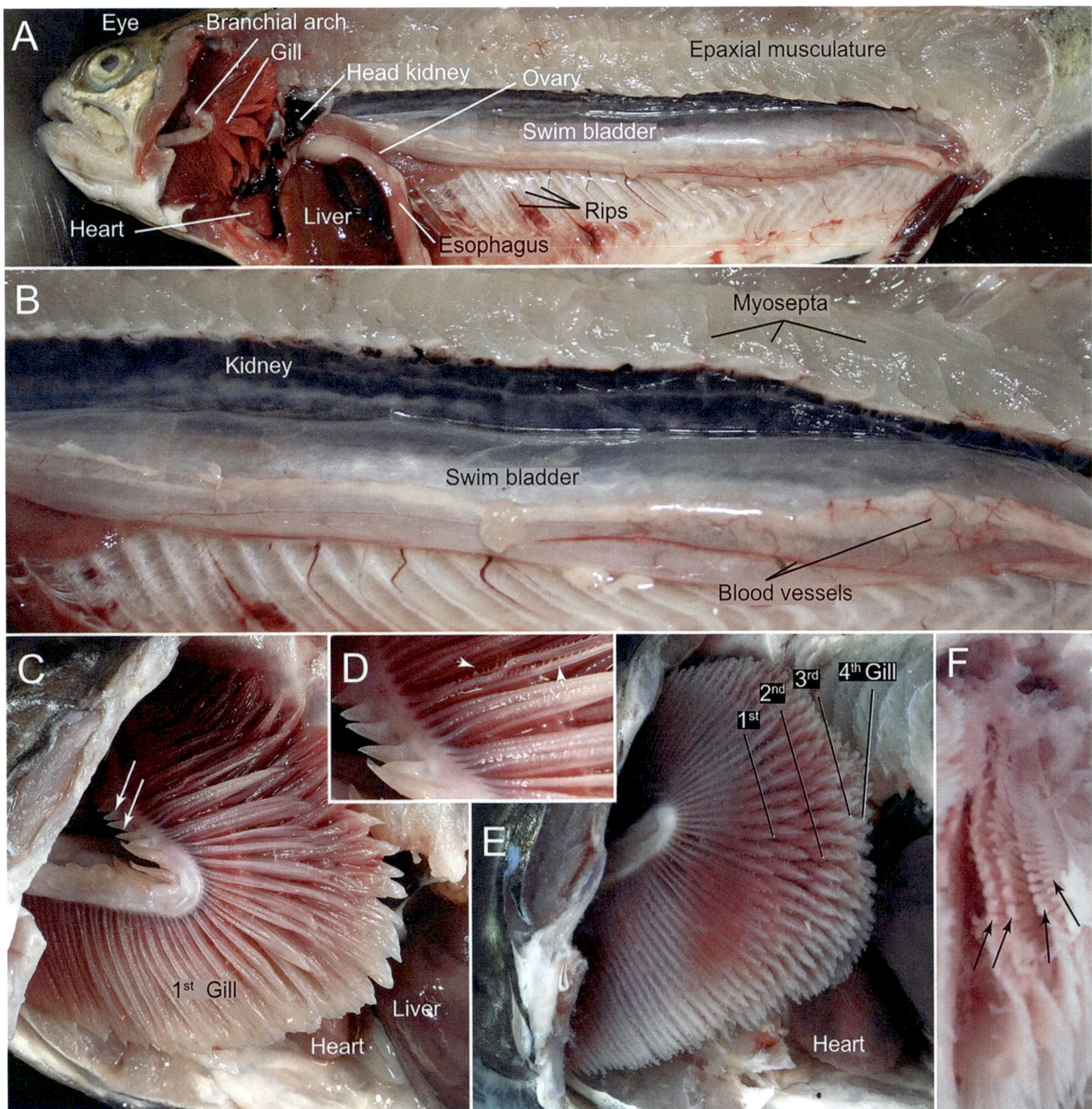

Fig. 12.7 *Oncorhynchus mykiss*. Swim bladder and gills. (**a**) Swim bladder in natural position; intestine laid ventrally. (**b**) Swim bladder with blood vessels and kidney (Ductus pneumaticus; see Fig. 12.8a). (**c**) First, the gill arch with gill raker and gill filaments, behind it, the heart and the liver. Arrows: inward-directed short raker spines. (**d**) Magnification of a part of the gill; arrowheads point to the gill lamellae, which are perpendicular to the gill filaments. (**e, f**) Representation of the four gills. Each gill consists of two rows of flat gill filaments (arrows), which, in turn, carry the gill lamellae, not visible here but in Figure 12.7d

makes it particularly suitable for breeding in fishponds, where these parameters often cannot be maintained precisely.

Since the species has now spread outside of fish farms, this leads to the known problems of all neobiota (aliens). In this particular case, the population of the European native river trout (*Salmo trutta fario*) is particularly endangered by displacement. Usually, the rainbow trout reach an age of about 10 years, a length of 35–70 cm, and a weight of 1.4–5.4 kg. The name rainbow trout refers to the characteristic coloration of the animals (Fig. 12.3a). Like most fish, the back is darker colored than the almost white belly. Dorsally, a green pigmentation with numerous black spots predominates, and the animals give off a silver shimmer on the flanks,

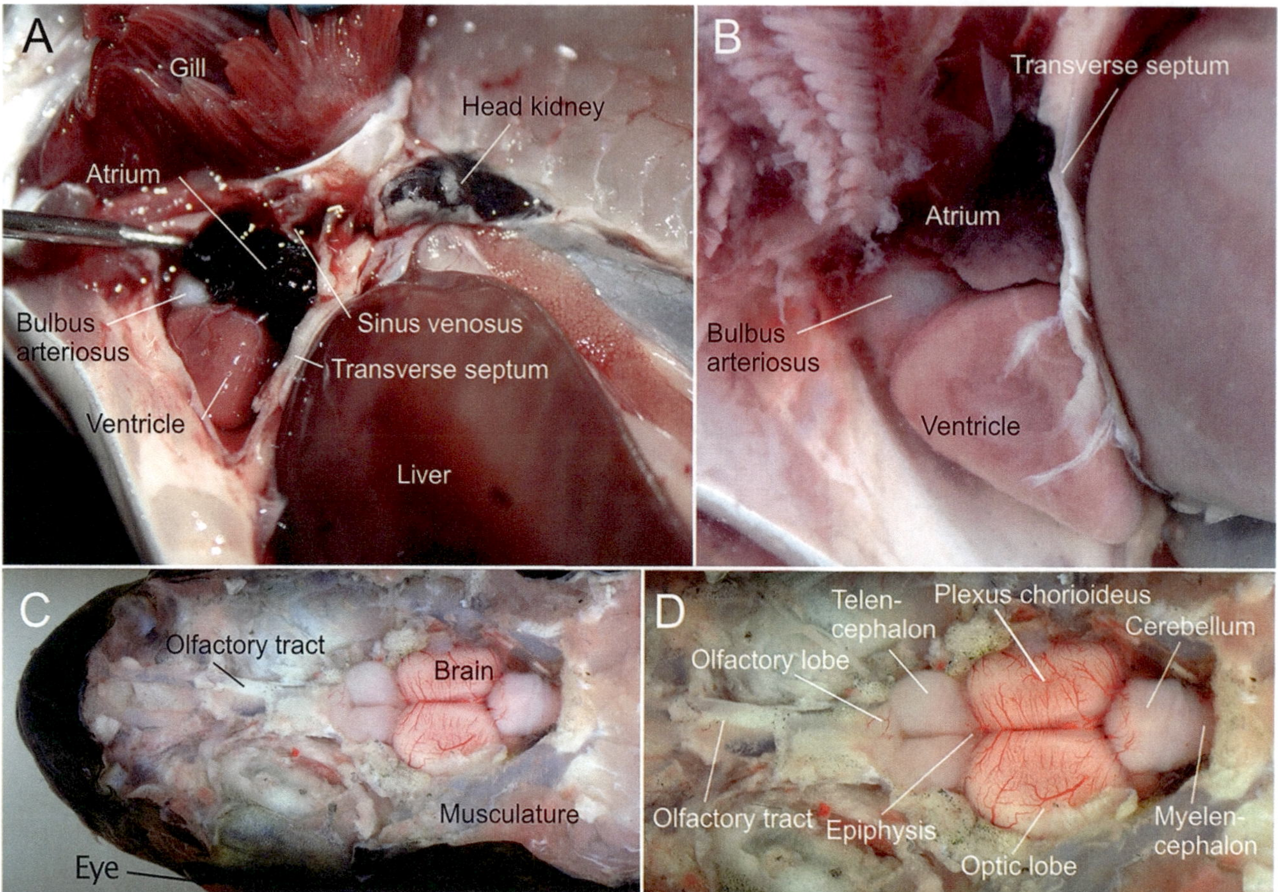

Fig. 12.8 *Oncorhynchus mykiss.* Heart and brain. (**a**) Heart after opening the pericardium. The large head kidney lies dorsally to the heart. (**b**) Atrium (auricle), ventricle and bulbus arteriosus. (**c**) Brain in the dorsally opened skull. (**d**) Brain with telencephalon (forebrain, cerebrum), mesencephalon (midbrain) with paired tectum opticum, cerebellum, and myelencephalon (medulla oblongata, hindbrain); diencephalon (interbrain) not visible

which are also characterized by a reddish longitudinal band (Fig. 12.3a, b). Thus, the animals are not easy to recognize from above or below, which is advantageous for hunting and protection against predators.

External Morphology

Except the head, the skin carries regularly arranged, overlapping scales (Fig. 12.3a–d). These are relatively small, soft, and located beneath the epidermis within the cutis. The variable number of scales fluctuates around an average value. For example, their number ranges between 135 and 150 along the lateral line. Scales are thin, cell-free bone plates, which are considered remnants of an original bony armor made of relatively thick and heavy enamel scales (each consisting of an enamel, dentin, and bone layer). Such scales still occur, for example, in bowfin fish. Due to the ring-shaped growth stripes, the scales of the rainbow trout and many carp, salmons and other fish, are referred to as cycloid scales. For microscopy, these scales can be easily removed with tweezers.

Recommended Material

Depending on availability, various teleost species are suitable for a basic zoology dissection course. Some can be obtained from local fish farms, fish sellers or from angling shops. Marine fish are best dissected as part of marine biology excursions. Often only gutted animals are sold in fish markets, making it more difficult to obtain marine fish for a zoological course. In addition to the rainbow trout presented here (*Oncorhynchus mykiss* [= *Salmo gairdneri*]), we recommend the following species for a basic zoological course: roach (*Rutilus rutilus*), common rudd (*Scardinius erythrophthalmus*) or carp (*Cyprinus carpio*). One should ensure that the acquired specimens do not exceed a body length of 20–30 cm, so they can still be examined in a standard dissecting dish. However, dissection can often also be unsatisfactory with too small specimens. In addition, some older sex-

(continued)

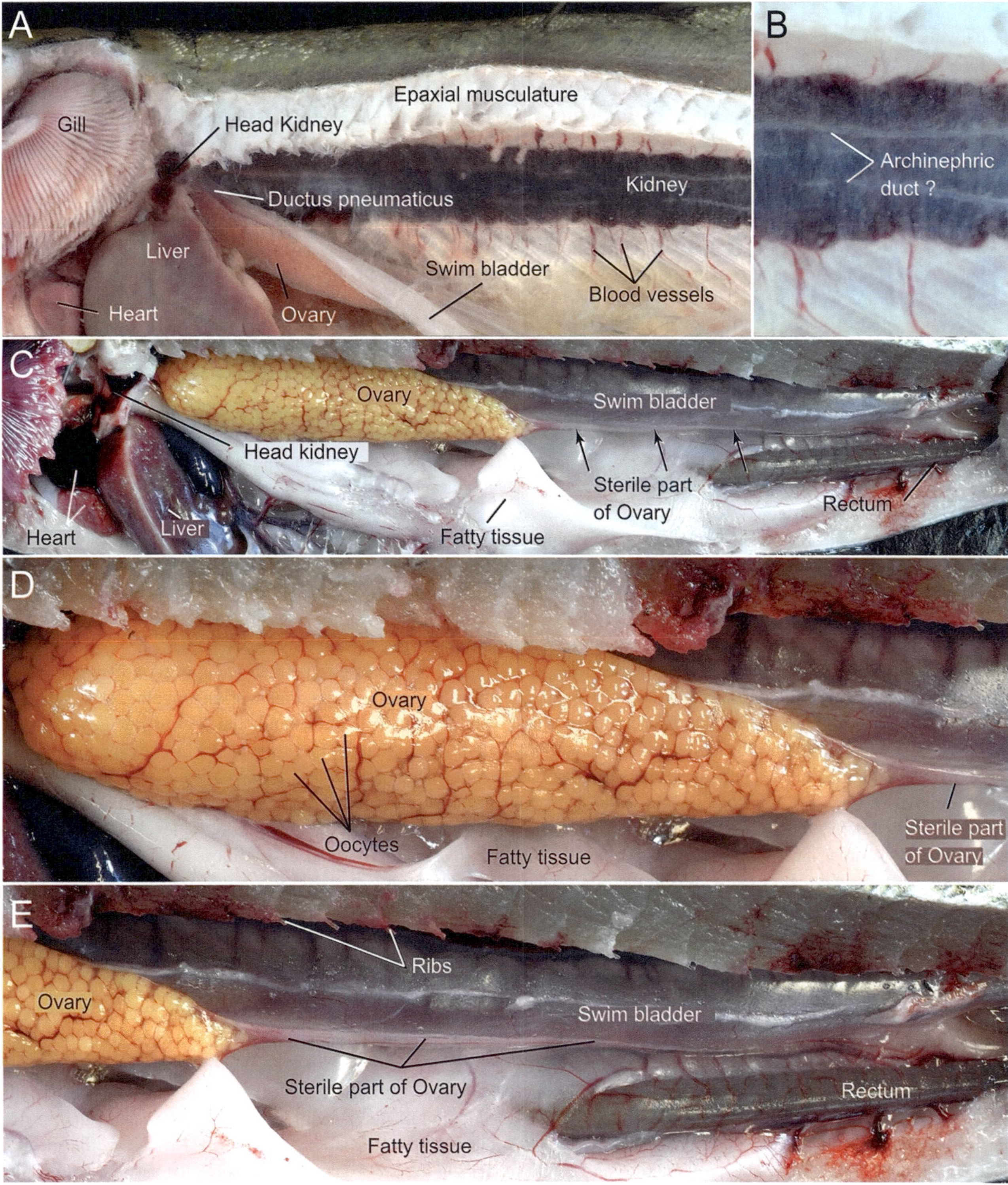

Fig. 12.9 *Oncorhynchus mykiss*. Kidney and reproductive organs. (**a**) Position of the kidney as a compact, dark red band beneath the spine. (**b**) Close-up. (**c**) Left ovary with posterior sterile section (arrows) beneath the swim bladder after removal of the intestine. (**d**) Enlarged ovary with orange-colored round oocytes and supplying blood vessels in the ovary's wall. (**e**) Sterile part of the ovary beneath the swim bladder (**d** and **e** Close-ups from **c**)

ually mature animals should be at hand to demonstrate fully developed gonads. The animals can be anesthetized and then killed with MS222 and should be examined fresh. Unfortunately, tissue preservation suffers greatly from frozen animals. Fish farms also provide freshly killed animals—however, these animals are usually killed by a blow to the head.

A vertebrate's nervous system and brain can probably be prepared most easily in a cartilaginous fish; here, one could use the small-spotted catshark (*Scyliorhinus canicula*) or a ray. Although classified as not endangered, one should limit oneself to a few specimens. From the Biologische Anstalt Helgoland (BAH, Heligoland, Germany), material supply, various bony fish, but also cartilaginous fish such as small-spotted catsharks (*Scyliorhinus canicula*), and rays (*Raja* spec.) can eventually be obtained.

The rainbow trout *Oncorhynchus mykiss* has a typical streamlined body adapted to aquatic life, in which the head, trunk, and tail merge without clear boundaries (Fig. 12.3a). The body is relatively slender, the ratio of length to height is about 5:1, and the largest diameter is reached after about one-third of the body length. The body is slightly compressed laterally, so it is somewhat taller than it is wide. The caudal fin is large and provides propulsion with the help of the muscular tail. This body structure shows us that the trout is a fast swimmer. In addition, the tail anatomy indicates that the trout accelerates very well. On the other hand, a mackerel (*Scomber scombrus*) swims also very fast but with great endurance. Mackerels have a stalk-shaped caudal fin distinctly set off from the rest of the body. On the other hand, very agile swimmers or maneuvering specialists have short and laterally compressed, disc-shaped bodies and various forms of caudal fins, as is the case with many inhabitants of coral reefs.

Due to their body structure, trout can also move against the current in river systems or accelerate very quickly and thus overpower their prey organisms. During the subsequent dissection, the predominant proportion of the musculature on the entire body becomes visible. The body musculature is basically still structured like that of Acrania (*Branchiostoma*) and consists of muscle segments (myomeres) and tendon plates (myosepta) (Fig. 12.7b). The three-dimensional arrangement is, however, much more complex and essentially has the shape of a horizontal W. The musculature consists predominantly of white fibers; only above the horizontal septum does a thin band of darker (gray–red) musculature extend. The myoglobin content and the mitochondria cause

this coloration; these are, therefore, aerobic muscle fibers. This dark musculature is responsible for slow and continuous performances. So, when we observe a slowly swimming fish in an aquarium, it achieves this movement with only a small proportion of its musculature. The white (anaerobic) musculature is intended for short-term peak performances (escape reactions, prey capture, jumping movements of salmon). A movie showing a swimming trout can be found at figshare: sn.pub/xr9qhi.

The dorsal fin is relatively short and lies approximately in the middle of the body. The fin rays become visible if you spread the fin (Fig. 12.3e). If the skeleton is visualized with the help of an Alzian/Alizarin stain, this becomes particularly clear (Fig. 12.4a, c). The fin rays in the rainbow trout are soft, distally branched rays. However, the name is misleading, as the fin rays consist of hard bony elements, each of which comprises several segments (Fig. 12.4c). By the way, the distribution, number, and occurrence of the fin rays are important identification features. The caudal fin is externally symmetrical (homocercal) and contains a complex system of modified caudal vertebrae (Fig. 12.4a, c). The spine bends dorsally, but in this region, it consists only of two tail vertebrae. The symmetrical shape of the caudal fin is achieved by an enlargement of the ventral arch elements of the spine, the so-called hypurals (Fig. 12.4c). A homocercal caudal fin generates neither an upward nor a downward movement component, thus working together with the swim bladder. On the other hand, the typical dorsally enlarged (hypocercal) caudal fin of many sharks, which do not have a swim bladder, generates an upward movement component. The pectoral fins are located directly behind the operculum, and the pelvic fins are roughly at the level of the dorsal fin (Fig. 12.3e). The anal and adipose fins are opposite each other (Figs. 12.3a and 12.4a).

On external examination of the head, the first thing one notices is the very large eyes (Fig. 12.3a–c). Due to their lateral position, the field of vision is quite vast. Another externally visible sensory organ is the nose, which has openings at the front and back, called nares (Fig. 12.3c, g). The olfactory epithelium is located in between. During swimming, water passes the receptor cells and is continuously analyzed. The front nostrils are precisely aligned in the direction of movement (Fig. 12.3g). The lateral line is also clearly visible. Some ray-finned fish also have tactile appendages in the lower jaw area, the so-called barbels, with which they can feel and chemically examine the ground or make their way through bodies of water with low visibility. The operculum, which reveals a view of the gills when spread apart, forms the rear end of the head region (Fig. 12.3h). The gills show a complex structure of the flat gill filaments protruding laterally from the gill arches, on which the thinner gill lamellae are then located. To see these structures clearly during dissection, the trout must be covered with water in the dissecting dish. Otherwise, the fine lamellae will stick together (Fig. 12.3h). In the skeletal staining, not only do the gill arch elements (red) become visible but so do the blue-colored filamentous cartilage elements that support the individual gill filaments (Fig. 12.4b). The operculum itself is a complex structure consisting of several bone elements.

If you open the mouth of the fish, the teeth on the upper and lower jaws and the large tongue become visible. The teeth are relatively large, pointed, and located on three paired bone elements, premaxilla and maxilla in the upper jaw and dentary in the lower jaw (Fig. 12.4b). In addition, the upper jaw features an unpaired facial bone, the so-called vomer, which carries two longitudinal and one transverse row of teeth. Unlike mammals, in teleosts, the premaxilla and maxilla are movable relative to the rest of the skull. This shows that the trout is an active predator that feeds on various invertebrates, but also smaller fish and amphibians. Insects are hunted both under and above water. Jumping of trouts is especially observable in the evening.

Skeleton

To examine the skeleton of the rainbow trout more closely, the animal can be stained as a whole in a relatively simple but lengthy preparation. Soft body parts are made transparent, while cartilage is stained blue and bones red (Fig. 12.4a–c). This preparation method has the advantage that the individual skeletal elements largely remain in their natural arrangement and can be examined. This shows that the skeleton of the teleosts is largely ossified. The most noticeable skeletal structures are the skull and the spine. The skull is a relatively complex formation, consisting of numerous bone elements. These can only be partially assigned in a total preparation and will not be discussed in detail here (Fig. 12.4b). Notable are, for example, the above-mentioned tooth-bearing elements, the circumorbitalia surrounding the eye socket, and the operculum composed of several elements. The spine has about 60 vertebrae and runs straight through the body; only the last two vertebrae bend dorsally. The vertebral bodies are biconcave and unipartite. They dorsally carry a neural arch, which encloses and protects the spinal cord and extends obliquely backward in a neural process. Ventral to the vertebrae in the tail are corresponding processes, known as hemal arches and processes, which accommodate the caudal aorta and caudal vein. In the trunk area, there are lateral processes instead of the hemal arches, the so-called parapophyses, where the ribs insert. The ribs support the body cavity and are developed in teleosts over the entire trunk region up to the anus. In front of the dorsal fin are additional skeletal elements above the spine, the supraneuralia, which each begin between two successive neural processes. The unpaired fins rest on multipart fin supports (pterygiophores), which lie in the median connective tissue septum. These are followed by the fin rays, which are at least partially branched. In the caudal fin, the fin rays follow the arch elements of the spine; one distinguishes the larger branched middle main rays, which are preceded dorsally and ventrally by the so-called procurrent rays. The fin rays of the paired extremities continue into radials, which are articulated at the extremity girdles. The pelvic girdle consists of only one rod-like element each and is not connected to the rest of the skeleton (Fig. 12.4a). On the other hand, the pectoral girdle is connected to the skull and represents a multipart formation that connects in an arc to the skull and extends far dorsally (Fig. 12.4a, b). In relatively primitive representatives of the teleosts like our course object, the pelvic fins are situated comparatively far back (abdominal). Still, the pelvic girdle can also be moved far forward under the pectoral fins. These skeletal elements are to be distinguished from the so-called fish bones occurring in many teleosts in various frequencies and positions. These are ossified tendons that can occur in the segmental connective tissue apparatus of the myosepta and in the horizontal septum. In the trout, ossified tendons are limited to the epineural fish bones above the spine in the trunk area.

Preparation Steps

After an in-depth external examination, the dissection and preparation of the internal organs can be started.

Oncorhynchus: **Preparation Step 1**
First, use a sharp pair of scissors to carefully puncture the body cavity on the ventral midline immediately in front of the anus without damaging the intestine (Fig. 12.5a). Then, continue the cut to the tip of the lower jaw; the pectoral girdle must be severed (Fig. 12.5b). To open the body cavity, place the animal on its right side and make an anterior and a posterior cut from ventral to dorsal up to the lateral line. The posterior cut is made at the level of the anus, the anterior cut immediately behind the pectoral fin (Fig. 12.5c). A fourth cut is made at the level of the lateral line from back to front. The resulting flap is now opened from back to front, and the remaining trunk muscles and the ribs are cut (Fig. 12.5d). Afterward, the internal organs are exposed. In the example shown, the air-filled, silvery-white shiny swim bladder was not visible and thus was accidentally damaged during the preparation, causing the gas to escape (Fig. 12.5e).

Oncorhynchus: **Preparation Step 2**

After first inspection, the next cuts can be made (Fig. 12.5f–h): Lift the operculum with tweezers and cut as close as possible to the skull. For this, the bones of the operculum must be cut. Afterward, the gills are exposed. Initially, only the first gill is visible, as it covers the following posterior ones (Fig. 12.5f). Before further preparation, observe the anterior boundary of the abdominal cavity, the transverse septum, which separates the abdominal body cavity from the pericard (Fig. 12.5g). Next, extend the dorsal cut to the gills and remove the pectoral girdle with the remaining parts of the body wall. The heart and the head kidney are now exposed (Fig. 12.5h). The pericardium is usually damaged in this preparation step and is no longer recognizable. After these cuts, the whole specimen should be covered with water; this always results in oil droplets, mainly from the more or less extensive adipose tissue (Figs. 12.5e and 12.7a, b). Therefore, from time to time, the water must be changed.

Oncorhynchus: **Preparation Step 3**

Once the described preparation steps have been carried out and the corresponding organs have been observed, the specimen can be placed on its ventral side and the dorsal musculature removed. Once the spine with the spinous processes is reached, carefully puncture the skull capsule from the back and cut it up to the tip of the upper jaw. Either make two lateral cuts and then fold the skull cap off, or make a median cut and only remove the skull side wall on one side. The brain becomes visible in the skullcap (calvarium) (Fig. 12.8c).

Digestive System and Swim Bladder

The intestine is only partially visible in a freshly opened body cavity (Fig. 12.6a). Anteriorly, the relatively large liver is noticeable first, followed by the pyloric ceca. In the posterior part of the body cavity, the lobed, differently developed adipose tissue dominates, depending on the nutritional state. The intestine is spread out for further examination, and the connective tissue suspensory ligaments (mesenteries) are cut. In addition to the digestive organs, the dark red spleen, located behind the stomach and embedded in adipose tissue, is visible (Fig. 12.6b). The intestine continues behind the septum transversum with the esophagus, which merges without a clear boundary into the U-shaped stomach, bending forward (Fig. 12.6b, c). Between the stomach and the adjoin-

ing midgut is the valve-like pylorus, which is not visible from the outside. The tubular pyloric ceca, which may be involved in digestive processes, open directly into the first section of the midgut. (Fig. 12.6a–c). Their number can vary significantly within teleosts, and in some species, these ceca and the stomach can be completely absent. For instance, this is the case in carp fish, Cyprinidae. The often in zoology courses used roach, *Rutilus rutilus*, also belongs to this group. In contrast, up to 200 ceca are present in Salmonidae, including trouts. Their function is not fully clarified.

In addition to the secretion of digestive enzymes, participation in absorption is discussed. In front of this region, the gallbladder, which is more or less hidden between the liver lobes, opens into the midgut (Fig. 12.6e). In some teleosts, the bile secretion is discharged directly into the intestine without a gallbladder being present. Due to the breakdown products of the red blood pigment acting as fat emulsifiers, it is green. The pancreas, which produces protein-degrading enzymes and insulin, is present in teleosts, but is very diffusely developed. It is located in the mesenteries and the connective tissue of the intestinal canal from finely branched canaliculi, which remain undiscovered in our preparation. The liver, as the central metabolic organ, receives the nutrients absorbed from the intestine via the portal vein and, in turn, supplies the body with nutrients via the hepatic vein. The hepatic vein can be displayed with careful preparation; it is quite short, penetrates the transverse septum, and opens into the sinus venosus. From the anterior part of the midgut with the pyloric ceca, this bends backward and runs as a more or less straight tube to the anus. The rectum is hardly separated from the rest of the intestine and is difficult to distinguish.

The swim bladder is located dorsally in the body cavity, above the body's center of gravity (Figs. 12.6b and 12.7a, b). This dorsal position prevents the body from turning over on its back due to the buoyancy of the swim bladder. During ontogenesis it develops as a ventral outgrowth of the foregut at the transition from the pharynx to the esophagus. The wall of the swim bladder is thin and can be easily damaged during preparation, so its structure becomes difficult to recognize. Due to the gas filling, the swim bladder wall, which histologically strongly resembles the intestinal wall, appears silvery-white. The blood vessels are, therefore, easily recognizable (Fig. 12.7b). In the rainbow trout, the swim bladder is a simple, elongated, sac-like structure, which maintains an open connection to the foregut throughout life. This connection is called the pneumatic duct (physostome state; Fig. 12.9a). In addition to the swim bladder epithelium, gas can also be absorbed or released via this duct. In carp-like fish, the swim bladder is divided into an anterior and a posterior chamber; in each lies a part of the gas-releasing epithelium (gas gland), while the pneumatic duct opens into the posterior chamber. This system allows for relatively quick pressure compensation, while, in other teleosts, there is no connection between the intestine and the swim bladder after ontogenesis. In these organisms, gas exchange occurs via the blood and the swim bladder epithelium. This system is somewhat slower, so if the fish rises too quickly to the surface, for example, in a fishing net, the swim bladder can burst—always a fatal injury. This regularly happens when catching cod and other fish in trawl nets, so throwing back small fish or by-catch into the sea often does not result in these animals surviving.

Gills and Circulatory Organs

The respiratory organs of teleosts are the gills, usually covered by the operculum, a plate composed of several bones (Fig. 12.3a, b). The operculum protects the sensitive gills and its movements maintain a water stream for respiration. After spreading the operculum, the gills become visible, colored red by the hemoglobin shining through from the blood vessels (Fig. 12.3h). After removing the operculum with a strong pair of scissors, the individual gill arches with the gill filaments can be seen more clearly (Figs. 12.5f and 12.7c–e). Teleosts have four gill arches, each carrying a double row of gill filaments (Fig. 12.7c–e), basally connected at different lengths, depending on the species. Each gill filament carries the perpendicular attached gill lamellae, which are often not recognizable during preparation and stick to the filaments (Fig. 12.7d). The gill arches carry inward-facing fin rakers in many Teleostei; in *Oncorhynchus mykiss*, these are relatively short and cone-shaped (Fig. 12.7c). They play an important role in food intake.

The heart is the most prominent part of the blood-vascular system (Fig. 12.8a, b). It is located in a separate coelomic space, the pericardium, whose whitish anterior and lateral walls are connected with the septum transversum. The pericardium ensures that the heart's pumping activity is not disturbed by the movement of the animal or its internal organs. The vertebrate heart is always composed of several—primarily consecutive—chambers. In the rainbow trout, like in all teleosts, these chambers are slightly shifted one on top of the other. The deoxygenated blood passes from the body via the sinus venosus, located in front of the heart and sitting dorsally on the pericardium, into the atrium. The atrium is usually dark red and relatively large. From there, the blood enters the main heart chamber (ventricle), which is significantly more muscular and lighter colored (Fig. 12.8a, b). From the ventricle, the blood is pumped into the whitish bulbus arteriosus, which leads into the ventral aorta. The bulbus prevents backflow of the blood into the ventricle. The ventral aorta carries the oxygen-poor blood to the gills and directs the blood via the afferent (supplying) gill arch arteries. The foremost gill arch arteries arise by bifurcation of the aorta. Four pairs of gill arch arteries correspond to the gill arches, numbered with Roman numerals from anterior to posterior

with III, IV, V, and VI. The arches I and II are assigned to the mandibular and hyoid arches. Since these gill arches are incorporated into the jaw apparatus, the corresponding gills are missing in gnathostomes. With careful preparation, the aorta can be displayed in the dorsal position of the specimen.

Kidneys

The excretory organs of teleosts, the kidneys, are located as a pair of elongated organs dorsally on both sides of the spine, between the peritoneum and body musculature (Fig. 12.9a, b). The kidney consists of two sections: the roundish head kidney, which lies in front of the septum transversum, and the trunk kidney, which extends through the entire length of the body cavity as a dark red-violet structure. The trunk kidney only becomes visible when the swim bladder has been removed, and the gonads have been slightly shifted ventrally.

The head kidney corresponds to the so-called pronephros, which is only formed from two nephrons. The head kidney regulates excretion and osmoregulation during the larval phase of the fish. In adults, the head kidney is no longer actively excretory and serves as a lymphoid organ, which presumably plays a significant role in blood formation.

After the larval phase, the functional kidney is the so-called opisthonephros. During development, a pair of nephrons (renal corpuscles) is formed in each body segment. Their ducts merge into the primary ureter (Wolffian duct), which also forms the definitive ureter of the teleosts. Especially in the posterior part, these can be recognized as parallel running, bright ducts on the kidney (Fig. 12.9b). Posteriorly, the two ureters unite into a single unpaired urinary bladder. The bladder opens separately from the reproductive organs to the outside. Depending on the habitat (sea or freshwater), the kidneys show corresponding adaptations to excretion and osmoregulation. The blood supply of the kidney in teleosts is complex: Arteries branch off from the dorsal aorta into the kidneys, and in addition, the kidneys receive venous blood from the renal portal vein coming from the tail. The draining vessels merge into the paired posterior cardinal vein (posterior cardinal veins), which opens into the sinus venosus via the Cuvier ducts.

Reproductive Organs

Teleosts are usually gonochoristic—as is the rainbow trout. However, the sexes cannot be distinguished externally. Hermaphrodites are rare in teleosts; these are then, like the clownfish (*Amphiprion* spec.), usually consecutive hermaphrodites, so the two sexes are formed one after the other. The reproductive organs of teleosts are relatively simple and have no connection to the kidney. Such a connection is common in many vertebrates, including the basal ray-finned fish (Actinopterygii), such as the sturgeons (Chondrostei), gars

(Ginglymodi), and bowfins (Halecomorphi) (Fig. 12.1f–h). The reason for this lies in the modified reproductive biology: Teleosts produce large amounts of gametes, which are released into the open water; the absence of internal fertilization and brood care is a derived feature within teleosts. Thousands of gametes are required for successful reproduction, as the loss of unfertilized eggs and early developmental stages is high. The ovaries are transparent, and the spherical oocytes inside are usually clearly visible (Figs. 12.6a and 12.9a, c–e). Rainbow trout become sexually mature at around 2–3 years, with a life expectancy of around 10–11 years. As we usually buy younger trout for our dissection course, only a few animals will have mature ovaries or testes. However, the paired ovaries of the rainbow trout are suspended dorsally in the coelom by a peritoneal fold. In small and not yet fully sexually mature specimens, the ovaries are relatively small, and eggs are found mainly in the anterior section (Figs. 12.7a and 12.9a, c, d). Some older and sexually mature fish should be at hand to demonstrate the gonads.

In trout, the oocytes are first released into the body cavity via the posterior sterile section of the ovary before they are taken up by the genital funnel, which opens to the outside via a short passage. The latter is usually invisible when dissecting a trout. The testes can be recognized by their milky white color caused by the sperm. Otherwise, the structure of the male sexual organs is similar; however, they open to the outside via paired secondary sperm ducts (vasa deferentia), which unite shortly before the genital opening. In addition to the gametes, the gonads in both sexes also produce sexual hormones.

Nervous System, Sensory Organs

After the described preparation steps and observations have been carried out, the brain can be prepared (Fig. 12.8c). This shows the characteristic serial structure of basal craniotes: The anteriormost part is the relatively small telencephalon with the two hemispheres (Fig. 12.8b). In front of the telencephalon and slightly lower lie the two olfactory bulbs. A small stalk-shaped dorsal appendage, the pineal organ, unfortunately usually lost during dissection, extends from the diencephalon. This is followed dorsally by the mesencephalon with the two large lobes of the tectum opticum, the largest area of the brain in trout. The optic nerves converge here, where the superior brain center is located in all teleosts. The metencephalon with the dorsal cerebellum then follows, and the myelencephalon makes up the last part, consisting of two rather inconspicuous spherical areas (Fig. 12.8d). The myelencephalon continues into the spinal cord without a clear boundary. The cerebellum is mainly responsible for the coordination of position in space; here, sensations from the vestibular system and the lateral line converge. If you now carefully cut into the brain, it becomes clear that it is a hollow structure, and you can open the brain ventricles. Ontogenetically these arise during the folding of the nervous

system from the dorsal ectoderm as the neural tube. In the brain, this cavity persists as a system of ventricles. The nose, eye, and labyrinth organ have already been mentioned above. The spherical lenses can be prepared from the large eyes for a detailed microscopic inspection. Unlike in terrestrial species, accommodation is not achieved by deformation of the lens, but by its displacement. The finer eye structure would then have to be examined in histological preparations. The vestibular organ, which lies next to the cerebellum, is usually damaged during preparation and more or less destroyed.

12.2 The Laboratory Rat (*Rattus norvegicus*, Mammalia, Mammals)

Less than 0.5% of all animal species living on Earth are mammals. Nevertheless, they are one of the most successful groups of animals to have colonized all habitats. Furthermore, the majority of domesticated animals are mammals, and, therefore, very important to human history. With only a few exceptions, all "large" animals living on the Earth are mammals. When animals, in general, are mentioned in everyday life, it is usually in reference to mammals, and they play a particularly important role for us, because our own species belongs to this group.

One of the most meaningful events in the evolution of Craniota was the conquest of land in the Devonian period, which began about 365 million years ago. This allowed vertebrates to colonize various new habitats and tap into new food sources. This was accompanied by numerous changes in the vertebrate body plan and the acquisition of new properties. As an example of a mammal, we present the laboratory rat, also known as brown rat (*Rattus norvegicus*).

The primarily aquatic vertebrates most closely related to land vertebrates are all extinct. We only know them through the fossil record. According to current knowledge, the lungfish (Dipnoi) are the sister group of all land vertebrates among the recent fishes. (Fig. 12.2). The conquest of land required numerous changes in anatomy and physiology to colonize this entirely new and different habitat successfully. Some adaptations, such as lungs, were already present in the aquatic ancestors of Tetrapoda, while others have newly emerged during this phase. The transition to terrestrial life required a restructuring of the locomotor system, respiration, excretion, reproduction, and the development of effective evaporation protection, since water was no longer unlimitedly available in the new habitat. Not all adaptations are perfect in the same way in all terrestrial vertebrates. For example, only amniotes (Amniota) reproduce completely independently of water and are the only tetrapods possessing effective evaporation protection.

For this reason, they are also referred to as the "true" terrestrial vertebrates. Amniota include Sauropsida (= Reptilia), comprising   Chelonia   (turtles),   Lepidosauria   with

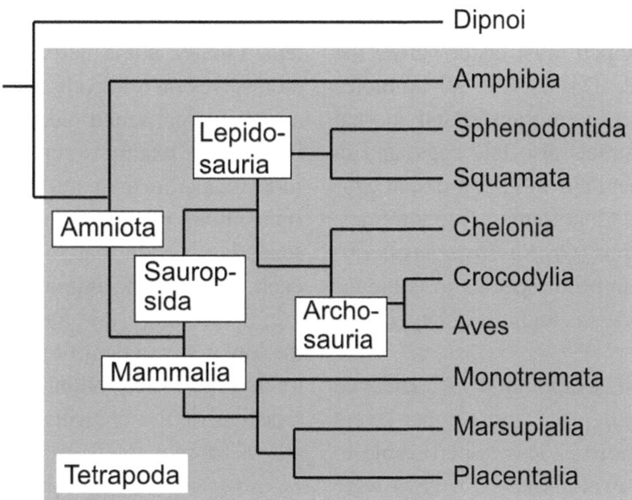

Fig. 12.10 Phylogenetic relationships of terrestrial vertebrates, Tetrapoda. Tetrapoda is monophyletic and comprises Amphibia and Amniota as the highest ranked sister groups. Sauropsida = Reptilia and Mammalia are the highest-ranked sister groups within Amniota; note that Reptilia also includes Aves (birds). Squamata (lizards and snakes) is the largest lepidosaurian subtaxon, with only snakes (Serpentes) being monophyletic. The three mammalian subgroups, Monotremata (echidnas and platypus), Marsupialia (marsupials), and Placentalia (placental mammals), form a grade. Only major tetrapod groups were included, and extinct taxa were omitted; after morphological and molecular data and combined from various sources

Rhynchocephalia (beak-heads, lizard-like animals with only one living species, the Tuatara (*Sphenodon punctatus*) of New Zealand), and Squamata (lizards and snakes), Archosauria with Crocodylia (crocodylians) and Aves (birds), as well as the sauropsid sister group Mammalia (mammals) (Fig. 12.10). It should be noted that the "classical" taxon Reptilia is paraphyletic, and in modern (phylogenetic) systematics, Aves, birds, is an ingroup of the former. To avoid confusion, many authors instead prefer the name Sauropsida and to abandon the misleading term Reptilia.

Mammals and birds, especially, have reached a comparably high level of adaptation to terrestrial life, allowing for colonization of all habitats. Furthermore, amniotes have repeatedly succeeded in conquering the sky (†pterosaurs, birds, and bats), as well as in returning to the water (e.g., †ichthyosaurs, †plesiosaurs, sea turtles, penguins, whales, manatees, and seals).

The ancestors of Tetrapoda had a typical streamlined fish-like shape, and were thus optimally adapted for life in water. In the course of evolution and adaptation to terrestrial life, a series of significant changes in body structure occurred:

1. Transformation of the paired extremities (fins) into legs, structures that can at least temporarily lift the body off the ground and carry it.
2. Separation of the pectoral girdle from the skull. This allowed to move head and extremities independently.
3. Regionalization of the spine and its conversion into the "supporting bar" of the body. This includes the formation of vertebrae with compact (one-piece) vertebral bodies (centra) and the differentiation of a neck region, allowing for independent head movement relative to the trunk.
4. Dorsal extension of the pelvic girdle and its connection with the spine.
5. Adaptation of the acoustic organ to air. The hyomandibular bone undergoes a functional change and becomes the only auditory ossicle initially present (now called columella). The hyomandibular bone is the uppermost component in a set of bones that forms the second jaw arch. The columella now transmits sounds from the air to the fluid-filled labyrinth (cochlea); thus, it works as an impedance transformer.
6. An increasing separation of the body's circulation from that of the lungs in the circulatory system. Birds and mammals are the only representatives with a complete separation of oxygenated from de-oxygenated blood, whereas, in all other tetrapods, there is a certain degree of mixing of the oxygen-rich blood from the lungs with oxygen-poor blood from the body.
7. A complete reduction of the gills and sole conversion to lung respiration.

Only the evolution of the so-called amniotic egg, with its four extra-embryonic organs—amnion, allantois, chorion, and yolk sac, as well as a heavily keratinized, and therefore water-impermeable, epidermis—made it possible to separate all aspects of life from the water. In amniotes, the eggs laid by females are primarily enclosed by a rigid shell.

Consequently, embryonic development takes place outside the female's body. This reproductive mode inevitably requires internal fertilization, which must occur before the oocyte is enclosed in the shell. This is why all amniotes reproduce, to this day, with direct sperm transfer and internal fertilization. Even the first mammals also laid eggs, and in contrast to the popular belief, there are still a few extant egg-laying mammal species, namely, Monotremata (monotremes, cloacal animals), with the platypus (*Ornithorhynchus anatinus*), the duck-billed platypus and four species of echidnas (spiny anteaters, with three *Zaglossus* and one *Tachyglossus* species).

Effective protection against evaporation is achieved through a heavily keratinized epidermis, whose upper layers consist of dead cells that are more or less impermeable to water. Only through these two evolutionary novelties were all habitats on earth open to amniotes. Within birds and mammals, particularly adapted species have survived even in very cold regions. This was made possible by an insulating body covering of feathers or hair and an internal heat source (endothermy, homeothermy). Finally, some species, such as seals and whales, have found their way back to the sea, so mammals naturally occur in almost all habitats today. Only inner Antarctica, the deep sea, and some smaller islands have not been reached by mammals.

Birds and mammals are the only groups within amniotes that are homeothermic or, more precisely, endothermic; that is, their metabolism allows them to maintain a constant body temperature. Accordingly, they have an aerobic metabolism, especially in their musculature. Mammals have an insulating fur that protects them from excessive heat loss. The muscles also provide most of the heat required to compensate for heat lost to the environment. On the other hand, if ambient temperatures become too high, endothermy also requires adaptations to avoid overheating of the body. The constant and relatively high body temperature enables maximum physiological performance, but it also has disadvantages. Mammals and birds have energy requirements about ten times higher than those of similarly sized ectothermic vertebrates. Mammals, for example, can run long and enduringly, while this is impossible for squamates (for example, lizards), with their predominantly anaerobic muscles. Even though some textbooks talk about the "era of mammals," the number of their species is not as large as one might assume with respect to this statement. There are only about 5400 extant species, only about half the number of bird species. This reflects a strongly anthropomorphic view of biology and evolution.

Several evolutionary innovations were decisive for the success of mammals. A different reproductive strategy, the remodeling of the teeth, the jaws, and the palate, as well as a new form of locomotion that goes hand in hand with a change in the entire locomotor system. The changes in the feeding apparatus are indeed linked to endothermy, as the high energy requirement requires the intake of relatively large quantities of food, coupled with an effective digestive system. Finally, some herbivorous mammals can also sustain themselves on relatively poorly digestible food such as grass and have particular adaptations for its digestion. These include, for example, ruminants, which can use cellulose as food thanks to their multipart stomach and microbial help. Innovations affecting all mammals include, for example, a secondary palate that shifts the internal nasal openings far back, allowing the animals to breathe while eating and chewing; a secondary jaw joint with the extreme remodeling of the lower jaw; a dentition consisting of different specialized tooth types (heterodont dentition); and a comprehensive redesign of the chewing muscles (Fig. 12.11a–e). All this allows for movement of the lower jaw in three axes (x, y, and z), as opposed to just opening and closing of the jaw, as the primary joint allows for in sauropsids, so that the food in the mouth can be crushed.

The teeth can be assigned to four types and are firmly anchored in the jaw with one or more roots: incisors (dentes incisivi), canines (dentes canini), premolars (dentes praemolares), and molars (dentes molares) (Fig. 12.11a, c, e). The deciduous teeth (also known as baby or milk teeth) are formed only from the first three tooth types and are replaced by permanent teeth during the juvenile phase. This one-time tooth change is called diphyodont and is a unique feature of mammals. The primary dental formula is probably 3, 1, 4, 3 in each quadrant, for a total of 44 teeth. The teeth and their respective numbers are so specific that, often, a single molar is enough to identify the subgroup or sometimes even the species. In connection with this change, two of the former lower jaw bones are relocated as auditory ossicles into the middle ear. Together with the hyomandibular, quadratum, and articular form, the three auditory ossicles are now referred to as the stirrup, the anvil, and the hammer. The lower jaw is formed only by a single bone, the dentale, which protrudes with a long extension far up into the zygomatic arch (Fig. 12.11b–e). This extension, among other things, accommodates the chewing muscles, which have changed significantly compared to sauropsids.

Mammals show a strong regionalization of the body with a head, a trunk and a tail. The head carries the brain, protected by a skull, and major sensory systems such as eyes, ears and the nose. The head is distinctly separated from the trunk by the neck region, which has in mammals an important function allowing independent moving of head against the trunk. This is best seen, when we study a mammalian skeleton. The neck is always formed by exactly seven vertebrae. The first two vertebrae, the atlas and axis, enable the nodding and turning movements of the head (Fig. 12.11d–f). Thus, two joints are formed: one between the skull and the atlas and one between the atlas and axis. Sauropsids have these head movement possibilities combined in only one

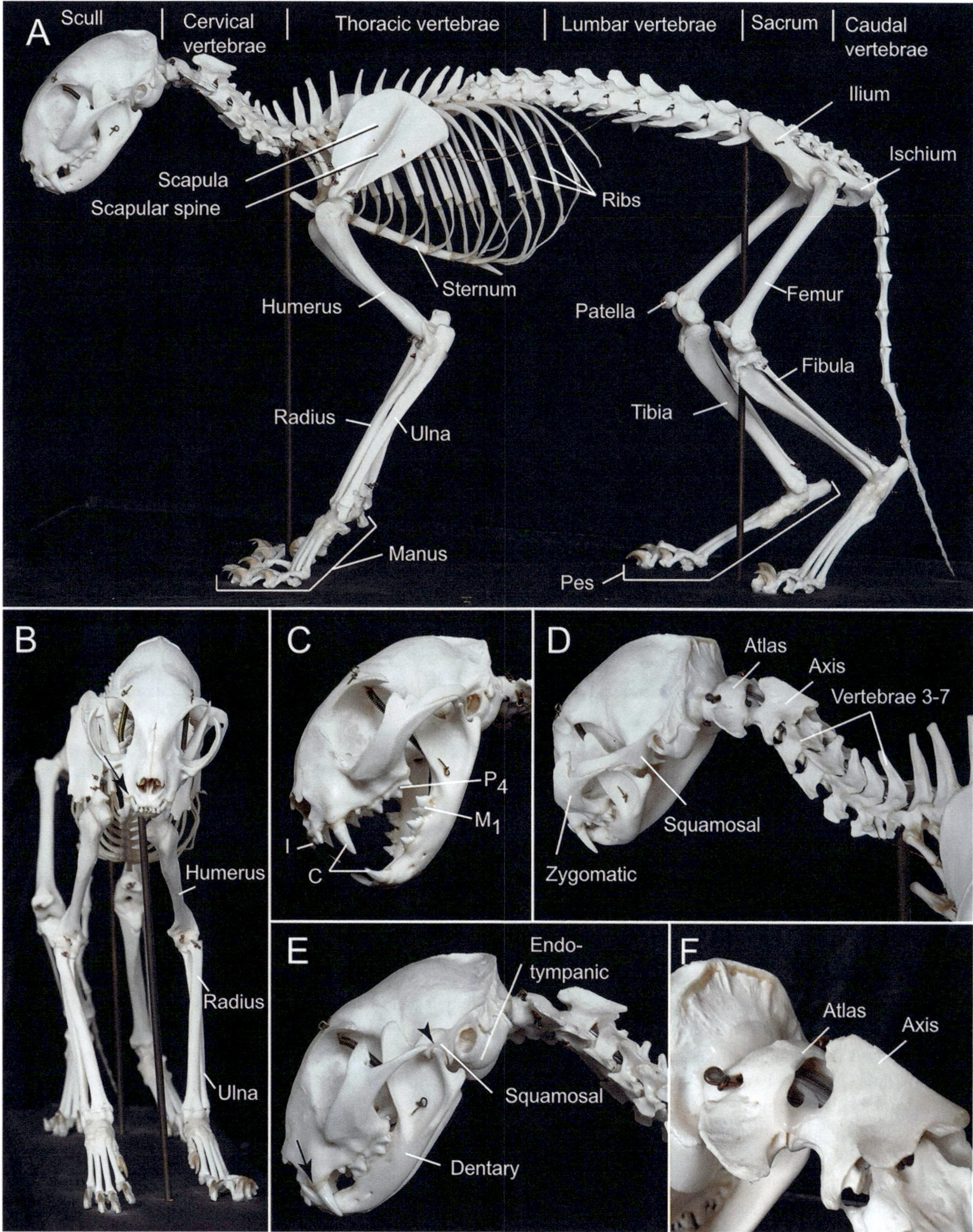

Fig. 12.11 *Felis silvestris* f. *catus*, Domestic cat. Example of a placental mammal skeleton. (**a**) Side view; subdivision of the spine, position, and posture of the extremities. (**b**) Front view, extremities are vertical and close to the body. Due to the rotation of the forelimb forward, radius, and ulna cross. (**c**) Head with open jaw, side view; characteristic carnivorous dentition with long canines (**c**) and large premolar 4 (P_4) and molar 1 (M_1), which form the carnassial pair. (**d**) Head and neck region obliquely from behind; atlas and axis form a double joint, each carrying muscle attachment surfaces offset by 90 degrees. Lower jaw (dental) with a long extension that reaches into the zygomatic arch (composed of jugale and squamosum). (**e**) Heterodont dentition with long canines that slide past each other when the jaw is closed (arrow). Secondary jaw joint (arrowhead). (**f**) Atlas-axis complex. The Department of Behavioral Biology at the University of Osnabrück kindly provided the skeleton

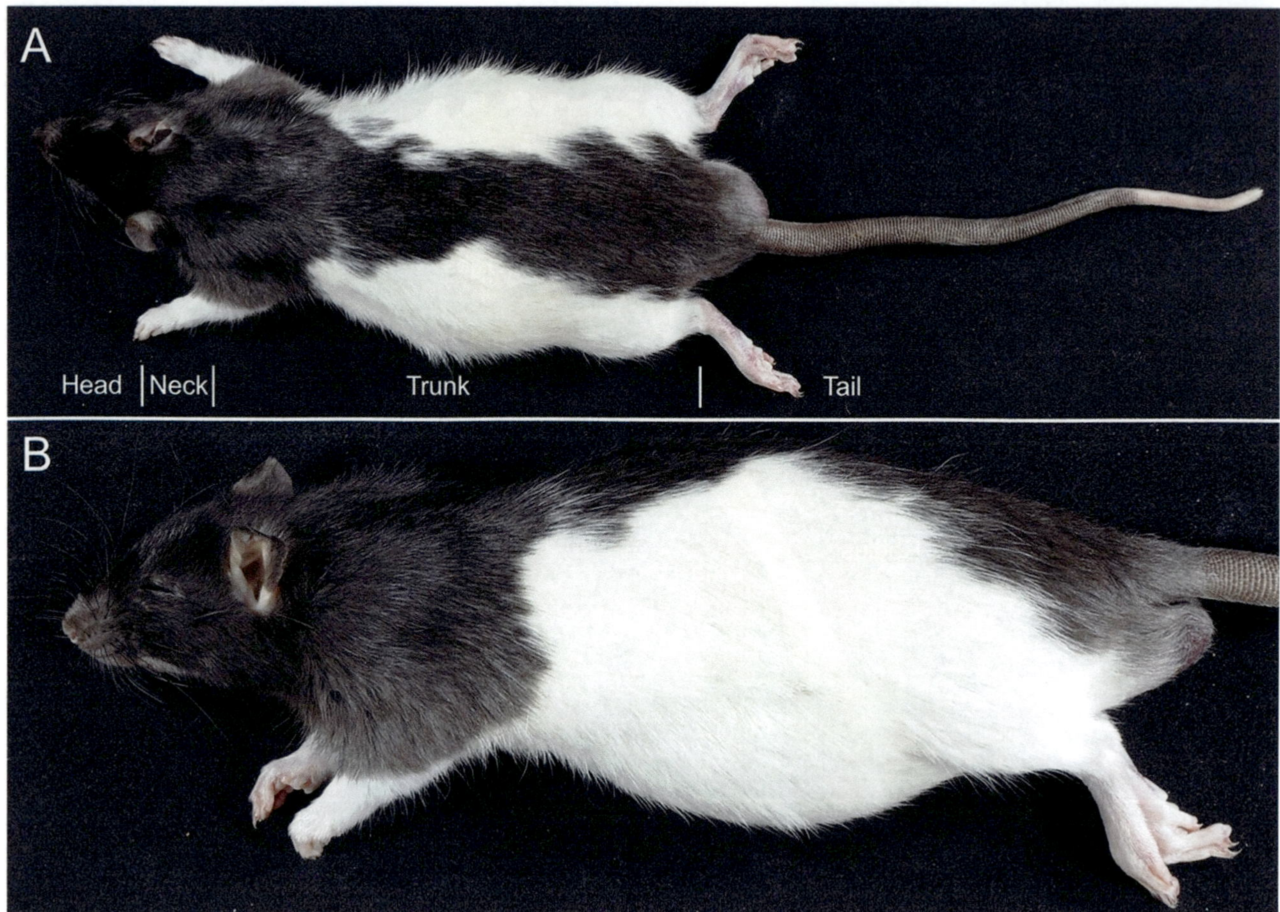

Fig. 12.12 *Rattus norvegicus*, Laboratory rat. Habitus and external morphology of a male animal. Except for the tail, the animal has a dense and complete coat of hair, which is white and black in this color variant. As inhabitants of caves and tunnels, the rats' limbs are relatively short. (**a**) Whole animal in dorsal view; typical body divisions of tetrapods into head, neck, trunk, and tail. Partition of the trunk into the thorax and abdomen is not recognizable. (**b**) Lateral view

joint between the skull and the atlas, with only one rod end at the atlas-occipital bone joint, instead of two found in mammals. By the way, the manatees (Trichechidae) and Hoffmann's two-toed sloth (*Choloepus hoffmanni*) are the only mammals with only six cervical vertebrae. Three-toed sloths (*Bradypus* spec.) have displaced thoracic vertebrae, previously mistakenly interpreted as additional cervical vertebrae. In birds, the number of cervical vertebrae is much higher, it varies between 10 and 31.

The trunk is divided into thoracic, abdominal, and pelvic regions (Fig. 12.12a); the diaphragm internally separates the first two. Ribs only occur in the thoracic region. The extremities are rotated laterally under the body so that they are closer to the center of gravity and can swing forward and backward (Fig. 12.11a, b). At the same time, the trunk muscles are relieved of generating a forward thrust component; that is, the spine no longer moves in the horizontal plane, as in lizards and snakes, but rather in the sagittal plane. These adaptations are essential for the fast and enduring running ability

of mammals. This rearrangement of the body musculature is why swimming movements (and the fluke's orientation) in whales are perpendicular to that of fish.

The skin of mammals is generally rich in glands; essential glands are the mammary glands, which are only developed in females and whose secretion serves to nourish the young. In the ground pattern of sauropsids, significantly highly developed young animals hatch from large eggs and can often feed independently. This applies, for example, to lizards, geese, and ducks. In contrast, the mammalian hatchlings are small and underdeveloped. They, therefore, have to be nourished by their mother's milk for a comparatively long time. In steppe and open land dwellers, but also marine mammals, precocial young emerged secondarily, for example, in elephants, antelopes, perissodactyls (horses, tapirs, rhinos), and whales.

The cardiovascular system also changed; mammals achieved these changes independently of birds. Associated with a complete separation of the pulmonary and systemic circulation, the heart consists of four separated chambers, two atria, and two ventricles. The right atrium receives the blood from the body and pumps it via the right ventricle into the lungs. From there, oxygen-rich blood flows into the body via the left atrium and left ventricle. In addition, mammals have erythrocytes without cell nuclei, thus creating more space for hemoglobin in the blood cells.

Dissection and Observations

As an example of a mammal, the laboratory rat from the taxon rodents (Rodentia) will be presented here. With about 2300 species, rodents are the largest and most widespread group of recent mammals. Most species are between 5 cm (mouse) and 25 cm (rat) in size, measured from head to body. *Hydrochoerus hydrochaeris* (capybara) is by far the largest rodent, with a size of 130 cm and 60 kg, closely followed by the European beaver (*Castor fiber*), which can weigh up to 30 kg and, in exceptional cases, even up to 45 kg.

> **Recommended Material**
>
> The laboratory rat (*Rattus norvegicus*) is suggested for studying Mammalia for a basic zoology course. Other species from breeding farms, such as guinea pigs, could also be used, depending on availability. Smaller species, like laboratory mice, are not as well suited for such a course, due to their body size. Breeding farms deliver frozen rats. They were usually anesthetized and killed by introducing CO_2 or ether. The preparation should take place in large dissecting dishes.
>
> In principle, any medium-sized species is suitable for examining a skeleton; one should proceed according to availability.

The laboratory rat originated from selective breeding of the brown rat (*Rattus norvegicus*), and goes back to the beginning of the twentieth century. It thus represents one of the youngest animals domesticated by humans. Compared to the wild form, laboratory rats are usually a bit smaller. Besides purely white animals, there are various piebald forms, with white usually accompanied by brown, gray, or black (Abb. 12.12). Rats are often used as experimental animals, as feed animals for other mammals or snakes, but also—like golden hamsters—as real pets. In these rats, special color variants have been bred. The fur of the wild form has a gray-brown to brown-black, slightly mottled color. From Asia, brown rats colonized all continents, except Antarctica, as synanthropic species with humans after the similar black rat (*Rattus rattus*). In some places outside their natural distribution area, rats often cause massive problems in the local ecosystems. Although rodents and brown rats are omnivores, problems for the native fauna in Australia and New Zealand have arisen. The spread of the brown rat occurred mostly passively, resulting from "stowaways" on ships; by now, it has largely replaced and displaced the black rat worldwide. Rats are highly social animals that live in groups of up to 60 individuals; the individual group members recognize each other primarily by smell. Black rats and brown rats were primarily involved in the spread of devastating plague epidemics, as the disease-causing bacteria (*Yersinia pestis*) are transmitted from rats to humans by various flea species.

External Morphology

Rats are relatively small and short-legged, quadruped animals that use burrows or other preexisting cave systems, which explains the typical habitus (Fig. 12.12a, b). The animals can climb exceptionally well and move through narrow spaces and crevices. The head, neck, trunk, and tail regions of the body can be easily distinguished. In contrast, the further division of the trunk into thoracic and abdominal regions is externally somewhat unclear. The entire body is covered with hair. One distinguishes the longer, finer bristle or contour hairs from the shorter, finer wool hairs, which can be seen upon closer examination. In addition, the rat also has whiskers (vibrissae), which form an important mechanical sensory system and are especially relevant in the close range (Fig. 12.13a–c). Most vibrissae are located in the upper lip region and arranged in a specific pattern (Fig. 12.13a), but they can also be found on the forelimbs (Fig. 12.13c). Only the soles of the feet are hairless (Fig. 12.13c, d). The tail also carries short hairs next to the characteristic scales (Fig. 12.13e). The vibrissae are found in all placental mammals (Placentalia, Eutheria) and marsupials (Marsupialia), but are absent in monotremes.

On the head, besides the vibrissae, the sensory organs, nose, eyes, and ears are noticeable (Figs. 12.12a, b and 12.13a, b). The large, movable funnel-shaped auricles, with their more or less numerous inner folds, allow for very precise spatial hearing. For example, foxes and coyotes can precisely locate a mouse's movements through thick snow cover, and thus successfully prey on these animals in winter. This spatial hearing has largely been lost in humans. The same applies to the olfactory sense, which in rats, as in most mammals, is essential for orientation and intraspecific communication and is only rudimentarily present in humans. Looking at the head from the front or below, the usually yellowish incisors become visible (Figs. 12.13b and 12.14a). There is only one incisor per quadrant (tooth formula for all quadrants: 1, 0, 0, 3). Behind the incisors is a distinct tooth gap; this structure, called a diastema, is characteristic of many herbivores and has evolved convergently several times within mammals. When rodents use their incisors, they push the lower jaw forward. Behind the incisors, the hairy upper

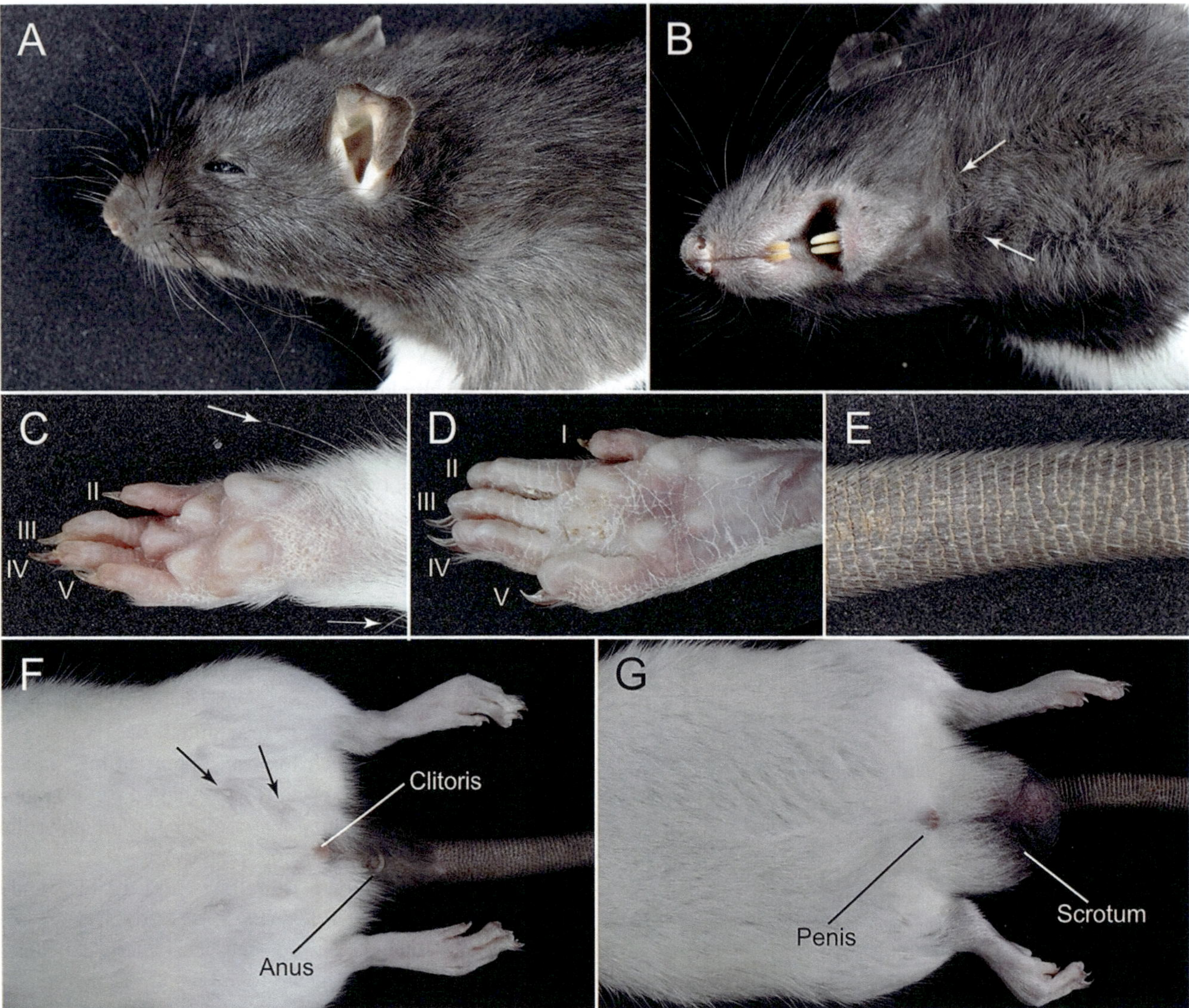

Fig. 12.13 *Rattus norvegicus*, Laboratory rat. Habitus and external morphology. (**a**) Side view, head, sensory organs: nostril, vibrissae (whiskers), eyes, and ears with movable auricles. The vibrissae are of different lengths and are particularly common in the nasal area (upper lip) (about 50–60 on both sides of the nasal area). (**b**) Mouth with incisors in the upper and lower jaw, a clear gap between the two teeth; on the lower jaw, a small group of vibrissae are recognizable (arrows). The rhinarium, the usually moist, bare surface of the tip of the nose, is reduced, a median groove (philtrum) runs from the rhinarium to the oral cavity; the hairy skin of the upper lip extends into the mouth. (**c**) Hand (forepaw) from below, tread surface hairless with sensory pads; roman numbers denote the corresponding finger rays, the first (thumb) is missing, functionally replaced by the prepollex (II); further up, there are whiskers on the hand (arrows). (**d**) Foot with five toe rays (I–V). (**e**) Close-up of the tail with scales and short hairs in between. (**f**) Female, genital region from the ventral; only the two posterior pairs of teats are visible (arrows). (**g**) Male, genital region

lip folds inward (typical for rodents). The incisors are characterized by two special features: (1) They have an open root and grow continuously to compensate for wear and tear. (2) Only the front side is covered by enamel (the hardest substance that exists in vertebrates), while the back side consists of the softer dentin. Thus, the incisors wear down more at the back than the front and remain permanently sharp. If rodents (rats, guinea pigs) are kept as pets, one must provide enough hard materials to avoid vet visits to trim the incisors. Canines mainly serve to hold the prey and, during the forced closure

of the jaws, ensure the precise interaction of the upper and lower jaw teeth in predators and omnivores. Their absence in rodents is probably mainly due to the grinding movements required for food crushing. Accordingly, the enamel cusps on the molars are connected to ridge-like structures (the so-called lophodont teeth), which one can confirm by carefully opening the mouth. In many Muridoidea (mouse-like rodents), the rodent subgroup to which the rat belongs, the molars also have open roots and continuous growth so that wear and growth balance each other out. Through a compli-

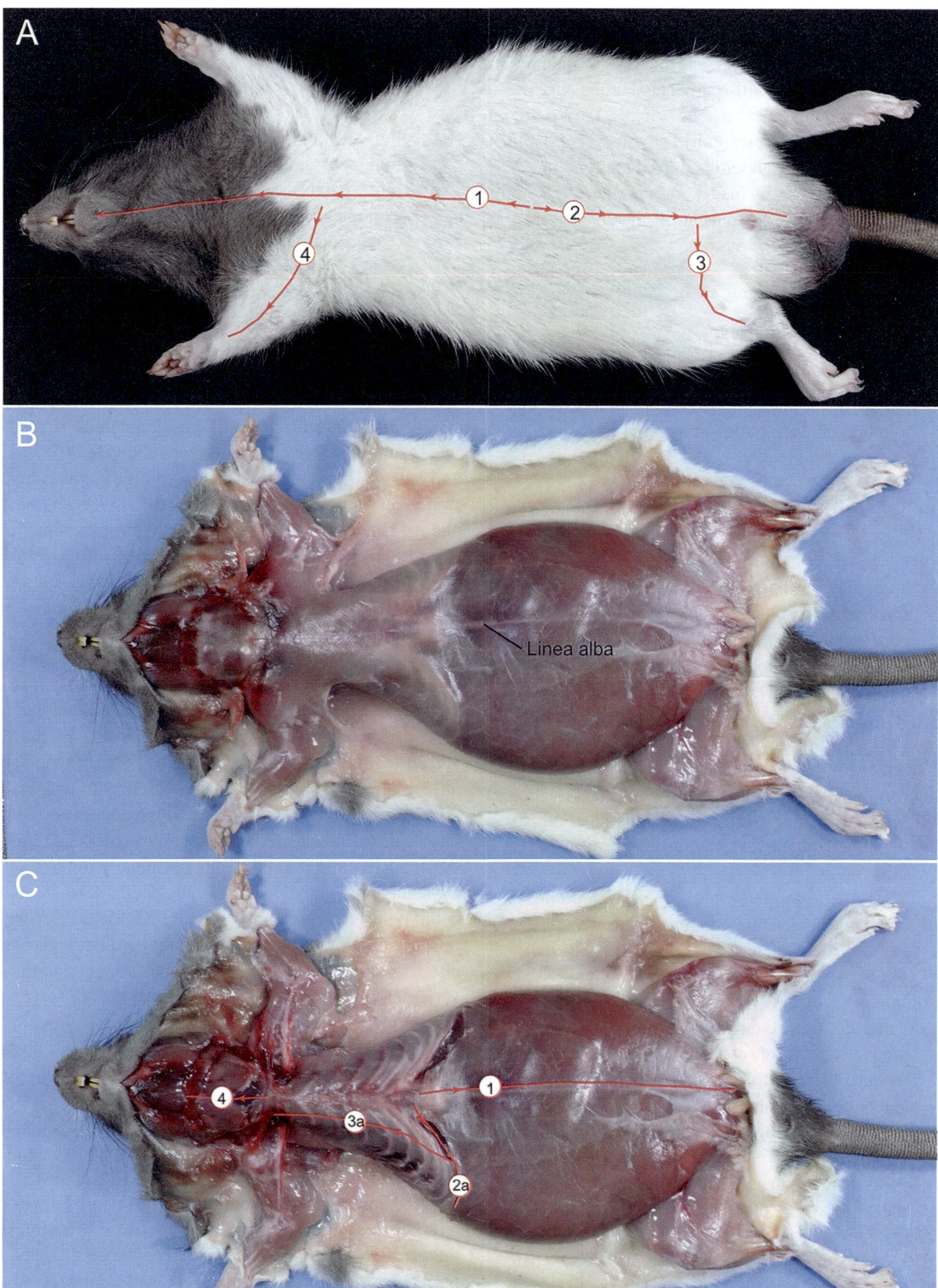

Fig. 12.14 *Rattus norvegicus.* Preparation steps. (**a**) Ventral view and position of the animal before dissection (male). For the initial cuts 1 and 2, lines, arrows, and numbers mark their location, direction, and sequence. Cuts 3 and 4 are to be performed similarly on the left side of the body. (**b**) Female after removal of the fur according to the dissection in A. (**c**) The same specimen after the removal of the musculature on the chest, forelimbs, and neck. The internal organs can now be exposed. First, open the abdomen along the tendinous linea alba (cut 1); through cut 2a (also performed analogously on the other side of the body) along the last rib, the abdominal wall can then be opened, and the organs of the abdomen are exposed. To open the thorax, the ribs are cut according to cut 3a on both sides, so that the sternum with a part of the ribs can be lifted. Only perform cut 4 after examining the salivary glands. This exposes the trachea with appendages

cated pattern of enamel edges and dentin, the molars are equipped with numerous self-sharpening cutting edges.

The forelimb only has four toes (Fig. 12.13c); here, the first ray has been lost. The hindlimb has five toes (Fig. 12.13d), the original number for all tetrapods. All toes end in a horn claw. The contact surfaces (pads) are more heavily keratinized, thereby thickened, and thus mechanically protected.

Dissection of the Rat
Rattus: **First dissection step**

(a) To remove the fur, lift it slightly in the middle of the belly and pierce the skin with sharp scissors without damaging the underlying musculature. The first cut (1) is then made up to the tip of the snout (Fig. 12.14a).

(b) The second cut (2) runs from the puncture site backward. Make the cut on the side of the penis or clitoris (Fig. 12.14a). Now, you can carefully separate the fur from the underlying musculature (e.g., with the blunt side of the scalpel). To proceed, cut along the limbs (cuts 3 and 4). After that, the skin can be separated from the rest of the body and pinned to the bottom of the dissection tray (Fig. 12.14b). The musculature of the ventral side is now fully visible. The muscles are carefully removed in the thorax and neck area to make the ribs and salivary glands visible (Fig. 12.14c). Proceed carefully to avoid damage to the salivary glands and lymph nodes in the lower jaw and neck area.

(c) Then, the abdominal wall is opened according to cut 1 along the ventral midline (linea alba); the cut is made along the tendinous linea alba and continues to the pubic symphysis.

(d) Initially, cut only up to the tip of the sternum. Then, cut open the abdominal wall along the rib arch (also perform cut 2a on the other side); do not damage the diaphragm. Now, the abdominal wall can be folded open on both sides, and the internal organs of the abdomen become visible (Fig. 12.15a, b).

(e) After that, the rib cage can be opened with a strong pair of scissors. A lateral cut is made from back to front on each side (perform cut 3a on both sides of the body), until you reach the collarbones. Then, connect the two cuts at the back close to the rib arch and cut through the diaphragm. Then, the ventral tissue can be folded up and completely separated at the front. Finally, cut through the col-

larbones and carefully expose the neck organs according to cut 4. The result is shown in Fig. 12.17c. You can separate these two preparation steps as described in time or perform them immediately, one after the other. Others prefer to examine the organs of the thorax first, as the smell that becomes noticeable when opening the abdominal cavity does not seem so strong and does not linger quite as long.

Digestive System

The digestive tract of mammals consists of the oral cavity, esophagus, stomach, small intestine, large intestine, and rectum, some of which can be further subdivided (Figs. 12.15 and 12.16). The salivary glands open into the oral cavity, and the two most important glands of the digestive tract, the liver, and the pancreas, open into the front part of the small intestine, the duodenum. The sac-like cecum is at the transition from the small intestine to the large intestine. During the preparation of the anterior body, the powerful masticatory (chewing) muscles that insert on the lower jaw are noticeable. The jaw opener, the digastric muscle or musculus digastricus, is located medially (Fig. 12.15a, c). On the outside, we see the masseter muscle, or cheek muscle, the musculus masseter. The masseter muscle is one of the three adductor muscles of the lower jaw. The musculus masseter is an evolutionary novelty of mammals and is essential for sucking, among other things. The lymph nodes (median), the three pairs of salivary glands, the sublingual gland (glandula sublingualis), the parotid gland (glandula parotis), and the submandibular gland (glandula submandibularis) are located behind the chewing muscles (Fig. 12.15c, d). After the removal of the muscles, the ducts of the salivary glands can also be seen (Fig. 12.15d). The mucous secretion of the salivary glands moistens the food and makes it slippery; at the same time, the secretion also contains digestive enzymes. The esophagus crosses the larynx with the trachea and runs from there dorsally, so it remains initially invisible during dissection from the abdominal side (Fig. 12.15c, d). Behind the diaphragm, the esophagus then opens into the stomach.

After the opening of the abdominal cavity, the liver, located directly behind the diaphragm, and the intestine, with its numerous loops held in their position by the mesentery, become noticeable (Fig. 12.15a, b). Beside the intestine, the pancreas becomes visible, extending across the abdominal body side behind the liver. The small intestine mainly occupies the remaining visible part. Depending on the dissection, the stomach, parts of the large intestine, colon, or the cecum, which is quite large in rats, can also be seen (Fig. 12.15b).

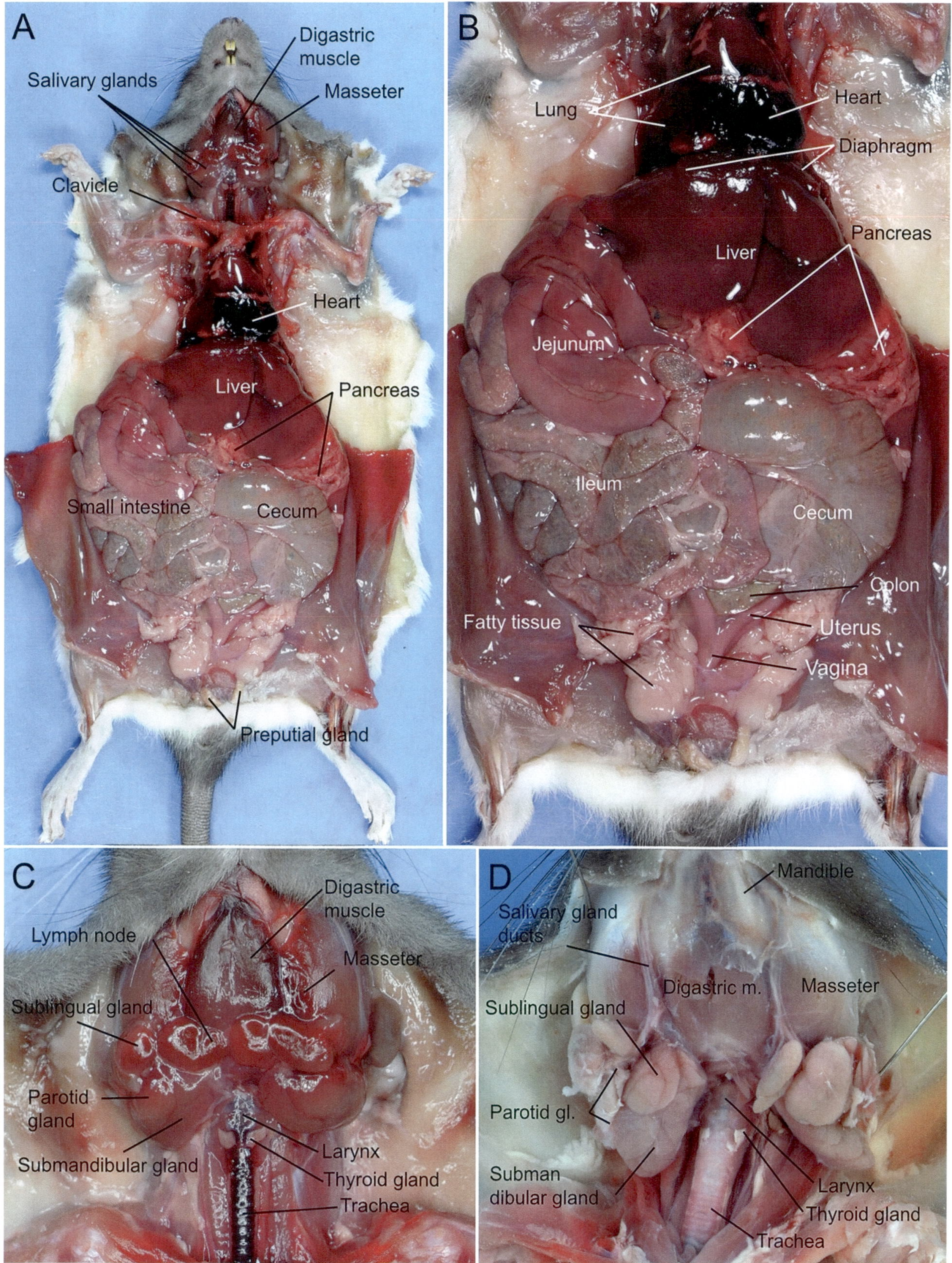

Fig. 12.15 *Rattus norvegicus*. Preparation of the internal organs, female. (**a**) Position of the organs after opening the abdominal cavity and the thorax. (**b**) Close-up of the abdomen; the intestine with numerous loops, held in position by the mesenteries. (**c**) Masticatory muscles and salivary glands. (**d**) Salivary glands, larynx, and trachea after removal of the neck muscles

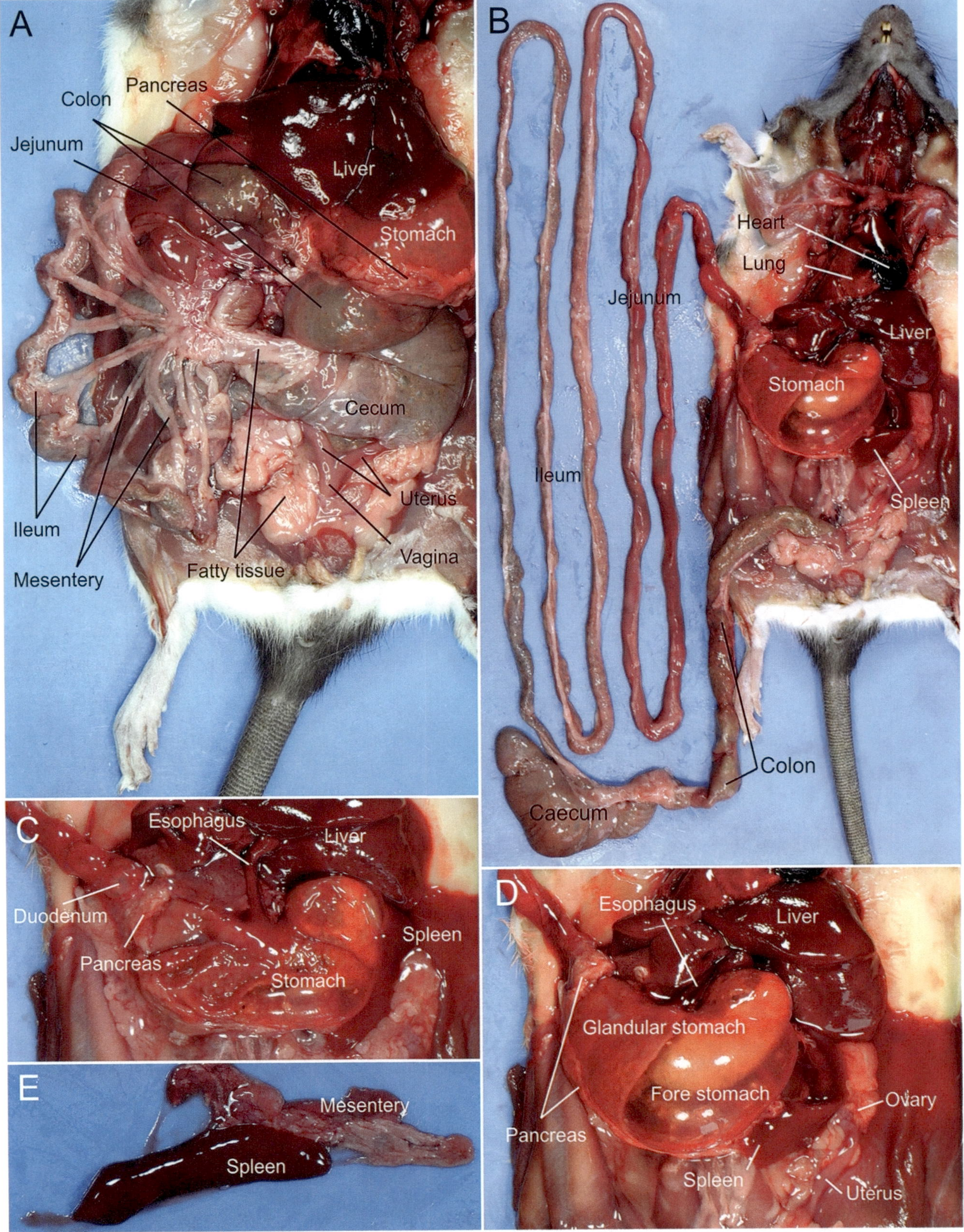

Fig. 12.16 *Rattus norvegicus*. Digestive tract, female. (**a**) Digestive tract and abdominal organs. The mesentery with blood vessels and adipose tissue are clearly visible in the ileum. (**b**) Digestive tract after cutting the mesenteries; the small intestine reaches several times the body length. (**c**) Opening of the esophagus into the stomach. (**d**) Stomach region, duodenum, and appendicular organs. (**e**) Dissected spleen and mesentery

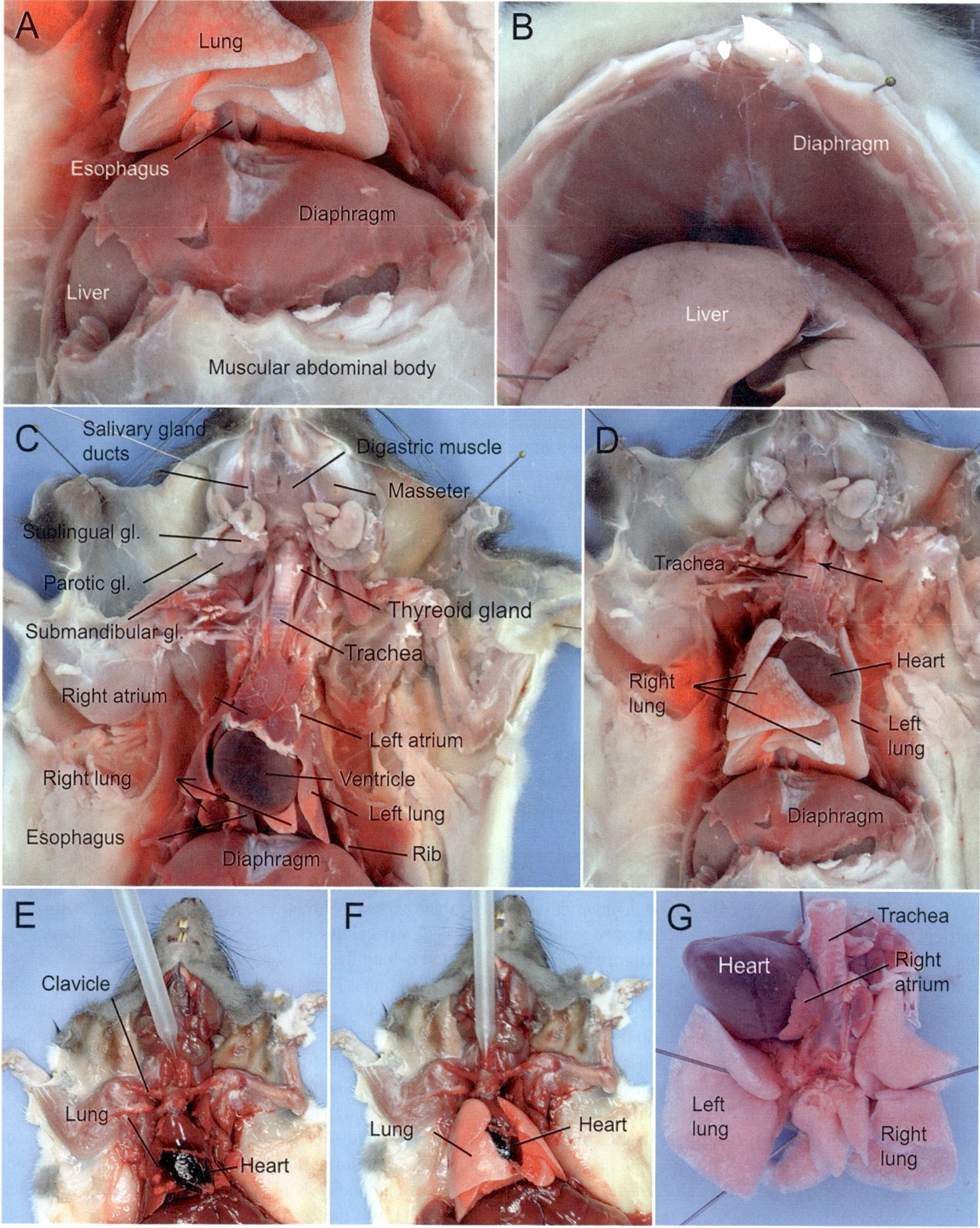

Fig. 12.17 *Rattus norvegicus*. Respiratory and circulatory organs. (**a, c, d, g**) Male, (**b, e, f**) Female. (**a**) Diaphragm from anterior, abdominal wall closed, lung partially unfolded. (**b**) Diaphragm from posterior, thorax still closed. (**c**) Position of salivary glands, larynx, trachea, lungs, and heart. (**d**) Close-up of C, arrow points to cut to open the trachea. (**e**) Insertion of the Pasteur pipette into the trachea. (**f**) Unfolded lungs after air supply through the pipette. (**g**) Heart and lungs separated, view from the dorsal

Rattus: **Second Dissection Step**

(a) For further dissection, the small intestine is now carefully exposed, initially without cutting through the mesenteries (Fig. 12.16a).

(b) Later, the intestine is completely removed from the body cavity (Fig. 12.16b). Below the stomach, the spleen becomes visible, which can then also be removed, together with the intestine (Fig. 12.16b, d, e).

The esophagus opens approximately at the level of the body's center into the bean-shaped stomach, and this section is referred to as the cardia. To the left (from the viewer's perspective to the right) is the sac-like cecum (Fig. 12.16a–d). The cardia is an intermediate region and mainly contains mucus-producing glands. At the same time, the glands of the corpus and fundus (the uppermost portion of the stomach) regions of the stomach also produce gastric acid and proteolytic enzymes. Gastric acid has a pH between 1 and 3 and plays a key role in the digestion of proteins. It activates proteolytic enzymes, which together break down the proteins. The epithelium of the stomach contains no cells that absorb nutrients; rather, the stomach cells protect it from self-digestion. The stomach exit (pylorus) is equipped with a sphincter, a circular muscles surrounding the gut tube. Sphincters function as valves. Depending on the diet, the stomach of other mammals can have further specializations.

Three sections are distinguished in the subsequent small intestine that merge into each other without clear boundaries (Figs. 12.15a, b and 12.16a–d). The small intestine begins with the duodenum, into which the pancreas and liver open. The large brown-red liver is divided into several lobes, while the pancreas is light red and branched. Like other vertebrates, the liver is also the central metabolic organ in mammals. The nutrients absorbed from the intestine are supplied to the liver via the portal vein. The hepatic vein, in turn, supplies the body with nutrients. In addition, the liver produces bile, which plays an important role as an emulsifier in fat digestion. It also synthesizes the cholesterol, lipids, and lipoproteins circulating in the blood plasma and the nitrogenous degradation products of various metabolic pathways. The liver stores iron, glycogen, and various vitamins. The glucose, fat, and protein metabolism essentially take place in the liver. Finally, the liver also plays a major role in detoxifying endogenous and exogenous substances (conversion to non-toxic substances). Unlike many other mammals, the rat has no gallbladder for storing bile, which is continuously released into the intestinal lumen. As the second digestive gland, the pancreas is like the liver, an epithelial gland embryonically derived from the midgut epithelium. About 98% of the pancreas is acinar glands, which release enzymes for the digestion of carbohydrates, fats, and proteins into the small intestine. The remaining 2% is the islets of Langerhans or pancreatic islets. They produce hormones such as insulin, which is important for sugar metabolism.

The duodenum forms a posterior loop and then merges into the digesting and absorbing section (Fig. 12.16c). The mesenteries ensure the intestine's mobility and connect it, via blood, lymph vessels, and nerves, with the rest of the body (Fig. 12.16a). Depending on the nutritional state, more or less fat is stored in the mesenteries. In the specimen shown here, it mainly accompanies the blood vessels (Fig. 12.16a). The small intestine is the longest section of the digestive tract and reaches several times the body length in rats (Fig. 12.16b). The anterior section of the small intestine, the jejunum, is somewhat reddish; it eventually merges into the ileum (Figs. 12.15a and 12.16a, b). In the mature rat a boundary between the jejunum and the ileum is not clearly visible. The long small intestine is essential for the digestion of food and leads to an enlargement of the digesting and absorbing surface; in the finer anatomy and histology not discussed here, one would recognize that this is further enhanced by the formation of villi of the epithelium and the dense microvilli space of the epithelial cells.

The cecum is located at the transition from the small to the large intestine (Figs. 12.15a, b and 12.16a). As in all mammals, the cecum in rats is ventral, which is why it is already recognizable in the early stage of dissection (Figs. 12.15b and 12.16a). Depending on the nutritional physiology of the various mammalian species, this cecum is of different sizes. While it is quite large in rodents (Fig. 12.16a), it is reduced to a small appendix in humans, for example. But even in these, this section is not functionless; ceca always house lymphatic tissue. In more recent textbooks, the cecum is attributed to the immune system. Rodents like the rat are the so-called hindgut fermenters, that is, they harbor symbiotic bacteria in their cecum that can break down cellulose. Since this fermentation chamber is located behind the absorbing section of the intestine, a special adaptation is necessary so that the animals can access the nutrients contained in the cellulose (and the bacteria). Rodents and lagomorphs (hares, rabbits, and picas) produce two different forms of feces. In addition to the actual feces, cecal feces are also formed, released through the anus, then re-ingested through the mouth, and thus subjected to another passage through the intestine. In addition to this cecotrophy—a certain form of rumination—there are also examples in other mammals where retrograde transport exists in the intestine. From the cecum, the large intestine (colon) rises upward and leads in a U-shaped loop to the rectum. Here in the rectum, water is removed from the feces. As all terrestrial vertebrates have to live with as little water loss as possible, this is an extremely important function.

Lungs and Circulatory Organs

The body cavity (coelom) of mammals is separated into the thoracic and abdominal cavities by the diaphragm (Fig. 12.17a, b). The two pleural cavities each house a lung lobe and form a sliding displacement layer for the lungs during breathing. The pericardium lies between the right and left pleural cavities. The diaphragm plays an essential role in breathing. At rest, it is domed forward (Fig. 12.17a, b); a contraction leads to a flattening, thus enlarging the thoracic cavity and creating a vacuum so that the lungs fill with breathing air. The intercostal muscles support the diaphragm. Exhalation at rest is largely passive. Thus, the diaphragm is not only one of the most active muscles in mammals, but also a unique structure.

The respiratory organs begin with the nose. The tip of the nose, the rhinarium, is a moist, furless skin area surrounding the openings of the respiratory system. It helps to sense smell. However, the rhinarium in rats is reduced and almost absent. The breathing air passes through the nostrils, the nasal cavity and pharynx to the larynx, a complex structure that directs air ventrally into the trachea and prevents food pulp from entering it. For this purpose, the trachea is closed by the epiglottis during swallowing. This is followed by the tubular trachea itself, whose lumen is kept stable by cartilage rings in its wall (Figs. 12.15c, d and 12.17c, d, g). One of the most important hormone glands, the thyroid gland, is associated with the trachea (Fig. 12.17c). It stores iodine and produces iodine-containing hormones, including thyroxine and calcitonin, which are important for numerous metabolic processes and for regulating bone resorption and formation.

The trachea branches into the two main bronchi below the heart, approximately in its center. These can be seen if the heart is flipped upward after cutting the ligaments with which it is connected to the diaphragm, or if the heart and lungs are removed from the thorax and viewed from the back (Fig. 12.17g). The lungs of dead animals are only filled with a little air and, therefore, relatively inconspicuous. The lung lobes only appear as two flat structures on the left and right next to the heart, so the thoracic cavity is dominated by the heart (Figs. 12.15a, b, 12.16b, and 12.17c, e). The right lung lobe consists of several parts, a cranial, a middle, and a pos-

terior caudal lobe, while the left lung is unilobed (Fig. 12.17d). By making a small, transverse cut through the trachea (Fig. 12.17d; arrow), you can carefully insert the tip of a Pasteur pipette into it (Fig. 12.17e) and fill the lung with air (Fig. 12.17f). The lung lobes are unfolded and appear bright red (Fig. 12.17a, f). A movie showing lung filling with a Pasteur pipette can be found at figshare: sn.pub/xr9qhi.

The heart of mammals consists of two completely separate atria and ventricles (Fig. 12.17c–g). In the course of evolution, this allowed the separation of the pulmonary and the systemic circulation from each other. Thus, mixing of oxygenated and deoxygenated blood is prevented. The deoxygenated blood coming from the body enters the right atrium via the two anterior and the posterior vena cava through the sinus venosus, which is not separated from the atrium, and from there into the right ventricle, which pumps the blood through the pulmonary arteries into the lungs. From the lungs, the oxygenated blood enters the left atrium, the left ventricle, and then proceeds through the left aortic arch into the body. From the aortic arch, the arteries supply the anterior body, starting with the right subclavian artery originating from a common branch (truncus brachiocephalicus) and the right common carotid artery. This then divides into the external and internal carotid arteries. Then follow the arteries of the left side of the body: the left common carotid artery, followed by the left subclavian artery. Subsequently, the aortic arch continues ventrally as the ventral aorta and supplies the rest of the body with oxygenated blood. The head arteries and anterior vena cava can be seen in the dissection of the neck region (Fig. 12.17c), mainly arranged parallel to the trachea. Beginners may have difficulties presenting these vessels—successful dissection requires some practice.

Since the spleen can be described as a blood lymph organ, it should be briefly discussed here, even though it belongs to the abdomen in terms of its location. The spleen is located in the dorsal mesentery dorsolaterally and to the stomach's left. It is, therefore, also removed from the abdominal cavity when the mesenteries are removed (Fig. 12.16b–e). The brown-red, elongated organ plays a central role in the breakdown of red blood cells, is a site of lymphocyte formation, and produces antibodies. As a highly vascularized organ, it also serves as a blood reservoir, releasing blood into circulation when oxygen demand increases. Rodents have a spleen that primarily serves for immune defense, while a storage spleen is found, for example, in seals. Of the structures of the lymphatic system, only the lymph nodes of the head-neck area located next to the salivary glands are noticeable in the dissection (Fig. 12.15c), so the lymphatic system, for all its complexity, is only treated here in a rudimentary way.

Kidneys

After the removal of the intestine, the kidneys and reproductive organs are accessible. To see them, however, extensive

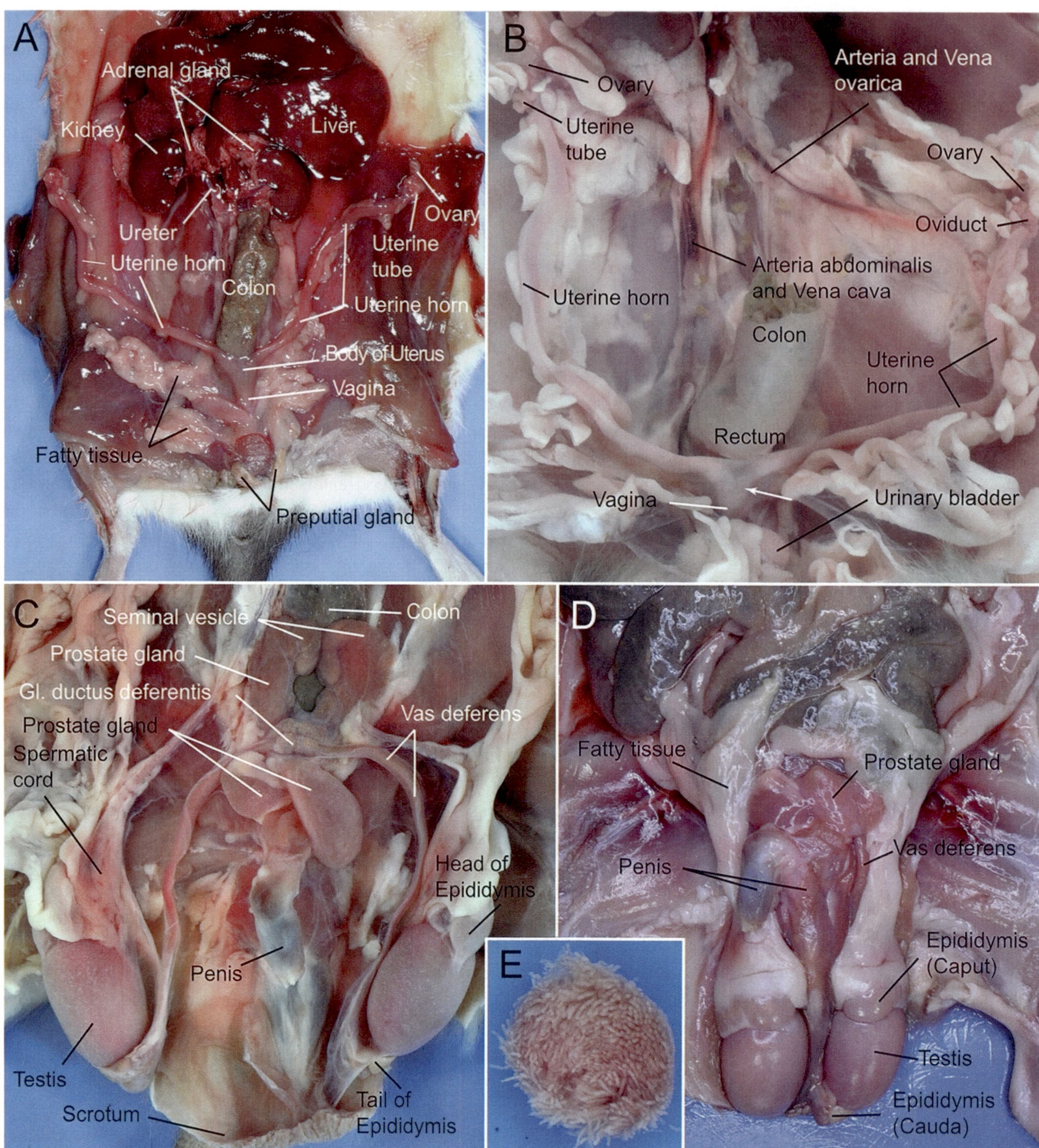

Fig. 12.18 *Rattus norvegicus*. Reproductive organs. (**a, b**) Female. (**a**) Overview; uteri slightly moved to the side. (**b**) Close-up, natural position. (**c–e**) Male. (**c**) Male organs after opening the scrotum and exposing the testes. (**d**) Testes and penis in the natural position. (**e**) Exposed testis with numerous testicular tubules

fat deposits in the suspensory ligaments of the female reproductive organs, the testicles, and around the kidneys must be removed (Fig. 12.18a–e). The functional kidney of mammals is the so-called metanephros; the segmental primitive kidney (mesonephros) is embryonically formed, undergoes a functional change, and plays an important role in the excretion of gametes as the epididymis and vas deferens. The paired kidney is located dorsally in the abdominal cavity, slightly behind the liver (Fig. 12.18a, b). As in many mammals, the rat's kidneys are bean-shaped and arranged slightly asymmetrically. The right kidney is positioned slightly further forward than the left (Fig. 12.18a). In the inward-facing indentation, a ureter originates in each case. This is also where the blood vessels that supply the kidney run. The slightly widened origin of the ureter is also called the renal pelvis; this is also where the only renal papilla in the rat is located (cut kidney lengthwise). The collecting ducts of the nephrons open onto the renal papilla, and the distal sections of the Henle's loops are also located here. In mammals adapted to desert environment and similar dry habitats, the Henle loops are particularly long, as they concentrate the urine. The ureters open caudally into the urinary bladder and, from there, to the outside via the urethra. In females, this is short and opens in front of the vagina at the base of the clitoris; in males, it is long and runs through the penis. In males, the accessory genital glands and the vas deferens also open into the urethra (Fig. 12.18c).

The adrenal gland is located cranially on the kidney and is an endocrine gland that has nothing to do with excretion. It produces important hormones of intermediate metabolism, including adrenaline and noradrenaline. In addition to the adrenal glands, there are also the renal lymph nodes, which lie further ventrally in the kidney's indentation.

Reproductive Organs

The female reproductive organs of placental mammals consist of the unpaired clitoris and the vagina, as well as the paired uteri (uterus duplex), oviducts, and ovaries (Fig. 12.18a, b). The uteri in mammals are very differently designed; the morphology of the uteri is always correlated with the size of the litter. They can, as in humans, be fused into a single uterus simplex. In the rat, the uteri are separated from each other until they open into the vagina. Even if the last sections seem connected, the lumina remain separate (Fig. 12.18a, b, arrow). The uteri extend far forward; in pregnant females, embryos are located in both uteri. After a gestation period of 22 days, an average of eight to nine blind and naked young are born; they are weaned after another 22 days and reach sexual maturity after about 3 months. The uteri are followed by the coiled fallopian tubes and oviducts, whose funnels each attach to the grape-shaped ovary. The ovaries are located approximately at the level of the kidneys.

Rattus: **Fourth Dissection Step**

To display the male reproductive organs, the scrotum should be carefully cut open. To trace the course of the vasa deferentia (singular: vas deferens), it is best to start at the testicles. The vas deferens must still be completely exposed, and the testicles placed to one side (Fig. 12.18c, d). Finally, the connective tissue sheath of the testes can be completely removed, so that the testicular tubules become visible (Fig. 12.18e).

The male reproductive organs consist of the paired, egg-shaped testicles, on which the epididymis is located. In addition, the vasa deferentia, sperm ducts, pass through the inguinal canal into the abdominal cavity and then turn medially, cross the urethra, and enter the urethra from the ventral side (Fig. 12.16c, d). From there, the common duct from the two vasa deferentia and the urethra runs through the penis to the outside. In mammals, the accessory genital glands are relatively diverse and essential for the activation and transport of sperm. Their arrangement, function, and structure vary within mammals. The testes are suspended by a muscular ligament and, in most mammals, are located outside the abdominal cavity in the scrotum (scrotal sac; in Placentalia, located at the base of the penis). The biological significance is not completely clarified, especially because several mammals have internal testes. The outer layer of the testes consists of a fibrous connective tissue layer that forms an organ capsule. The testes consist of individual testicular tubules (Fig. 12.18e). Spermatogenesis takes place in these tubules from the outside to the inside. The differentiated, immobile sperm are released into the lumen of these tubules and transported to the epididymis. This is located anteriodorsally and medially to the testes; the dorsal part of the epididymis is referred to as the caput, from which it runs medially and ends in the cauda, subsequently merging into the vas deferens and entering the abdominal cavity (Fig. 12.18c, d). These structures originate in ontogenesis from the mesonephros. At the junction of the vasa deferentia into the urethra, we also find the accessory genital glands: the prostate, which consists of several lobes, the glandula ductus deferentis, and the large glandula vesiculosa (seminal vesicle). A special feature in rats and other rodents is the so-called coagulation gland, which is part of the prostate. Its secretion forms a plug after copulation in the vagina, and thus prevents the fertilization of the eggs with sperm of other males. The penis of the rat is in a non-erect state in a backward bend, forming a U shape (Fig. 12.18d).

Nervous System, Sensory Organs

A beginner's course can only represent the nervous system and sensory organs of the rat to a limited extent. Certain structures, such as those of the spinal cord, the brain, or the eyes, should be examined in histological sections.

Rattus: **Fifth Preparation Step**

From the nervous system, only the brain will be dissected and discussed here. Due to its protected position in the skull and its soft structure, dissection is only possible very roughly without further treatment or fixation.

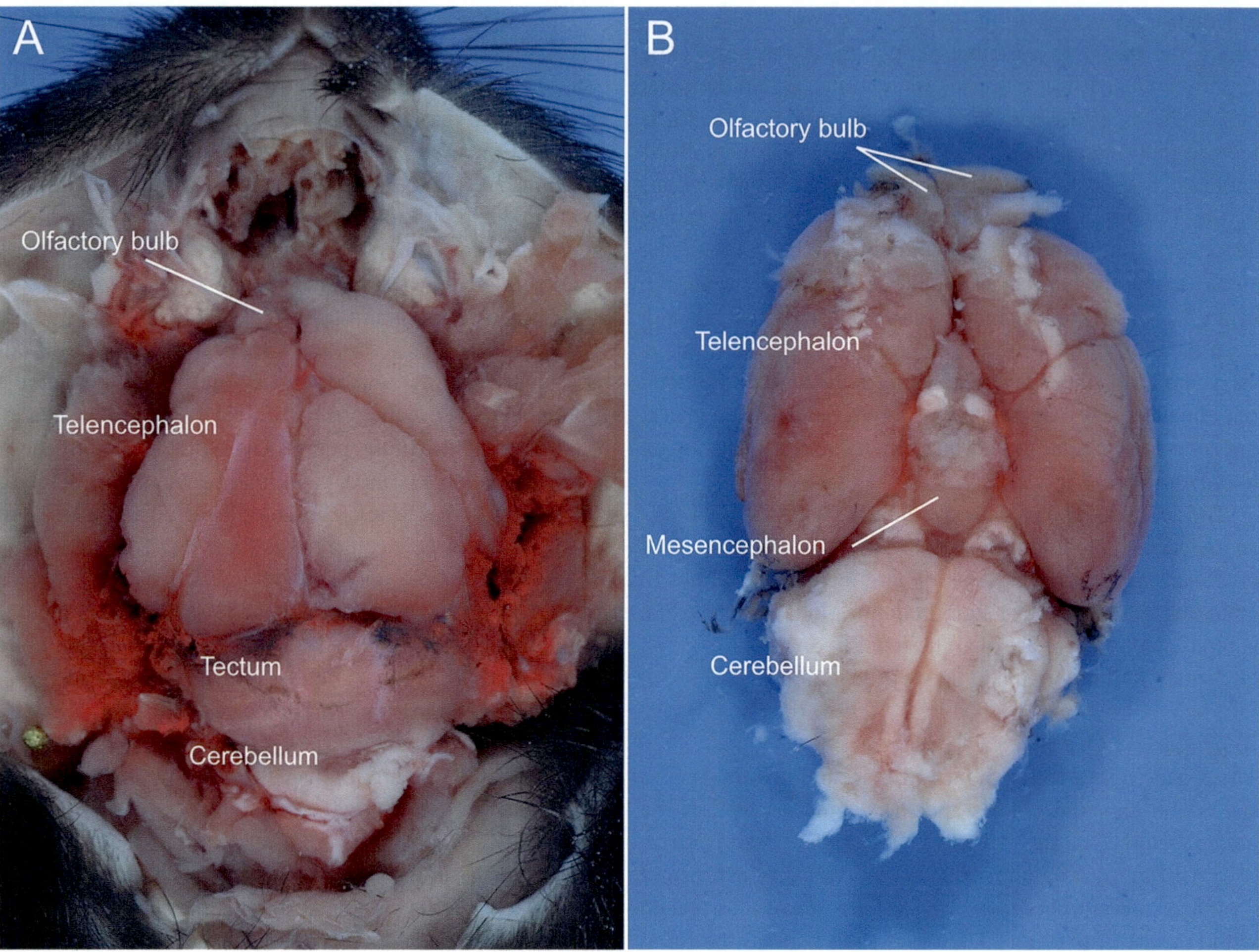

Fig. 12.19 *Rattus norvegicus*. Brain. (**a**) Position in the opened skull roof from the dorsal. (**b**) Prepared, ventral; Medulla oblongata (Myelencephalon) not preserved

The head is detached from the body at the level of the first cervical vertebra, the dorsal skin and musculature are removed with a median incision, and two perpendicular cuts are made in it, so the skull can be exposed. The skull capsule is then opened with a fine saw to the left and right of the occipital foramen above the eye socket. Then, the skull cap can be lifted off and the brain is accessible. At this point, a fixation with formaldehyde would be advantageous, as the tissue becomes somewhat firmer. Starting from the back, one can carefully lift the medulla oblongata and then cut the cranial nerves as far as possible from their origin from back to front. After cutting the optic and olfactory nerves, the brain can be removed and fixed in ethanol. If time is short, one can also try to open the skull capsule by making a median puncture with the tip of a pair of fine scissors and then opening the skull capsule by unfolding the scissors.

When viewed from the dorsal, the two hemispheres of the telencephalon (endbrain), which make up the largest part of the brain, become noticeable first (Fig. 12.19a, b). In rodents, the cerebral cortex is not furrowed. In front of it are the two olfactory bulbs, into each of which the neurites of the olfactory nerve enter. The nose is one of the most important sensory organs of many mammals; in humans it plays a rather minor role. The diencephalon is not recognizable from the dorsal and is covered by the forebrain hemispheres. This is followed posteriorly by the midbrain, of which the tectum is visible, followed by the metencephalon. This is followed by the narrow myelencephalon, which merges into the spinal cord. The latter structures are more or less well preserved, depending on how carefully they are prepared. The 12 cranial nerves can only be demonstrated in a very good preparation.

Further Readings

Guides to Dissection

Berman W (1984) How to dissect. Exploring with probe and scalpel. Special projects for advanced study, 4th edn. A Fireside Book. Simon & Schuster. ISBN 978-0-6717-6342-8

Bullough WS (1958) Practical invertebrate anatomy, 2nd edn. MacMillan, London. ISBN: 978-0-333313-206

Dales RP (ed) (1981) Practical invertebrate zoology. A laboratory manual, 2nd edn. University of Washington Press, Seattle. ISBN 978-0-632007-554

De Iuliis G, Pulera D (2019) The dissection of vertebrates. A laboratory manual, 3rd edn. Academic Press, London. ISBN 978-0-12-410460-0

Holyoak AR (2016) Laboratory exercises in invertebrate zoology, 2nd edn ISBN 978-1-539392-408

LaDouceur ESB (ed) (2021) Invertebrate histology. Wiley Blackwell. ISBN 978-1-119507-659

Lytle CF, Meyer JR (2005) General zoology. Laboratory guide, 14th edn. McGraw-Hill Higher Education. ISBN 0-07 234900-X

Löw P, Molnar K, Kriska G (2016) Atlas of animal anatomy and histology. Springer, Heidelberg. ASIN B01F4THHKC

Rogers E (1986) Looking at vertebrates. A practical guide to vertebrate adaptations. Longman Group Limited. ISBN 0-582-45086-1

Rowett HGQ (1962) Guide to dissection. John Murray Publishers Ltd.. ISBN 978-0719511875

Storch V, Welsch U (2014) Kükenthal. Zoologisches Praktikum, 27th edn. Springer, Heidelberg. ISBN 978-3642419362

Wallace RL, Taylor WK (1996) Invertebrate zoology. A laboratory manual, 5th edn. Prentice Hall. ISBN 9-780 132 700269

Zalisko EJ, Kardong KV (2008) Comparative vertebrate anatomy. A laboratory dissection guide, 4th edn. McGraw Hill Education. ISBN 978-0072528312

Zboray G, Kovacs Z, Kriska G, Molnar K, Palfia Z (2010) Atlas of comparative sectional anatomy. Springer, Heidelberg. ISBN 978-3-211-99762-8

Textbooks on Animal Morphology

Brusca RC, Giribet G, Moore W (2022) Invertebrates, 4th edn. Oxford University Press Inc. ISBN 978-0197554418

Long JA (2011) The rise of fishes. 500 million years of evolution, 2nd edn. John Hopkins University Press. ISBN 978-0801896958

Ruppert EE, Fox RS, Barnes RD (2004) Invertebrate zoology. A functional evolutionary approach, 7th edn. Books/Cole—Thomson Learning, Belmont. ASIN B009CPW2LO

Urry LA, Cain ML, Wasserman SA, Minorsky PV, Reece JB (2019) Campbell Biologie (11. Aufl.). Pearson Studium

Holidays Readings

Westheide W, Rieger G (2013) Spezielle Zoologie. Teil 1: Einzeller und Wirbellose Tiere, 3th edn. Springer. ISBN 978-3642346958

Westheide W, Rieger G (eds) (2015) Spezielle Zoologie. Teil 2: Wirbel- oder Schädeltiere: Wirbel- Oder Schädeltiere, 3th edn. Springer. ISBN 978-3642554353

Holidays Readings

Attenborough D (1979) Life on earth, 1st edn. William Collins & Sons. ISBN 978-0002190916

Godfrey-Smith P (2020) Metazoa: animal minds and the birth of consciousness, 1st edn. William Collins, Glasgow. ISBN 978-0008321192

Godfrey-Smith P (2018) Other minds: the octopus and the evolution of intelligent life, 1st edn. William Collins, Glasgow. ISBN 978-0008226299

Halliday T (2023) Otherlands: a world in the making. Penguin. ISBN 978-0141991146

Scales H (2021) The Brillant abyss: true Tales of exploring the Deep Sea, discovering hidden life and selling the seabed. Bloomsbury Publishing Plc. ASIN B08YP26KTQ

Wallace AR (2010) A narrative of travels on the Amazon and Rio Negro: with an account of the native tribes, and observations on the climate, geology, and natural history of the Amazon Valley. Cambridge University Press, Cambridge. ISBN 978-1108007290

Yong E (2023) An immense world: how animal senses reveal the hidden realms around us. Vintage. ISBN 978-1529112115

Field Guides

Fischer A (ed) (2013) The Helgoland manual of animal development: notes and laboratory protocols on marine invertebrates, 1st edn. Dr. Friedrich Pfeil, Munich. ISBN 978-3899371611

Hayward PJ, Ryland JS (eds) (1990a) The marine Fauna of the British Isles and North-West Europe. Volume 1 introduction and protozoans to arthropods. Clarendon Press, Oxford. ISBN 0-19-857356-1

Hayward PJ, Ryland JS (eds) (1990b) The marine Fauna of the British Isles and North-West Europe. Volume 2 molluscs to chordates. Celarendon Press, Oxford. ISBN 0-19-857515-7

Hayward PJ (2017) Handbook of the marine Fauna of North-West Europe, 2nd edn. Oxford University Press, Oxford. ISBN 9780199549443

Hayward P, Nelson-Smith T, Shields C (1999) Sea shore of Britain & Europe (Collins pocket guide), 4th edn. GB Gardners Books. ISBN 978-0002199551

Larink O, Westheide W (2011) Coastal plankton. Photo guide for European seas, 2nd edn. Dr. Friedrich Pfeil, München. I SBN 978-3-89937-127-7

Muus BJ, Nielsen JG (2013) Die Meeresfische Europas in Nordsee, Ostsee und Atlantik. Franck Kosmos, Stuttgart. ISBN-13 978-3440135150

Newell GE, Newell RC Marine Plankton – A Practical Guide. Pisces Conservation Ltd; Facsimile Edition (12. Mai 2023). ISBN-13:978-1904690429

Riedl S, Schweder-Riedl B (eds) (2011) Rupert Riedl Fauna und Flora des Mittelmeeres. Seifert, Wien. ISBN 978-3-902406-60-6

Schmidt-Rhaesa A (ed) (2020) Guide to the identification of Marine Meiofauna. Dr. Friedrich Pfeil, München. ISBN 978-3-89937-244-1

Synopses of the British Fauna. Field Studies Council. This is a series of great books presenting the fauna of the Atlantic coast

Todd CD, Laverack MS, Boxshall GA Coastal Marine Zooplankton: A practical manual for Students. Cambridge University Press; 2. Edition (1. Juni 2009), ISBN-13: 978-0521555333

Phylogeny of Metazoa

Bleidorn C (2019) Recent progress in reconstructing lophotrochozoan (spiralian) phylogeny. Org Divers Evol 19:557–566. https://doi.org/10.1007/s13127-019-00412-4

Edgecombe GD, Giribet G, Dunn CW, Hejnol A, Kristensen RM, Neves RC, Rouse GW, Worsaae K, Sörensen MV (2011) Higher-level metazoan relationships: recent progress and remaining questions. Org Divers Evol 11:151–172. https://doi.org/10.1007/s13127-011-0044-4

Giribet G, Edgecombe GD (2020) The invertebrate tree of life. Princeton University Press, Princeton/Oxford. ISBN 978-0691170251

Laumer CE, Fernández R, Lemer S, Combosch D, Kocot KM, Riesgo A, Andrade SCS, Sterrer W, Sørensen MV, Giribet G (2019) Revisiting metazoan phylogeny with genomic sampling of all phyla. Proc R Soc Lond B 286:20190831. https://doi.org/10.1098/rspb.2019.0831

Manni L, Penati R (2016) Tunicata. In: Schmidt-Rhaesa A, Harzsch S, Purschke G (eds) Structure and evolution of invertebrate nervous systems. Oxford University, Oxford, pp 699–718. ISBN 978-0199682201

Marlétaz F, Peijnenburg KTCA, Goto T, Satoh N, Rokhsar DS (2019) A new spiralian phylogeny places the enigmatic arrow worms among gnathiferans. Curr Biol 29:312–318. https://doi.org/10.1016/j.cub.2018.11.042

Schierwater B, DeSalle R (2022) Invertebrate zoology. A tree of life approach. Taylor and Francis Group/CRC Press, Boca Raton/Abingdon/Oxon. ISBN 78-0367685676

Simion P, Philippe H, Baurain D, Jager M, Richter DJ, Di Franco A, Roure B, Satoh N, Quéinnec E, Ereskovsky A, Lapébie P, Corre E, Delsuc F, King N, Wörheide G, Manuel MI (2017) A large and consistent phylogenomic dataset supports sponges as the sister group to all other animals. Curr Biol 27:958–967. https://doi.org/10.1016/j.cub.2017.02.031

Telford M, Budd GE, Philippe H (2015) Phylogenomic insights into animal evolution. Curr Biol 25:R876–R887. https://doi.org/10.1016/j.cub.2015.07.060

Whelan NV, Kocot KM, Moroz LL, Halanych KM (2015) Error, signal, and the placement of Ctenophora sister to all other animals. Proc Natl Acad Sci USA 112:5773–5778. https://doi.org/10.1073/pnas.1503453112

Porifera

Adams EDM, Goss GG, Leys SP (2010) Freshwater sponges have functional, sealing epithelia with high transepithelial resistance and negative transepithelial potential. PLoS One 5(11):e15040. https://doi.org/10.1371/journal.pone.0015040

Müller WEG (ed) (2012) Sponges. Progress in molecular and subcellular biology. Springer, Berlin/Heidelberg. ISBN 978-3642624711

Simpson TL (1984) Gamete, embryo, larval development In: The Cell Biology of Sponges. New York: Springer; p. 341–413

Vernale A, Prünster MM, Marchiano F, Debost H, Brouilly N, Rocher C, Massey-Harroche D, Renard E, Le Bivic A, Habermann BH, Borchiellini C (2021) Evolution of mechanisms controlling epithelial morphogenesis across animals: new insights from dissociation-reaggregation experiments in the sponge Oscarella lobularis. BMC Ecol Evol 21:160. https://doi.org/10.1186/s12862-021-01866-x

Cnidaria

Dupre C, Yuste R (2017) Non-overlapping neural networks in Hydra vulgaris. Curr Biol 27:1085–1097. https://doi.org/10.1016/j.cub.2017.02.049

Holstein T, Tardent P (1984) Ultrahigh-speed analysis of exocytosis: nematocyst discharge. Science 223:830–833. https://doi.org/10.1126/science.6695186

Holz O, Apel D, Hassel M (2020) Alternative pathways control actomyosin contractility in epitheliomuscle cells during morphogenesis and body contraction. Dev Biol 463(1):88–98. https://doi.org/10.1016/j.ydbio.2020.04.001

Tardent P (1978) Cnidaria. In: Seidel F (ed) Morphogenese der Tiere. Lieferung 1: A1 Einleitung, Cnidaria. Gustav Fischer, Jena

Platyhelminthes

Lofty WM, Brant SV, DeJong R, Hoa Le T, Demiaszkiewicz A, Jayanthe Rajapakse RPV, Perera VBVP, Laursen JR, Loker ES (2008) Evolutionary origins, diversification, and biogeography of liver flukes (Digenea, Fasciolidae). Am J Trop Med Hyg 79(2):248–255. PMCID: PMC2577557

Mehlhorn H (2023) Human parasites. Diagnosis, treatment, prevention, 2nd edn. Springer Spektrum, Heidelberg. ISBN 978-3031417047

Mehlhorn H (2016) Animal parasites. Diagnosis, treatment, prevention. Springer Spektrum, Heidelberg. ISBN 978-3319328010

Annelida

Bartolomaeus T, Purschke G (eds) (2005) Morphology, molecules, evolution and phylogeny in Polychaeta and related taxa. In: Developments in hydrobiology, vol 179. Springer, Dordrecht. ISBN 978-1402029516

Bely AE, Zattara EE, Sikes JM (2014) Regeneration in spiralians: evolutionary patterns and developmental processes. Int J Dev Biol 58:623–634. https://doi.org/10.1387/ijdb.140142ab

Edwards CA, Arancon NQ (2022) Biology and ecology of earthworms, 4th edn. Springer, Heidelberg

Harrison FW, Gardiner SC (1992) Annelida. In: Harrison FW (ed) Microscopic anatomy of invertebrates, vol 7. Wiley Liss, New York. ISBN 978-0471561149

Hauenschild C, Fischer A (1969) *Platynereis dumerilii*. In: Siewing R (ed) Großes Zoologisches Praktikum. Gustav Fischer, Jena

Purschke G (2016) Annelida: basal groups and Pleistoannelida. In: Schmidt-Rhaesa A, Harzsch S, Purschke G (eds) Structure and evolution of invertebrate nervous systems. Oxford University Press, Oxford. ISBN 978-0199682201

Rouse G, Tilic E, Pleijel F (2022) Annelida. Oxford University, Oxford. ISBN 978-0199692309

Mollusca

Chase R, Blanchard KC (2006) The snail's love-dart delivers mucus to increase paternity. Proc R Soc B 73(1593):1471–1475. https://doi.org/10.1098/rspb.2006.3474

Dimitriadis VK, Domouhtsidou GP, Cajaraville MP (2004) Cytochemical and histochemical aspects of the digestive gland cells of the mussel *Mytilus galloprovincialis* (L.) in relation to function. J Mol Histol 35(5):501–509. https://doi.org/10.1023/b:hijo.0000045952.87268.76

Hinzmann M, Lopez-Lima M, Teixeira A, Varandas S, Sousa R, Lopes A, Froufe E, Machado J (2013) Reproductive cycle and strategy of *Anodonta anatina* (L., 1758): notes on hermaphroditism. J Exp Zool 319A:378–390. https://doi.org/10.1002/jez.1801

Lodi M, Koene JM (2016) The love-darts of land snails: integrating physiology, morphology and behavior. J Molluscan Stud 82:1–10. https://doi.org/10.1093/mollus/eyv046

Nematoda

Anderson TJC (1995) *Ascaris* infections in humans from North America: molecular evidence for cross-infection. Parasitology 110(2):215–219. https://doi.org/10.1017/s0031182000063988

Leles D, Gardner SL, Reinhard K, Iniguez A, Araujo A (2012) Are *Ascaris lumbricoides* and *Ascaris suum* a single species? Parasit Vectors 5:42. https://doi.org/10.1186/1756-3305-5-42

Arthropoda-Crustacea

Hessen DO, Taugbøl T, Fjeld E, Skurdal J (1987) Egg development and lifestyle timing in the noble crayfish (*Astacus astacus*). Aquaculture 64:77–82. https://doi.org/10.1016/0044-8486(87)90207-9

Ingle R (1997) Crayfishes, lobsters and crabs of Europe: an illustrated guide to common and traded species. Springer, Heidelberg. ISBN 978-0412710605

Kawai T, Kouba A (2020) A description of postembryonic development of *Astacus astacus* and *Pontastacus leptodactylus*. Freshwater Crayfish 25(1):103–116. https://doi.org/10.5869/fc.2020.v25-1.103

Longshaw M, Stebbing P (eds) (2021) Biology and ecology of crayfish. CRC Press Taylor/Francis Group, Boca Raton/London/New York. ISBN 978-0367782986

Richter S (2002) Evolution of optical design in the malacostraca (Crustacea). In: Wiese K (ed) The crustacean nervous system. Springer, Berlin/Heidelberg. ISBN 978-3642086182

Arthropoda-Insecta

Berry R, van Kleef J, Stange G (2007) The mapping of visual space by dragonfly lateral ocelli. J Comp Physiol A 193:495–513. https://doi.org/10.1007/s00359-006-0204-8

Chapman RF, Simpson SJ, Douglas AE (2013) The insects. Structure and function, 5th edn. Cambridge University Press, Cambridge. ISBN 978-1107624801

Honkanen A et al (2018) The role of ocelli in cockroach optomotor performance. J Comp Physiol A 204(2):231–243. https://doi.org/10.1007/s00359-017-1235-z

Vilcinskas A (ed) (2013) Insect biotechnology, 1st edn. Springer, Heidelberg/London/New York. ISBN 978-9400733886

Echinodermata

Brundu G, Cannavacciuolo A, Nannini M, Somma E, Munari, M, Zup V, Farina S (2023) Development of an efficient, noninvasive method for identifying gender year-round in the sea urchin *Paracentrotus lividus*. Aquaculture 564: 739082. https://doi.org/10.1016/j.aquaculture.2022.739082

Reich A, Dunn C, Akasaka K, Wessel G (2015) Phylogenomic analyses of Echinodermata support the sister groups of Asterozoa and Echinozoa. PLoS One 10(3):e0119627. https://doi.org/10.1371/journal.pone.0119627

Sigl R, Steibl S, Laforsch C (2016) The role of vision for navigation in the crown-of-thorns starfish, *Acanthaster planci*. Sci Rep 6:30834. https://doi.org/10.1038/srep30834

Acrania

Holland ND, Venkatesh TV, Holland LZ, Jacobs KK, Bodmer R (2003) *AmphiNk2-tin*, an *Amphioxus* homeobox gene expressed in myocardial progenitors: insights into evolution of the vertebrate heart. Dev Biol 255:128–137. https://doi.org/10.1016/s0012-1606(02)00050-7

Lacalli T, Stach T (2016) Acrania (Cephalocordata). In: Schmidt-Rhaesa A, Harzsch S, Purschke G (eds) Structure and evolution of invertebrate nervous systems. Oxford University Press, Oxford. ISBN 978-0199682201

Urochordata-Tunicata

Anderson HE, Christiaen L (2016) *Ciona* as a simple chordate model for heart development and regeneration. J Cardiovasc Dev Dis 3:25. https://doi.org/10.3390/jcdd3030025

Aoyama M, Kawada T, Satake H (2012) Localization and enzymatic activity profiles of the proteases responsible for tachykinin-directed oocyte growth in the protochordate, *Ciona intestinalis*. Peptides 34:186–192. https://doi.org/10.1016/j.peptides.2011.07.019

Brunetti R, Gissi C, Pennati R, Caicci F, Gasparini F, Manni L (2015) Morphological evidence that the molecularly determined *Ciona intestinalis* type A and type B are different species: *Ciona robusta* and *Ciona intestinalis*. J Zool Syst Evol Res 53:186–193. https://doi.org/10.1111/jzs.12101

Cima F, Peronato A, Ballarin L (2017) The haemocytes of the colonial aplousobranch ascidian *Diplosoma listerianum*: structural, cytochemical and functional analyses. Micron 102:51–64. https://doi.org/10.1016/j.micron.2017.08.007

Ebner B, Burmester T, Hankeln T (2003) Globin genes are present in *Ciona intestinalis*. Mol Biol Evol 20(9):1521–1525. https://doi.org/10.1093/molbev/msg164

Horstmann E (1977) Ciliary movement for particle transport (Food Intake) – Pond Mussel. IWF Film K85, IWF, Göttingen. https://av.tib.eu/media/14768

Jeffrey WR (2015) The tunicate *Ciona*: a model system for understanding the relationship between regeneration and aging. Invertebr Reprod Dev 59(sup1):17–22. https://doi.org/10.1080/07924259.2014.925515

Kusakabe T, Motoyuki T (2007) Photoreceptive systems in ascidians. Photochem Photobiol 83:248–252. https://doi.org/10.1562/2006-07-11-IR-965

Satoh N (1994) Developmental biology of ascidians. Cambridge University Press, Cambridge. ISBN 978-0521352215

Satoh N (2016) Chordate origins and evolution. The molecular evolutionary road to vertebrates. Academic Press, Amsterdam. ISBN 978-0521352215

Sawada H, Yamamoto K, Yamaguchi A, Higuchi A, Nukaya H, Fukoka M, Sakuma T, Yamamoto T, Sasakura Y, Shirae-Kurabayashi M (2020) Three multi-allelic gene pairs are responsible for self-sterility in the ascidian *Ciona intestinalis*. Sci Rep 10:2514. https://doi.org/10.1038/s41598-020-59147-4

Craniota

King GM, Custance DRN (1982) Colour atlas of vertebrate anatomy: an integrated text and dissection guide. Blackwell Scientific Publications, Oxford. ISBN 978-0950767604

Kramer B (1996) Electroreception and communication in fishes. Progr Zool 42:1–119. ISBN 3-437-25038-8

GPSR Compliance

*The European Union's (EU) General Product Safety Regulation (GPSR)
is a set of rules that requires consumer products to be safe and our
obligations to ensure this.*

*If you have any concerns about our products, you can contact us on
ProductSafety@springernature.com*

In case Publisher is established outside the EU, the EU authorized
representative is:

Springer Nature Customer Service Center GmbH
Europaplatz 3
69115 Heidelberg, Germany

Batch number: 10121907

Printed by Printforce, the Netherlands